THE INDUSTRIAL OPERATOR'S HANDBOOK

Petroleum & Chemical Industries Edition

Gulf Publishing Company
Houston, Texas

THE INDUSTRIAL OPERATOR'S HANDBOOK

Petroleum & Chemical Industries Edition

by
H. C. Howlett II

Foreword by Trevor Kletz

The Industrial Operator's Handbook:
A Systematic Approach to Safety and Reliability in Industrial Operations

Copyright © 1996, 1995 by H. C. Howlett II.
All rights reserved. No part of this book may be reproduced or transmitted in any form or by any means, electronic, mechanical, photocopying, recording, or otherwise, without the prior written permission of the publisher.

Petroleum and Chemical Industries Edition, Published October, 1996.

Published by Gulf Publishing Company, Houston, Texas.

Gulf Publishing Company
Book Division
P.O. Box 2608 □ Houston, Texas 77252-2608

10 9 8 7 6 5 4 3 2

Library of Congress Cataloging in Publication Data

Howlett, H. C. (Hop C.)
    Industrial operator's handbook : a systematic approach to safety and reliability in industrial operations / H.C. Howlett II.
       p. cm.
     ISBN 0-88415-413-0 (alk. paper)
    1. Operations research—Handbooks, manuals, etc.   2. Industrial management—Handbooks, manuals, etc   3. Industrial safety—Handbooks, manuals, etc   I. Title.
T57.6.H72   1996
658.4'08—dc20   96-26232
                                                                       CIP

Printed on Acid-Free Paper (∞)

Printed in the United States of America

Foreword

The many fascinating case histories in this book show that accidents have many causes and that many people have the opportunity to prevent them. Every accident is due to a human failure: someone has to decide what to do, someone has to decide how to do it, and someone has to do it, and all of them can fail, through ignorance, deliberate decision, a slip or lapse of attention, and so on. This book concentrates on people in the third category—those who do it, the operators, and, of course, on those who should ensure that they are well trained and properly instructed and that adequate checks and inspections are carried out.

The case histories are a valuable part of the book, partly because we learn more from failure than from success but mainly because they grab our attention and give us the stimulation necessary to read, digest, and act on the detailed advice which follows. Advice alone, I fear, is usually glanced at and put aside to read when we have time and we all know what that means.

Prevention of accidents by good design is not the subject of this book, so I hope that those designers who read it will not feel that accident prevention is a job for operators and that designers can relax. It is true that about half the accidents that occur could not have been prevented by better design, only by better methods of operation or by following existing methods. But, nevertheless, we should design plants so that hazards and opportunities for error are minimal. Designers should assume that the plant will be operated by Murphy (we all have days when we act like Murphy) while operators should assume that the plant was designed by Rube Goldberg or his British cousin, Heath Robinson. Designers should read this book so that they know the sort of errors that operators make and can allow for them in their designs. The accident described in Chapter 25, for example, would not have occurred if the company's procedures had been followed. It would also not have occurred if the designers had not made an elementary error: using the same size and type of hose connection when interchange of them is hazardous.

Similarly, senior managers should not assume that because they have issued a policy statement and invested time and money in good training and procedures everything will be okay. All that effort will be negated if they suggest by their words or actions that production comes before safety, if they turn a blind eye to shortcuts or if, after an accident, their first question is, "When will we be back on line?" in short, if the culture is wrong. It was wrong in the accident described in Chapter 25 because everyone knows that the company's procedures were not being followed.

As I have said, the book shows that accidents have many causes. However, they are not due to unlikely coincidences, to the failure of different people or pieces of equipment at the same moment in time. Most of the failures are hidden or latent: equipment fails but testing does not detect it. People develop bad habits, but no one notices or says anything. Ultimately, the last line of defense gives way and an accident becomes inevitable.

Following the advice in this book calls for a lot of effort. Is it really necessary? A lot of the time it may not be. A lot of the time we can get by with a performance that is only just adequate. Only in key situations are our best efforts really necessary. But we never know until afterwards which were the key situations. We never know when the others have failed and everything depends on us, when we are the last line of defense. The only safe way is to follow Hop Howlett's advice all the time.

There are some things we cannot do because there are no known methods. Preventing accidents is not one of them. The information is available, much of it in this book. Whether or not we use it is up to us.

Trevor Kletz

Preface

What's This Book About?

The Industrial Operator's Handbook presents the elements vital to systematic industrial operation.

Safe, efficient industrial operation is predicated, in part, upon conservative design and well-executed construction. Ultimately, however, the responsibility for successful operation falls to skilled, alert humans systematically controlling the equipment that they are charged with operating.

Accordingly, this book is dedicated to the study of improved industrial performance through a team approach to operation. Its premise is that *there is no substitute for the alert, well-trained operator controlling equipment within specified operating bands in accordance with approved procedures.*

Who Is This Book For?

This book is written to educate operators, maintenance technicians, laboratory analysts, craftsmen, engineers, team leaders—anyone engaged in the operation or technical support of an industrial complex—in the principles and skills of systematic industrial operation.

This version, a special Petroleum and Chemical Industry Edition, has been expanded to include a unique section which analyzes OSHA Regulation 1910.119 (Process Safety Management of Highly Hazardous Chemicals), comparing its requirements with the tenets presented in the main text.

Although written from the perspective of the operator, the concepts presented herein apply equally to the maintenance technician, lab analyst, engineer, or team leader.

Why Read This Book?

Whether you're a company CEO or an entry level technician, **The Industrial Operator's Handbook** develops the strategy for systematic industrial operation in such a logical and fascinating way that you will be intrigued while you learn.

The philosophy of industrial operation, if properly developed and applied, is essentially the same for every technology. Whether you build microchips, manufacture automobiles, operate a nuclear power-generating station, fly commercially, or run the gravel crusher for your company's road construction operation, once you understand it, you can apply it wherever you go. As a bonus, you will soon find that the principles and skills presented within this text are as important for driving, flying, hobby woodcraft and metal-working, or farming as they are for commercial industrial operation.

The text is divided into five parts: *The Nature of Industrial Failure* (Part I), *A Strategy for Operating Success* (Part II), *Vital Operating Skills* (Part III), *Implementing the Strategy* (Part IV), and *Process Safety Supplement* (Part V). Through a series of case studies and the lessons drawn from them, you will:
- Probe the methods of failure by which most industrial accidents occur,
- Investigate nineteen common components of accidents,
- Explore a common-sense strategy for systematic industrial operations,
- Determine the purpose of operating limits and the safe operating envelope,
- Learn two simple, but immutable principles of operating success,
- Review how alert, well-trained operators are developed,
- Evaluate twelve vital operating skills that every operator and leader should master,
- Examine the importance of investigating abnormal events,
- Realize the worth of continuing training,
- Discover the role of self-assessment in implementing the strategy,
- Analyze a case study in implementing the systematic approach, and
- Discern how the principles and skills presented in the main text apply to process safety management of highly hazardous chemicals.

Principles with "Staying Power"

The principles and conventions presented in this book aren't "gimicks". Rather, they are time-proven tenets that, once learned, renew and confirm themselves daily to every member of an industrial team. Developed and refined over a four-year period, this text incorporates decades of learning, evaluating, and practicing the principles of successful industrial operations. *TECHSTAR* has provided training in these principles for industrial and nuclear facilities nationwide—principles that have become lifelong benchmarks for thousands of industrial professionals.

Consequently, one great feature of **The Industrial Operator's Handbook** is that it will never become outdated. In fact, the text can be used again and again as a basis for analyzing your own successes and failures as well as those from other industries. It provides a distinct and unchanging path to direct you and your team in improving industrial operations.

Use It for Training

At least as important as your own education is educating and training your team in the philosophy and skills of effective and efficient industrial operations. This text provides an extraordinary tool for accomplishing just that. The book progresses in a logical sequence intended to facilitate leading your team through a comprehensive study of industrial operations, one or two chapters at a time. The *Topic Summary and Questions to Consider* at the end of each chapter provides an excellent forum to compare your operations with those described in the text, prompting you to ask, "Are we making the same mistakes we just read about?"

Furthermore, after you've studied *The Nature of Industrial Failure* (Part I) and *The Strategy for Operating Success* (Part II), you can begin a sequential review of Part III's twelve *Vital Operating Skills*. And, since the skills build on one another, you have a wonderful opportunity to study and practice one skill at a time. In fact, by working through the book systematically in a "study one, practice one" sequence, you can build an effective one-year continuing training program that (rigorously applied) will dramatically improve your team's performance. Remember, though, there aren't any "free lunches". Don't expect much if you approach it with an attitude of "I'll let them read it and hope they get the point". If you don't believe it, teach it, and reinforce it by your own example, no one else will take it seriously or put much effort into it, either.

For Your Enjoyment

We wish you well on your reading journey. We think that you will find these case studies (and the lessons drawn from them) both fascinating and useful. Best of all, we believe that they will dramatically improve your industrial operation when intelligently applied!

Contents

Foreword v
Preface vii
List of Figures xv

Part I: The Nature of Industrial Failure

Chapter 1 — Anatomy of an Accident 3

Crash of Eastern Airlines Flight 401, December 1972 3
A Chain of Insignificant Events 4
Complex Failure Formation 5
Lessons of Eastern 401 7
Topic Summary and Questions to Consider 9

Chapter 2 — Common Components of Accidents 11

Grounding of the Exxon-Valdez, March 1989 12
Nineteen Common Accident Components 16
Topic Summary and Questions to Consider 29

Part II: A Strategy for Operating Success

Chapter 3 — Systematic Industrial Operations 33

Objective of the Strategy 33
Elements of the Strategy 33
An Underlying Operating Philosophy 34
Reliable Equipment and Facilities 36
Well-Defined Operating Boundaries 38
Valid Policies and Procedures 40
An Efficient Operating Structure 41
Alert, Well-Trained Operators 45
Superb Leaders 47
A Team Approach 49
Topic Summary and Questions to Consider 54

Chapter 4 — Boundaries of Safe Operation 55

Safety Analysis 55
Identifying the Hazards 56
Determining the Risk 56
Establishing Protection 57
Categories of Operating Limits 59
Developing Operating Limits 61
The Safe Operating Envelope 65
Topic Summary and Questions to Consider 67

Chapter 5 — Principles of Operation 69

Chernobyl Atomic Power Station Accident, April 1986 69
The Primary Cause 73
Principles of Operation 74
Topic Summary and Questions to Consider 75

Chapter 6

The Alert, Well-Trained Operator

77

Big Bayou Canot Bridge Accident, September 1993 77
Tenets of Training 83
Determining Job Requirements 83
Selection of Candidates 84
Developing the Training Program 84
Phases of Training 87
Initial Training Phase 88
Certification Phase 89
Continuing Training Phase 89
Fitness for Duty 89
Effective Management Oversight 90
Topic Summary and Questions to Consider 90

Part III: Vital Operating Skills

Chapter 7

Controlling Equipment and Processes

95

Fatal Gas Release at Bhopal, December 1984 95
Combined Accident Contributors 99
The Cost 101
Twelve Vital Operating Skills 101
Topic Summary and Questions to Consider 104

Chapter 8

Conducting Pre-Task Briefings

107

Utility Crew Electrocution, April 1994 107
Accident Prevention through Pre-Task Briefing 111
Elements of a Pre-Task Briefing 111
Pre-Task Briefing Guidelines 113
Common Pre-Task Briefing Errors 113
Post-Task Debriefing 114
Topic Summary and Questions to Consider 114

Chapter 9

Understanding and Using Procedures

115

Policy or Procedure? 115
Developing Procedures 115
Desirable Characteristics of Procedures 116
Procedure Format 116
Review and Approval 117
Controlling Procedures 117
Revising Procedures 117
Using Procedures 118
An Example of Procedure Use 119
Reporting Status Changes 120
Training to Procedures 120
Enforcing Procedural Compliance 121
Topic Summary and Questions to Consider 121

Chapter 10 — 123
Monitoring Critical Operating Parameters
- Why Monitor? ... 123
- When to Monitor ... 124
- How to Monitor ... 124
- Actions for Abnormal Values ... 125
- Special Concerns ... 125
- Topic Summary and Questions to Consider ... 126

Chapter 11 — 127
Independent Verification
- What Is It? ... 127
- For All Tasks? ... 127
- Who Decides? ... 128
- Process or Point Verification? ... 128
- Guidelines for Independent Verification ... 129
- Informal Verification ... 129
- Topic Summary and Questions to Consider ... 130

Chapter 12 — 131
Communicating Vital Information
- Sinking of R.M.S. Titanic, April 1912 ... 131
- Questions of Cause ... 133
- The Lesson of Communication ... 134
- Elements of Effective Communication ... 135
- Effective Verbal Communication ... 136
- Written Instructions ... 137
- Emergency Communications ... 138
- Topic Summary and Questions to Consider ... 139

Chapter 13 — 141
Keeping Logs and Recording Data
- Data Record Sheets ... 141
- Narrative Logs ... 143
- Specialized Formats ... 144
- Topic Summary and Questions to Consider ... 145

Chapter 14 — 147
Recognizing Abnormalities
- Crash of Air Florida Flight 90, January 1982 ... 147
- An Accumulation of Abnormalities ... 154
- Sources of Abnormalities ... 154
- Recognizing Abnormalities ... 155
- Performing Inspection Tours ... 157
- Prioritizing Abnormalities ... 157
- Responding to Abnormalities ... 159
- Topic Summary and Questions to Consider ... 160

Chapter 15 — 161
Combatting Emergencies and Casualties
- The TMI-2 Accident, March 1979 ... 161
- The Problems ... 164
- Responding to Industrial Crisis ... 166
- Planning for Emergencies and Casualties ... 168
- Training for Emergencies and Casualties ... 169

	Responding to Emergencies and Casualties 172
	Learning from Emergencies and Casualties 174
	Topic Summary and Questions to Consider................................ 176

Chapter 16 — 177
Overseeing Maintenance, Modification, and Testing

- Maintenance, Modification, or Testing? 177
- Who is Responsible? 177
- Tiers of Maintenance 179
- Operations/Maintenance Interface 179
- Guidelines for Maintenance Control 181
- Types of Testing 182
- Operations/Testing Interface 182
- Guidelines for Testing Control 183
- Topic Summary and Questions to Consider 184

Chapter 17 — 185
Isolating Energy Hazards

- Sources of Hazardous Energy 185
- Other Hazards 186
- Means of Isolation 187
- Lockout Isolation 187
- Tagout Isolation 189
- Using Danger and Caution Tags 189
- Determining Tag Locations 190
- Topic Summary and Questions to Consider 191

Chapter 18 — 193
Training On-the-Job

- The Purpose of OJT 193
- Elements of a Good OJT Program 193
- On-the-Job Training Controls 195
- Topic Summary and Questions to Consider 197

Chapter 19 — 199
Performing Shift Turnovers

- Continental Express Accident, September 1991 199
- The Problems 204
- Definition, Purpose, and Applicability 205
- Eight Principles of Shift Turnover 206
- The Process of Shift Turnover 207
- Preparing the Station for Turnover 208
- Pre-Shift Meeting 209
- Pre-Shift Tour 209
- Station Turnover 210
- Post-Turnover Meeting 211
- Guidance for Successful Turnovers 211
- Topic Summary and Questions to Consider 213

Part IV: Implementing the Strategy

Chapter 20 — 217
Investigating Abnormal Events

- The Challenger Accident, January 1986 217
- Cause of the Accident 221

Contributing Problems 221
An Investigative Model 223
Deciding to Investigate 223
Designating Investigators 224
Gathering and Preserving Evidence 224
Event Critique 225
Analyzing the Evidence 227
Determining Causes 228
Correcting Causes 228
Documenting the Investigation 229
Disseminating the Lessons 229
Topic Summary and Questions to Consider 230

Chapter 21 — 231

Conducting Continuing Training

What Is It? 231
Airline Industry Approach 232
Application to Industry 233
Choosing Continuing Training Topics 234
Integrating Continuing Training 235
Impediments to Continuing Training 236
Topic Summary and Questions to Consider 237

Chapter 22 — 239

Evaluating Operating Performance

What Is Evaluation? 239
Why Evaluate? 240
Levels of Evaluation 243
What Should Be Evaluated? 246
Cultivating the Self-Assessment Habit 249
Evaluation Strategy and Logistics 249
Responding to Evaluation 251
Factoring Evaluation Results Into Training 252
Topic Review and Questions to Consider 252

Chapter 23 — 255

A Case Study in Implementation

The Situation 255
Background 255
The Problems 256
Your Task 257
What Should You Do? 257
The Obstacles 257
The Approach 258
Preparing to Take Over 259
Controlling Immediate Safety Risks 260
Establishing Leadership Confidence 260
Identifying the Problem 261
Evaluating the Problem 262
Establishing Written Guidance 263
Teaching the Systematic Approach 264
Guidelines for Conducting the Training 265

Implementing and Evaluating.. 266
How Long Will It Take?.. 267
Measuring Improvement... 268
Topic Summary and Questions to Consider................................. 269

Chapter 24 — 271
Your Challenge
It Won't Happen by Itself.. 271
It Won't Happen Unless You Show Them.................................... 271
It Won't Happen Overnight.. 271
It Won't Happen Without Persistence.. 271
It Won't Happen Unless You Get Started!................................... 272

Part V: Process Safety Supplement

Chapter 25 — 275
A Study in Process Failure
Phillips 66 Chemical Complex Explosion and Fire, October 1989........... 276
A Preventable Accident?... 284

Chapter 26 — 287
Process Safety Management Overview
Definition and Purpose.. 287
The Law and Process Safety.. 288

Chapter 27 — 291
Dissecting the Law (OSHA Regulation 1910.119)
Application.. 291
Employee Participation.. 292
Process Safety Information... 294
Process Hazard Analysis.. 299
Operating Procedures... 303
Training... 307
Contractors... 311
Pre-Startup Safety Review.. 314
Mechanical Integrity.. 316
Hot Work Permit... 320
Management of Change... 321
Incident Investigation.. 323
Emergency Planning and Response.. 326
Compliance Audits... 328
Trade Secrets... 332
Conclusion.. 332

Part VI: Glossary/Index

Glossary/Index — 333

List of Figures

Figure 1-1:	Hand Grenade Cross-Section.	6
Figure 1-2:	Nature of Complex Failure.	7
Figure 2-1:	The Exxon-Valdez. [From NTSB Report]	11
Figure 2-2:	Exxon-Valdez Accident Location. [From NTSB Report]	13
Figure 2-3:	Inside the VTC. [From NTSB Report]	14
Figure 2-4:	Exxon-Valdez Tank Damage. [From NTSB Report]	15
Figure 3-1:	Successful Operating Strategy Elements.	34
Figure 3-2:	Margins of Safety.	39
Figure 3-3:	Triangle of Safe Operation.	43
Figure 3-4:	Operations Molecule.	44
Figure 4-1:	Risk Management Priorities.	57
Figure 4-2:	Model Operating Envelope.	60
Figure 4-3:	Prince William Sound Operating Envelope.	61
Figure 4-4:	Vehicle Safe Following Distance.	63
Figure 4-5:	Steam Boiler Operating Map.	66
Figure 5-1:	Chernobyl Units 3 and 4 (Cross-Section). [From NUREG 1250, Rev.1]	70
Figure 5-2:	Simplified RBMK-1000 Schematic. [From NUREG 1250, Rev. 1]	71
Figure 6-1:	Big Bayou Canot Bridge. [From NTSB Report]	77
Figure 6-2:	Mobile River Chart. [From NTSB Report]	79
Figure 6-3:	MAUVILLA Tow Configuration. [From NTSB Report]	81
Figure 7-1:	MIC Tank E-610 Details (Bhopal).	96
Figure 7-2:	Simplified MIC Tank and Header Arrangement (Bhopal).	97
Figure 7-3:	Twelve Vital Skills.	102
Figure 8-1:	Utility Crew Accident; Street Light Pole Comparison.	108
Figure 8-2:	Utility Crew Accident Location; Detail Map.	109
Figure 8-3:	Utility Crew Accident; Worker Locations.	110
Figure 12-1:	RMS Titanic. [From 1912 news photo]	131
Figure 12-2:	Communications Cycle.	135
Figure 14-1:	Air Florida Flight 90 Impact Attitude. [From NTSB Report]	148
Figure 14-2:	Engine Sensor Locations. [From NTSB Report]	150
Figure 14-3:	Boeing 737-222 Center Panel Instruments.	151
Figure 15-4:	Air Florida Flight 90 Flight Path. [From NTSB Report]	153
Figure 15-1:	TMI-2 Simplified Schematic.	162
Figure 15-2:	Typical Pressurized Water Reactor (PWR).	163
Figure 17-1:	Typical Danger (red) and Caution (yellow) Tags.	189
Figure 19-1:	Embraer 120 Brasilia (EMB-120).	200
Figure 19-2:	EMB-120 Tail Section (as found). [NTSB Photo, edited]	200
Figure 19-3:	EMB-120 Tail Components (Detail).	202

Figure 19-4:	Typical Shift Turnover Sequence.	206
Figure 20-1:	Challenger Launch (STS 61-A 22).	218
Figure 20-2:	Challenger Booster-Rocket Field Joint (Detail). [From Congressional Report]	219
Figure 20-3:	O-Ring Incidents vs. Temperature °F (Graph).	220
Figure 20-4:	Critique Guidelines.	226
Figure 22-1:	Levels of Evaluation.	242
Figure 25-1:	Aftermath of the Phillips 66 Explosion and Fire. [From OSHA Report]	275
Figure 25-2:	Partial Equipment Location Plan. [From OSHA Report]	277
Figure 25-3:	Typical Settling Leg Arrangement. [From OSHA Report]	278
Figure 25-4:	DEMCO® Valve. [From OSHA Report]	279

Part I

The Nature of Industrial Failure

○ Anatomy of an Accident
○ Common Components of Accidents

You may be wondering why a text about successful industrial operations begins with a whole section devoted to *failure*.

First, a very basic objective of successful operations is the avoidance of failure, and (hopefully) we all have an ability to learn from mistakes—preferably *other* people's mistakes. Chapter 2 is especially important to this objective by identifying nineteen common factors that are present with uncanny regularity in industrial accidents.

But, more importantly, Part I is intended to demonstrate the need for a systematic strategy for operating success. And the best way to establish that need is to show what happens when either there is no such strategy, or when the operating strategy fails or falls into disuse.

Chapter 1

Anatomy of an Accident

Seldom does a single problem alone lead to industrial disaster. More often, calamitous failure results from a combination of minor problems, errors, and flawed operating habits which have developed gradually over time. No one component problem seems to be significant; yet, when all of the components merge, disaster results.

Only through personal experience and studying past mistakes does the importance of accumulated small problems become apparent. Mature leaders, operators, technicians, and engineers have all observed and experienced complex operating problem development and, consequently, realize that simple problems must not be allowed to grow.

This chapter illustrates how "minor" problems in combination can quickly grow to unmanageable proportions and that conservative operating habits are the best safeguard against complex problem development.

Crash of Eastern Airlines Flight 401, December 1972

On the night of December 29, 1972, a Lockheed L-1011 operated by Eastern Airlines crashed into the Florida Everglades at 11:42 P.M. Of 178 passengers and crew on board, 99 died.

Eastern 401 departed Kennedy Airport in New York at 9:20 P.M. on a routine passenger flight to Florida's Miami International Airport. During the landing approach sequence, the nose wheel indicating light failed to come on when the cockpit crew placed the landing gear in the gear-down position. However, indicators for both left and right main landing gear did illuminate. Unsure of whether the nose gear had malfunctioned or the indicating lights were burned out, the captain recycled the gear. Again, the nose wheel indicator failed to light.

At 11:34:05 P.M., Eastern 401 called the tower and stated, *"Ah, tower this is Eastern, ah, four zero-one, it looks like we're gonna have to circle, we don't have a light on our nose gear yet."* At 11:34:14, the Miami tower replied, *"Eastern four oh one heavy, roger, pull up, climb straight ahead to two thousand, go back to approach control, one twenty eight six."*

Eastern 401 acknowledged the control tower instructions and returned to the approach control radio frequency. Contacting approach control, the captain stated, *"All right, ah, approach control, Eastern four zero one, we're right over the airport here and climbing to two thousand feet, in fact, we've just reached two thousand feet and we've got to get a green light on our nose gear."* Miami approach control acknowledged Eastern 401's transmission, provided them with a new course, and instructed them to remain at 2000 feet while they attempted to determine whether the nose wheel was down and locked in position. Eastern 401 acknowledged and complied.

At 11:36:04, the captain instructed the first officer to engage the autopilot. The first officer, who was flying the aircraft at the time, complied and then removed the nose wheel indicating light lens assembly to examine its two bulbs. While replacing the assembly, he jammed it in a position 90 degrees clockwise from its intended seat.

At 11:37:08, with the lighting assembly temporarily disabled, the captain ordered the second officer to descend into the forward electronics compartment where he could observe the wheel well compartment through a viewing port. (The sighting device is in a pressure tight wall looking aft into the wheel well compartment. With the wheel well illuminated, an index rod visible through the sight will indicate whether the nose gear is up, partially down, or fully down and locked.) Preoccupied with the indicator light problem, the second officer ignored the captain's instructions. Meanwhile, the captain and first officer continued efforts to free the jammed lighting assembly.

At 11:37:48, approach control radioed Eastern 401 and directed a new heading. Eastern 401 acknowledged and complied.

At 11:38:34, the captain again directed the second officer to enter the forward electronics compartment to view the nose wheel index rod. This time, the second officer complied.

Eastern 401's flight data recorder shows that, up to this point, the aircraft responded as if in altitude control mode on the autopilot. At this time, however, a downward pitch and acceleration transient indicate that the status of the automatic pilot was changed. (The L-1011 has two autopilots. One for the captain's position and the other for the first officer's position. When sufficient pressure is applied to either control yoke while the automatic pilot is in the altitude control mode, the system reverts to control wheel steering mode, and the aircraft will respond to light pressure on the yoke.)

At 11:40:38, a warning alarm known as a *C-chord* sounded, indicating an altitude deviation of 250 feet below the selected altitude of 2000 feet. Neither the cockpit voice recorder nor the flight data recorder indicate any response from the cockpit crew.

At 11:41, the second officer re-entered the cockpit and stated, *"I can't see it, it's pitch dark and I throw the little light, I get, ah, nothing."* Unknown to any of the cockpit crew, the wheel well light switch was above the captain's head rather than near the viewing port, below. The switch that the second officer operated was probably the control to uncover the lens on the wheel well side of the viewing port. Conversation recorded in the cockpit indicates that the crew believed the wheel well light automatically illuminated whenever the nose wheel was operated. After a brief discussion of the function of the wheel well light, a maintenance specialist riding with the crew in the forward observer seat accompanied the second officer back into the forward electronics bay.

At about the same time, an air traffic controller at Miami approach control noted that Eastern 401's altitude indicated 900 feet on his Automated Radar Terminal Service (ARTS III) equipment. At 11:41:40, the controller radioed and asked, *"Eastern, ah, four oh one how are things comin' along out there?"* Eastern 401, having flown out of approach control's airspace, was now nearing the controller's boundary. Eastern 401 responded, *"Okay, we'd like to turn around and come, come back in."* The controller acknowledged the communication and granted the request, stating, *"Eastern four oh one turn left heading one eight zero."* Eastern 401 acknowledged and began to comply.

During the investigation, the controller testified that he believed Eastern 401 was not in any trouble at the time. He had received a positive response from the flight crew and had returned his attention to five other aircraft in his area of responsibility. The ARTS III equipment sometimes requires up to three scans for an accurate altitude indication, so the controller was not concerned about an indicated altitude of 900 feet.

Eastern 401's flight data recorder showed that the aircraft was descending at a rate of approximately 3000 feet per minute. At 11:42:05, the first officer observed the dramatic decrease in altitude and stated, *"We did something to the altitude."* The captain replied, *"What?"* The first officer responded, *"We're still at two thousand, right?"* The captain then said, *"Hey, what's happening here?"*

At 11:42:10, the cockpit voice recorder registered the radio altimeter alarm warnings and the first sounds of impact as the aircraft's left wing tip struck the ground. Eastern 401 had been in a 28 degree left bank, coming to the last heading directed by Miami approach control while the crew thought they were still at 2000 feet.

A Chain of Insignificant Events

The crash of Eastern Airlines Flight 401 is yet another classic example of industrial failure. No single circumstance in the chain of events caused the disaster. Rather, the accident resulted from a coupling of abnormal circumstances and flawed operator responses.

In their accident report issued June 14, 1973, the National Transportation Safety Board wrote:

> *The Board is aware of the distractions that can interrupt the routine of flight. Such distractions usually do not affect other flight requirements because of their short duration or their routine integration into the flying task. However, the following took place in this accident:*
> *1. The approach and landing routine was interrupted by an abnormal gear indication.*
> *2. The nose gear position light lens assembly was removed and incorrectly reinstalled.*
> *3. The first officer became preoccupied with his attempts to remove the jammed light assembly.*
> *4. The aircraft was flown to a safe altitude, and the autopilot was engaged to reduce workload, but positive delegation of aircraft control was not accomplished.*
> *5. The captain divided his attention between attempts to help the first officer and orders to other crew members to try other approaches to the problem.*
> *6. The flight crew devoted approximately 4 minutes to the distraction, with minimal regard for other flight requirements.*

Individually, each circumstance and operator response might appear inconsequential. Indeed, when such *minor* problems are separately identified as potentially dangerous, other leaders or team members often resist corrective action as inappropriate or unnecessary. But experienced leaders and operators know that, in combination, such "insignificant" defects can have disastrous consequences.

Unfortunately, such a *chain of insignificant events* is characteristic of most accident patterns. No individual problem seems to be of any consequence. As a result, individual elements are often overlooked as harmless.

But, experienced professionals understand the subtle nature of accident condition formation. Whether leaders or workers, they have learned that **accidents are almost always the cumulative result** of a chain of apparently minor deviations from known standards, practices, and procedures. Consequently, they pay special attention to maintaining disciplined operations. To an outsider, systematic communications, religious use of procedures, and controlled exchanges of responsibility might *appear* to be overkill; but, to the skilled professional, they are effective barriers against "little problems".

Complex Failure Formation

As the Eastern 401 accident illustrates, the combined effects of small problems often result in an event of unanticipated proportion. A sort of *reverse synergism* occurs in which the final effects of a disaster are far greater than the sum of the individual events that fabricated it.

Multiple Contributors As with most accidents, this one was characterized by multiple contributing factors. After investigating the crash, the National Transportation Safety Board wrote:

> *It is obvious that this accident, as well as others, was not the final consequence of a single error, but was the cumulative result of several minor deviations from normal operating procedures which triggered a sequence of events with disastrous results.*

This was an event with many components—a *complex* accident.

Accident Initiation When non-conservative factors and circumstances align, all it takes to activate an accident sequence is an *initiating event*. In accident studies, the term **initiating event** denotes an action which, by itself, has little or no consequence, but (in the context of accident conditions) starts the accident chain.

To illustrate, consider the Figure 1-1 cross-sectional view of a hand grenade. In this simple analogy, no single component can cause the damage that all of the components together can cause. Assembled in the proper configuration, all that is necessary is to pull the pin and the effect of the device can be calamitous.

The grenade example is startlingly similar to the pattern of most complex accidents. No single problem is likely to result in disaster. Yet, all of the problems in combination, set in motion by an initiating event, are catastrophic.

The Eastern 401 accident was just such a chain of *insignificant* events. By itself, the failed nose wheel indicator was an insignificant problem. Yet, combined with other non-conservative circumstances, a failed indicator "pulled the pin" on this accident sequence.

Anatomy of an Accident

Composite of Simple Problems

Accidents are usually complex in nature. *Complex industrial failure* may be simply defined as an undesirable industrial event which has more than one causal component. When analyzed, complex failure is usually found to be a composite of simple problems that have accumulated over a long period. (Figure 1-2 illustrates this concept.) As each new problem is added, a nucleus of non-conservative circumstances develops and accumulates to create a larger, more serious problem. Moreover, the simple problems become more difficult to separate and root out.

Often, each of these "little" problems is nothing more than a poor habit established through minor deviation from prescribed practices and procedures. When allowed to go uncorrected, they take root and grow. When a leader attempts to correct such a habit, his effort is frequently met with a moan, a rolling of the eyes, and the complaint, *"But, we've always done it that way before!"*

Grown Over Time

Poor operating habits usually develop over long periods as they are allowed to stray unchecked. The habits and attitudes which engender such actions evolve with certainty as standards degrade in countless—often undocumented—previous events. As a result, the growth of a complex problem is usually a very subtle process. Since each of the simple problems seems insignificant, operating personnel and their supporting teams are not alarmed when they occur—they become accepted.

Industrial operations, since they are usually complex in nature, are particularly susceptible to forging these kinds of multiple links in accident chains. Since production or service schedules are subject to time and resource constraints, the drive for task completion may blind team members and their leaders to the formation of these dangerous conditions.

Detection of developing accident conditions is complicated by the characteristic failure of humans to sense slowly changing conditions in the environment. In fact, we are like the proverbial frog placed in a pan of slowly heating water. As a result, accident components can form undetected by all but the most experienced and vigilant of managers and employees.

Finally, these component problems frequently become inextricably interwoven as they are allowed to coexist. They become difficult (or impossible) to separate. In fact, as they grow, they seem to increase one another's severity. A communication failure, for example, can seriously compound other accident conditions. In the Eastern 401 tragedy, the air traffic controller's poorly worded inquiry regarding Eastern 401's altitude served only to complicate the developing accident conditions.

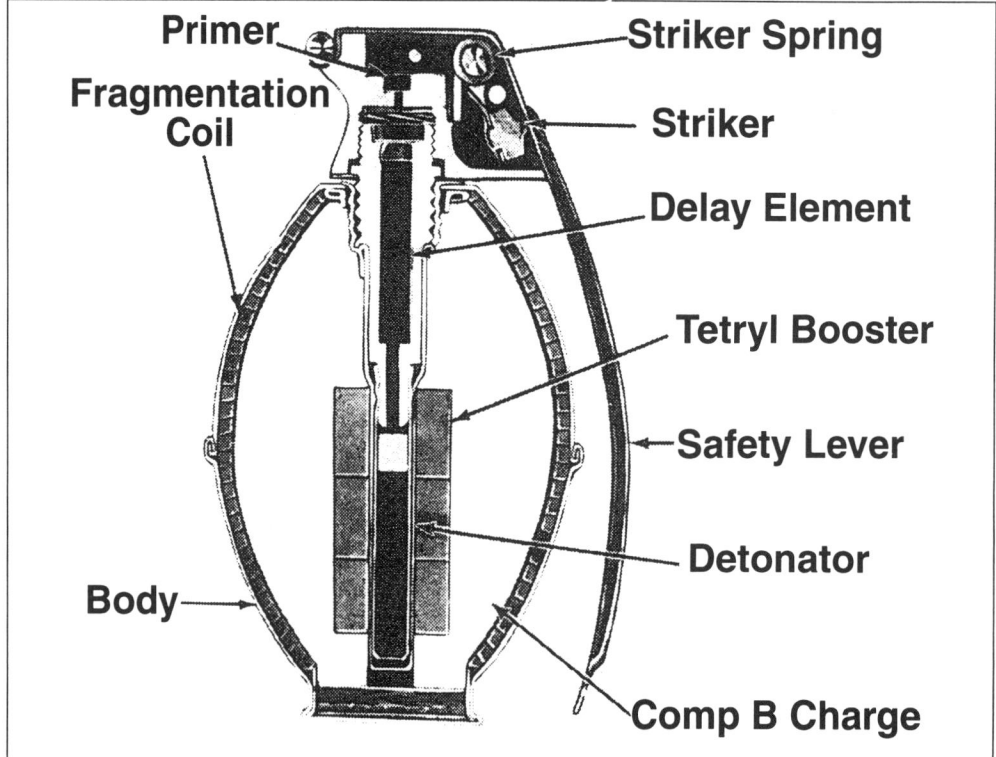

Figure 1-1: Cross section of a Hand Grenade.

Lessons of Eastern 401

During the Eastern 401 accident investigation, the National Transportation Safety Board determined that:

> *There was no failure or malfunction of the structure, powerplants, systems, or components of the aircraft before impact, except that both bulbs in the nose landing gear position indicating system were burned out.*

The aircraft had functioned as designed. The nose wheel was, in fact, down and locked. A prompt and properly conducted investigation—looking at the index rod—would have disclosed the condition of the nose wheel, and the aircraft would have landed safely. What, then, went wrong?

Like most accidents, the Eastern 401 misfortune was easily preventable. This was an accident purely attributable to human error. And, though a terrible tragedy, it holds for every industry a series of lessons which, if learned and applied, can save untold hours of anguish and many wasted dollars.

The Lesson of Training

With proper care and control, well-designed machines usually perform as intended. It follows that operators must thoroughly understand the functions, characteristics, and operating limits of their equipment if it is to be safely controlled.

In evaluating the Eastern 401 accident, the National Transportation Safety Board wrote:

> *The throttle reductions and control column force inputs which were made by the crew, and which caused the aircraft to descend, suggest that crew members were not aware of the low force gradient input required to effect a change in aircraft attitude while in CWS [control wheel steering mode]. The Board learned that this lack of knowledge about the capabilities of the new autopilot was not limited to the flight crew of Flight 401.... Although formal training provided adequate opportunity to become familiar with this new concept of aircraft control, operational experience with the autopilot was limited by company policy. Company operational procedures did not permit operation of the aircraft in CWS; they required all operations to be conducted in the command modes. This restriction might have compromised the ability of pilots to use and understand the unique CWS feature of the new autopilot.*

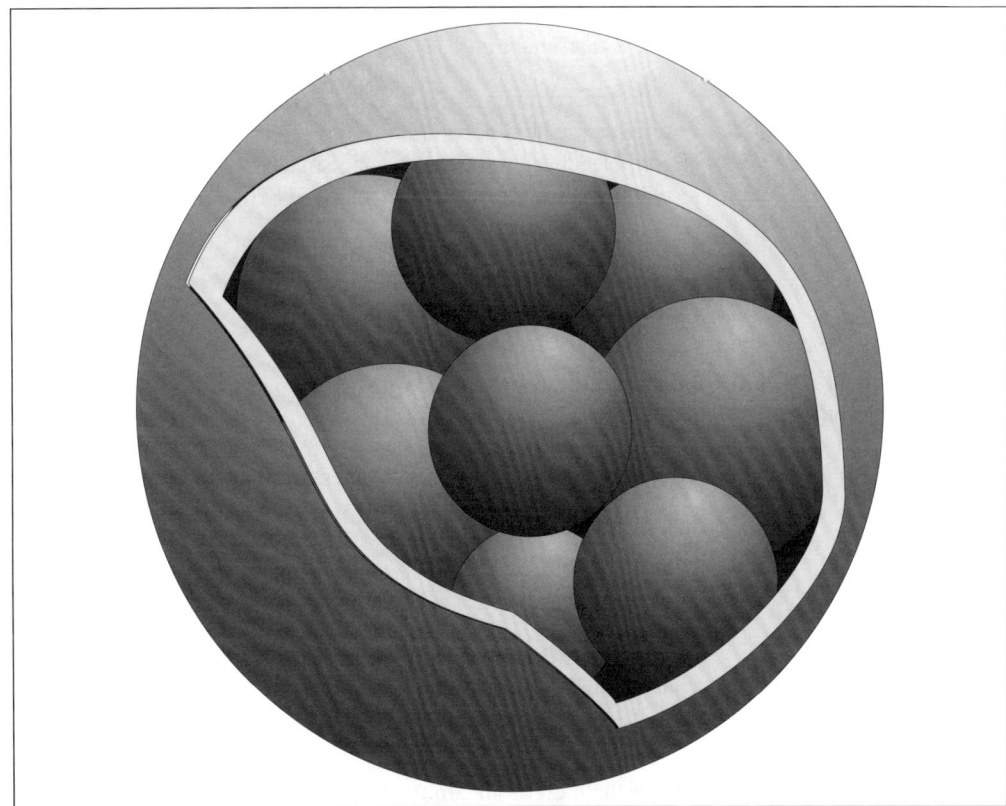

Figure 1-2: The Nature of Complex Failure—usually a composite of accumulated simple problems.

Anatomy of an Accident

The flight crew was also unaware of the means for illuminating the nose wheel compartment. The cockpit voice recorder transcripts reveal this foreboding series of communications between the maintenance specialist in the cockpit and the second officer:

 Maintenance Specialist: *Wheel well light's on?*

 Second Officer: *Pardon?*

 Maintenance Specialist: *Wheel well light's on?*

 Second Officer: *Yeah. Wheel well light's always on if the gear's down.*

Apparently, none of the flight crew knew about the light switch in the captain's overhead panel.

Talking about an equipment operating feature is seldom as valuable as hands-on experience with it. Though both equipment characteristics (the CWS autopilot mode and the nose wheel light) were part of the flight crew's training program, the effectiveness of the training was clearly deficient.

The Lesson of Communication

Confusion in high risk situations must be scrupulously avoided since the results of poor communication can be disastrous. Strict patterns of communication require identification of sender and receiver, transmission of clear messages, and responses which confirm or correct the original message.

Too often, however, experienced operators lapse into poor communication habits without realizing that these same poor habits may well become future components in accident chains. At 11:34:58 P.M., the first officer asked the captain, *"You want me to fly it, Bob?"* Rather than replying clearly to the first officer's important inquiry, the captain said, *"What frequency did he want us on, Bert?"* The first officer responded, *"One twenty-eight six."* The captain replied, *"I'll talk to 'em."* This opportunity to clearly designate who was to fly the aircraft was squandered through poor communication.

Similarly, when the air traffic controller noted an altitude of 900 feet for Eastern 401, his inquiry was ambiguous: *"Eastern, ah, four oh one, how are things comin' along out there?"* The National Transportation Safety Board, commented on this communication sequence:

> *The Board recognizes that the ARTS III system was not designed to provide terrain clearance information and that the FAA has no procedures which require the controller to provide such a service. However, it would appear that everyone in the overall aircraft control system has an inherent responsibility to alert others to apparent hazardous situations, even though it is not his primary duty to effect the corrective action.*

All effective communication is characterized by a few fundamental elements: a sender, a receiver, a message, a medium of communication, an environment which will support a successful communication, a feedback response mechanism, and as many confirmations or corrections as necessary to ensure that the message has been received as intended.

These communication steps may at first seem artificial and unwarranted. But, practicing and enforcing these disciplined, formal communication patterns is one of the most effective tools available for avoiding accident conditions.

The Lesson of Critical Parameter Awareness

Most machinery is equipped with critical parameter indications. A parameter is considered a *critical parameter* if an excursion outside its specified tolerances is likely to cause failure.

For safe operations, critical parameters must be frequently monitored. In this case, altitude was certainly one of the critical parameters. Yet, the National Transportation Safety Board found that *"The flight crew did not monitor the flight instruments during the final descent until seconds before impact."* In addition, the crew *"...did not hear the aural altitude alert which sounded as the aircraft descended through 1,750 feet m.s.l."*

Complacency is the enemy of awareness. Yet, awareness is not guaranteed by training alone. Only vigilant, professional attitudes on the part of team members can ensure such awareness. Those attitudes begin with the leader.

The Lesson of Independent Verification

Airline flight crews are among the best qualified and best trained personnel of any industry. Most have extensive prior experience in military or general aviation. Yet, anyone can make a mistake. That is precisely why independent verification is so important.

Independent verification is the process of separate observation to ensure that a high risk or critically important task is properly performed. It was absent, however, in the cockpit of Eastern 401. The National Transportation Safety Board wrote:

> *The captain failed to assure that a pilot was monitoring the progress of the aircraft at all times.... [G]ood pilot practices and company training dictate that one pilot will monitor the progress of the aircraft at all times and under all circumstances.*

Both the captain and the first officer should have frequently (and separately) verified autopilot status and altitude. Instead, they relinquished complete control to the autopilot, trusting it to perform the function of flying without monitoring its performance.

The Lesson of Leadership

Even when highly skilled people gather to accomplish a task, it is unlikely that the group will perform as a team without proper leadership. The NTSB concluded that:

> *...the probable cause of this accident was the failure of the flight crew to monitor the flight instruments during the final 4 minutes of flight, and to detect an unexpected descent soon enough to prevent impact with the ground. Preoccupation with a malfunction of the nose landing gear position indicating system distracted the crew's attention from the instruments and allowed the descent to go unnoticed.*

In essence, this cockpit crew forgot to fly the aircraft. They failed to function as a team.

Leadership is fundamental to successful teamwork. An effective leader is a catalyst for team development. The leader helps the team focus on a mission and set priorities. He or she defines and enforces the roles of the team members and the structure of the team. Most of all, the example set by the leader establishes the behavior standards for all the team members.

In the Eastern 401 accident, the captain became so preoccupied with the nose wheel indicator light that he failed to maintain full perspective of the problem. He was responsible to either fly the aircraft himself or delegate that responsibility to the first officer. Though team structure and team member responsibilities were well-known, the team failed, in great part, because the leader failed.

Leaders must continuously enforce the roles and the rules for the team to be successful. At 11:37:08 P.M., the captain of Eastern 401 directed the second officer to enter the forward electronics bay to check the nose wheel index:

> *Hey, hey, get down there and see if that (expletive deleted) nose wheel's down.*

The second officer failed to acknowledge the captain's directive and continued to examine the nose wheel indicating light. At 11:38:34 P.M., about a minute and a half later, the captain *again* ordered the second officer to check the index rod:

> *To (expletive deleted) with it, to (expletive deleted) with this. Go down and see if it's lined up with that red line. That's all we care. [Don't] (expletive deleted) around with that (expletive deleted) twenty cent piece of light equipment we got on this (expletive deleted).*

Finally, the second officer complied. But his delay, coupled with the clearly degraded atmosphere of professionalism in the cockpit, casts doubt on the teamwork of this cockpit crew and, necessarily, of their leader.

The importance of leaders in preventing complex problem formation must not be underrated. A team in name is not necessarily a team in deed. The difference is good leadership.

Topic Summary and Questions to Consider

The tragedy of Eastern Flight 401 reminds us that accidents simply must be prevented. The consequences of a failed mission are far too expensive and destructive to accept.

Simple, conservative design and well-executed construction are certainly fundamental to accident prevention. Ultimately, however, the responsibility for successful operation falls to skilled, alert humans systematically controlling the equipment that they are charged with operating.

Ask Yourself

Does your operating staff view accidents as "unavoidable tragedies"? If so, they probably aren't familiar with the mechanisms of accident formation—the anatomy of accidents.

Are your team members easily distracted by low-priority problems? Operating priorities must be ingrained through constant education and practice. Otherwise, when a light bulb burns out, your team may forget who is supposed to fly the plane.

Do you engage your team members in studying accidents and close calls—both their own and those of others—with the objective of team improvement? If not, you're missing an important opportunity to educate your team in the basics of accident prevention.

The overriding principle of the late Admiral Hyman Rickover's extraordinarily successful Naval Nuclear Propulsion Program is that **there is no substitute for the alert, well-trained operator controlling equipment and processes within specified parameter bands in accordance with approved procedures**.

Chapter 2

Common Components of Accidents

The components of an accident sequence are usually very apparent *after* the accident has occurred. In the Eastern 401 accident, poor communication, deficient leadership, and failure to monitor critical parameters combined to a calamitous end. Unfortunately, the mistakes in judgment that are so clear to the critic on Monday are often indistinct and difficult decisions that have to be made by the professional on Sunday.

Anyone can be a critic. Amateur criticism requires little competence. Lamentably, the ranks of industry, government, and the news media are filled with "experts" whose livelihoods depend upon criticizing the mistakes of others after the event.

The time to identify accident components is *before* calamity. Therefore, to truly master a profession, one must acquire the ability to *recognize* evolving accident conditions and act to interrupt them.

On first thought, this may seem a daunting task. But, accidents usually share several common features. In fact, the most common accident components are readily apparent after even cursory studies of industrial mishaps. One need not be a skilled industrialist to identify them. Common factors and accident patterns appear, regardless of the technology, with such uncanny regularity that they defy classification as coincidence.

This chapter introduces and discusses nineteen of the most common accident contributors. Through proper training, industrial team members can become adept at recognizing and interdicting them—*forestalling* the development of accident conditions.

Figure 2-1: The Exxon-Valdez. [From NTSB Report]

Grounding of the Exxon-Valdez, March 1989

Not all accidents result in injury or death. Some are just very expensive. Just after midnight on March 24, 1989, the supertanker *Exxon-Valdez* (Figure 2-1) ran aground on Bligh Reef, spilling 258,000 barrels of oil into the pristine waters of Prince William Sound. The tankship had departed the normal traffic lanes after passing through the Valdez Narrows. Because of heavy ice in the water, the master chose to navigate his Alaska-class tanker through a one mile stretch of open water to the north and northwest of the reef. At the time of the grounding, the vessel's third mate (not certified for navigation in the sound) was at the helm. Despite Exxon Shipping Company policies and federal law, the ship's master elected to delegate control of the bridge to the third mate and return to his quarters. Within 20 minutes of his departure from the bridge, the vessel was aground. The *Exxon-Valdez* sustained damage to 8 of her 11 cargo tanks, allowing over ten million gallons of her 50 million gallon load to leak. Within a few days, the spill had spread to the shores of the sound, complicating a mammoth and sometimes clumsy cleanup effort that would continue for months.

The costs would be great. Loss of cargo cost $3.4 million. Repair costs to the $125 million tankship were $25 million. Though environmental damage is still being assessed, spill cleanup costs in 1989 alone were $1.85 billion dollars. And in early 1992, a federal judge ordered Exxon to set aside $1.02 billion for civil and criminal litigation payments. Experts estimate that the total cost of the accident will eventually be in the range of $15 billion to $50 billion.

Tankship Description

The *Exxon-Valdez* (now *SeaRiver-Mediterranean*) was designed to transport oil from the oil pipeline terminus at the Port of Valdez to Panama where the oil is transferred to waiting tankships for transport up the eastern seaboard of the United States. It is a typical modern tankship of all-welded steel construction, utilizing a single one-inch thick hull. The vessel is 987 feet long, 166 feet wide, and 88 feet from deck to keel. (For comparison, it is longer than three football fields and wider than one.) Cargo capacity is 1.48 million barrels. When fully loaded, she drafts 64.5 feet.

The tankship is powered by a 31,650 horsepower, slow speed diesel engine directly coupled via the main shaft to a single-blade propeller. The diesel turns at a maximum 79 RPM giving a sea speed of over 16 knots. The engine can be reversed if necessary to back the ship, but it relies primarily upon tugs for maneuvering in port.

The tankship is equipped for navigation with a Sperry-Rand autopilot. When activated, the autopilot will lock in to either the current heading or to a computer controlled course, depending on the selected mode. For speed control, the ship is also equipped with a *loadup program* feature to sequentially load the driving shaft of the single-blade propeller. Over-torquing and shearing of the shaft is possible if load is not correctly controlled.

Vessel Crew

The Coast Guard certificate of manning required a minimum crew complement of 15 for the *Exxon-Valdez*. In charge of the vessel was a master or ship's captain. Working for the master in the operating crew were three officers: a chief mate, a second mate, and a third mate. In the maintenance crew were four more ship's officers: a chief engineer, a first assistant engineer, a second assistant engineer, and a third assistant engineer. A ninth officer, the radio-electronics officer, rounded out the officer staff. Also within the operations group, the crew was required to employ a minimum of three able-bodied seamen (ABs) whose duties ranged from lookout to helmsman. For maintenance, a minimum of three technicians—qualified members of the engineering department or QMEDs—were required for servicing the ship's equipment under the direction of the engineering staff.

Traffic Control

To control tanker traffic entering Prince William Sound and the Valdez Arm, the Coast Guard established the Vessel Traffic Service (VTS) in Port Valdez in 1977.

Tankship traffic in the sound posed new hazards to shipping and the environment. Prince William Sound is a haven for arctic wildlife and a major source of fishing and tourism income. A major oil spill in the region would thus have devastating effects. But hazards analysis for safe operation in the terminus and tanker navigation in the sound predicted that an oil spill in the range of several hundred thousand gallons would be likely only once in 241 years.

To protect against such a contingency, strict traffic controls were established in the Valdez Arm and Prince William Sound. The controls included a traffic separation scheme (TSS) consisting of 1,000 meter-wide inbound and outbound traffic lanes (Figure 2-2) for shipping operations. The lanes were divided by a 1,500 meter separation zone to prevent the possibility of collision. Regulations governing tankers prohibited deviation from the scheme without permission from the Officer of the Deck (OOD) in charge of the Coast Guard's Vessel Traffic Center (VTC) which controlled tanker traffic in the Sound. The OOD was further required to clear any such deviation with the Coast Guard's Commanding Officer or Executive Officer of the Marine Safety Office in the Port of Valdez.

Other safeguards included requirements for a licensed harbor pilot to control navigation from the time a tanker entered Prince William Sound, at Cape Hinchinbrook, until it re-entered open waters of the Gulf of Alaska. Tanker speed through the Valdez Narrows was limited to 6 knots. Special care was required to navigate around Middle Rock, an underwater rock outcropping in the middle of the narrows. Tanker traffic in the narrows was also limited to one ship at a time.

Radar and radio nets operated by the VTC also facilitated traffic control in the Port of Valdez area. Radio communication equipment, an important element of navigational control for the Valdez VTC, was specified for a 99.9 percent reliability level. However, due to inadequate maintenance, radio reliability had degraded to about 75 percent by the night of the accident.

The radar net, as originally conceived, was to blanket all of Prince William Sound. As built, however, the net covered only the vicinity of the Port of Valdez. Nevertheless, it had sufficient range to track vessels beyond Bligh Reef, through the Valdez Arm and Narrows, and into the Port of Valdez.

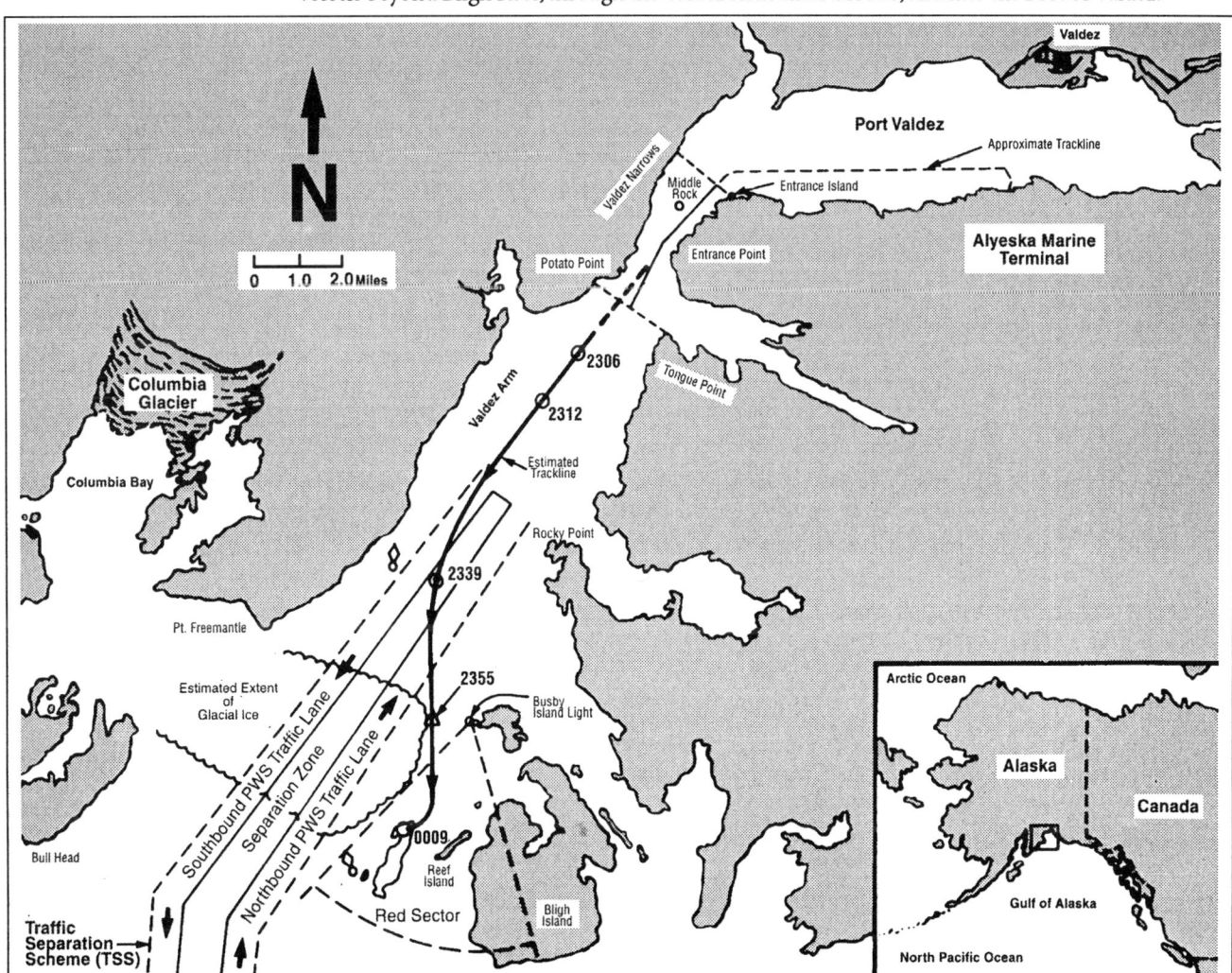

Figure 2-2: Traffic Separation Scheme showing Exxon-Valdez trackline and accident location. [From NTSB Report]

Common Components of Accidents

Each vessel, of course, was also equipped with radar and radio equipment for safe navigation, and navigation hazards were clearly marked in the Valdez Arm. Bligh Reef is indicated by a buoy marker and the danger area to the north of the reef is designated by the red sector of Busby Island Light.

Since opening the terminus to tanker traffic, over 8,800 safe tanker passages had been made through Prince William Sound. Fears of tanker grounding or collision seemed to be unfounded. The strict traffic controls, though still regulatory requirements, began to be relaxed in practice.

Staff Reductions Originally, VTC watchstanders monitored tanker traffic in the sound by plotting the location and course on navigation charts. But, as automated tracking equipment was incorporated into the VTC's control apparatus, the plotting process was abandoned. Watchstanders no longer monitored tanker traffic with the originally intended rigor. (Figure 2-3 shows the VTC.)

Staff reductions in the Port of Valdez Marine Safety Office also contributed to relaxation of standards. The VTC was originally staffed continuously with a supervisor, a certified radar specialist, and a communications specialist. The watch supervisor, a certified Coast Guard Deck Watch Officer, was charged with overseeing VTC operations and ensuring safe and professional operation of the station in accordance with the governing policies and regulations. This officer served, while on duty, as the OOD of the station and was required to be certified on the radar equipment.

Forced to accommodate staffing reductions, however, Coast Guard officials elected to eliminate the requirement for an officer to be in the VTC at all times. The position of OOD was relegated to an administrative post even though VTC requirements mandated that the OOD oversee safe operation of the traffic service. In fact, the OOD position was regularly filled by enlisted personnel who were not certified in VTC operations. The OOD was eventually only required to check the VTC periodically during each shift.

Ice in the Lanes As the original navigation safeguards deteriorated, hazards to navigation escalated. As early as 1975, ice in Valdez Arm had been identified as a navigational obstacle for tanker traffic. At times, ice calving off the Columbia Glacier would drift across the traffic lanes, requiring ships to slow down and pick their routes through the ice. By 1979, the retreat of Columbia Glacier had intensified the problem. Though most large icebergs were trapped at the terminal moraine of the glacier, smaller ice formations—growlers, bergy bits, and brash—often obstructed the traffic lanes. Though the ice problem was discussed by the Coast Guard and member companies of the Alyeska pipeline consortium, no distinct actions were taken to address the problem.

With deteriorated control in the VTC, tanker captains began to routinely deviate from the traffic lanes. By 1986, it had become common practice for captains, after checking with the VTC to determine if there was any traffic in the opposing lane, to swing out of their lane toward Bligh Reef where a stretch of open water often existed. Deviation from the lanes eliminated the need to slow down and navigate through the ice; but it also placed ships precariously close to the reef.

The practice of leaving the traffic lanes became even more common after pilotage requirements in the sound were compromised when the harbor pilot station, a vessel called the *Blue Moon* foundered. Rather than maintaining a pilot station at Cape Hinchinbrook, a decision was made to certify tanker captains and chief mates as pilots for navigation in the sound. The state harbor pilot station

Figure 2-3: Inside the Vessel Traffic Center (VTC). [From NTSB Report]

was moved back to a point northeast of Bligh Reef known as Rocky Point. When tankers reached Rocky Point, a state harbor pilot would board the tanker to guide it through the narrows and into the terminus. Federal pilotage requirements were met so long as a pilot-certified master or chief mate navigated the vessel while in Prince William Sound beyond Rocky Point.

Accident Sequence It was in this setting that the *Exxon-Valdez* entered the Port of Valdez just before midnight on March 22, 1989. Within fifteen minutes of docking, hookups had been made to off-load ballast, a process which began shortly before 1:00 A.M. Ballast transfer was complete by 4:15 A.M. and cargo loading began shortly thereafter.

Around 10:30 A.M., with cargo loading progressing well, the master of the *Exxon-Valdez*, his chief engineer, and the radio-electronics officer left the vessel for the town of Valdez. The master was scheduled to attend a short meeting with the Exxon Shipping Company representative and the trio intended to spend the remainder of the day shopping and attending to errands. Meanwhile, the crew continued cargo loading under direction of the chief mate into late evening. Most had been awake for over 20 hours.

Cargo loading was complete by 7:24 P.M. and the state harbor pilot came aboard at 8:20 P.M. in preparation for vessel departure. He learned that the master had not yet returned; so he went to the navigation bridge and began the obligatory navigation equipment checks. The master, chief engineer, and radio-electronics officer returned to the vessel at 8:30 P.M. The state pilot later testified that he smelled alcohol on the captain's breath.

The *Exxon-Valdez* cast off under navigational control of the pilot at 9:12 P.M. and was clear of its berth by 9:21 P.M. At 9:25 P.M., the third mate relieved the chief mate as watch officer on the bridge. The chief mate, after long hours without sleep, went to his quarters. At 9:40 P.M., the master left the bridge. By 11:24 P.M., the state harbor pilot had navigated the *Exxon-Valdez* out of port, through the narrows, and into the outbound lane of the TSS. At that time, the master returned to the bridge and accepted control of vessel navigation from the state harbor pilot. His job complete, the pilot disembarked to the pilot's vessel alongside the *Exxon-Valdez*.

Taking control of the ship, the master reported to the VTC that he was increasing sea speed to 16 knots. Seven minutes later, after confirming with the VTC that there was no conflicting on-coming traffic, he reported to the VTC that he was changing course to 200° to steer around ice in the channel and was reducing speed to 12 knots.

The third mate returned to the bridge at 11:36 P.M. The master then ordered another course change to 180° to facilitate navigation around the ice, and placed the vessel on autopilot. At 11:50 P.M., a watch relief occurred between the off-going and on-coming helmsmen. At 11:52, the master placed

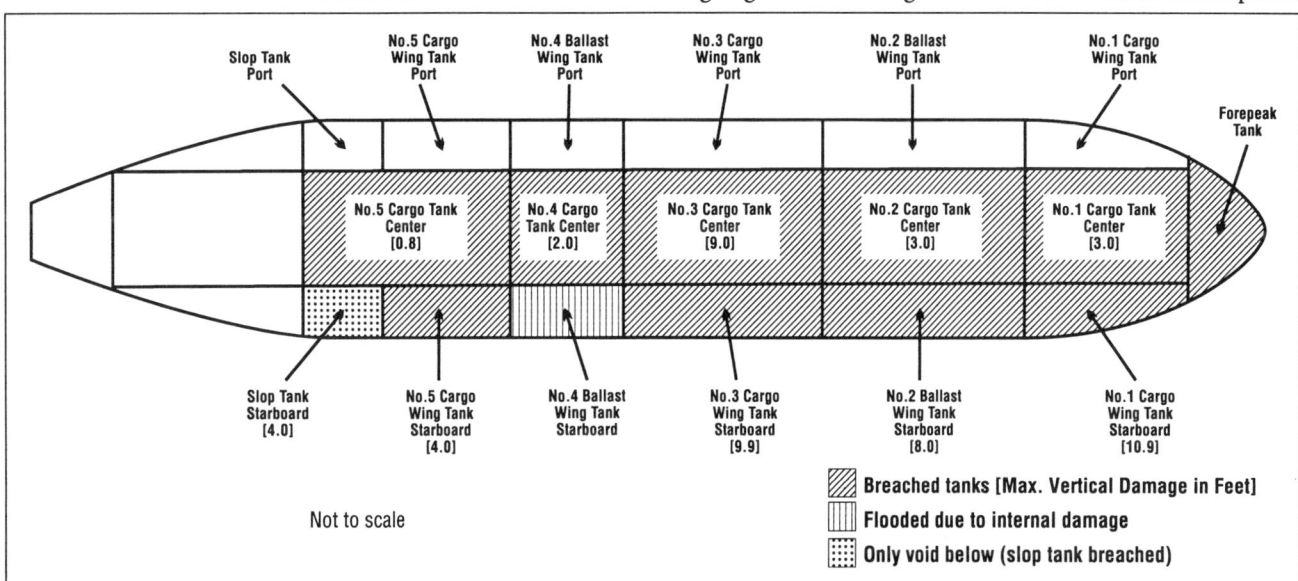

Figure 2-4: Exxon-Valdez tank damage. [From NTSB Report]

Common Components of Accidents

the main engine on loadup program and, violating both federal law and Exxon policy, turned navigational control over to the third mate. For navigation in confined waters, federal regulations and Exxon policy require three people to be on the bridge—a federally licensed pilot in charge of vessel navigation (in this case, the master); a watch officer to chart course and location, control the bridge, and assist the pilot as needed; and a helmsman to steer the ship.

Having turned control over to the third mate, the master gave instructions to return to the traffic lanes when he came abeam of Busby Island Light and then to call him back to the bridge. The master asked the third mate if he felt "comfortable" with what he was about to do and left the bridge.

The third mate testified that use of the autopilot and loadup program were unusual equipment conditions for steering in confined waters and that he was surprised to see such an equipment configuration. Monitoring course and location with radar, the third mate stated that he took the vessel off autopilot and noted that they reached the location specified by the master at 11:55 P.M., just three minutes after the master left the bridge. The third mate testified that, in accordance with his instructions, he ordered the helmsman to apply rudder for a 10 degree right turn. He then went back to monitoring course and position. Yet, neither the third mate nor the helmsman remembered monitoring rudder position to verify response to wheel command.

A little over six minutes after ordering the 10 degree right turn, the third mate noted with alarm that the vessel was not turning. To compensate, he ordered a 20 degree right turn. Further monitoring indicated that the ship was nearing Bligh Reef. The third mate directed the helm to hard right and called the master, stating, *"I think we're in serious trouble."* Within seconds, the *Exxon-Valdez* was aground on Bligh Reef and leaking cargo. (Figure 2-4 shows the extent of damage.)

A little over 20 minutes later, the master reported the event to the Coast Guard VTC. Adjusting the radar unit pulse frequency on the master radar unit, the VTC radar operator located the *Exxon-Valdez* as a stationary target aground on the reef. The radar pulse frequency had been improperly selected by the previous shift's radar operator, allowing the *Exxon-Valdez* to "fade" on the radar screen as it departed the TSS.

Nineteen Common Accident Components

The *Exxon-Valdez* mishap is a distinct example of a very costly, easily preventable accident. It clearly demonstrates how deterioration of organizational structure, degradation of leadership, and degeneration of training, maintenance, and performance standards can all combine to create adversity.

After an extensive investigation of the accident, the National Transportation Safety Board determined that the grounding of the *Exxon-Valdez* was attributable to five fundamental causes:

> *The National Transportation Safety Board determines that the probable cause of the grounding of the Exxon-Valdez was the failure of the third mate to properly maneuver the vessel because of fatigue and excessive workload; the failure of the master to provide a proper navigation watch because of impairment from alcohol; the failure of Exxon Shipping Company to provide a fit master and a rested and sufficient crew for the Exxon-Valdez; the lack of an effective Vessel Traffic Service because of inadequate equipment and manning levels, inadequate personnel training, and deficient management oversight; and the lack of effective pilotage services.*

Clearly, the ship's master bears great responsibility for this event; but, almost every element of the team failed. Factors contributing to the grounding included:

- Unfitness for duty of most team members and leaders aboard the *Exxon-Valdez*
- Inadequate staffing in both the VTC and aboard the *Exxon-Valdez*
- Inadequately trained watchstanders in the VTC
- Failure of both the VTC watchstanders and the crew of the *Exxon-Valdez* to follow established regulations and procedures
- Failure of team members within the VTC and aboard the *Exxon-Valdez* to properly monitor critical equipment and system parameters
- Deficient management oversight in the VTC, aboard the *Exxon-Valdez*, and within the Exxon Shipping Company
- Poor maintenance of equipment vital to sustaining safe navigation within Prince William Sound
- Failure to detect and investigate the warning signs of impending disaster

Yet, it is unlikely that any single contributor could have caused the grounding. As for most accidents, a series of unfavorable events and non-conservative actions aligned to create the event. This was an accident that had been forming for years. Many contributing factors were the result of "minor" procedural and performance deficiencies which were allowed to grow untended and unchecked. Yet, it is just such a chain of insignificant events that leads to development of accident conditions.

Perhaps no event in recent history better illustrates the characteristic alignment of common accident components better than the *Exxon-Valdez* grounding. This mishap illustrates nineteen of the most common contributors. They are:

1. Defective night operations
2. Unfitness for duty
3. Incomplete communication
4. Inadequate turnovers
5. Ignorance of equipment operating characteristics
6. Non-compliance with procedural requirements
7. Failure to monitor equipment status
8. Deficient equipment maintenance
9. Failure to independently verify critical tasks
10. Sense of invulnerability
11. Acceptance of abnormalities
12. Ignorance of warning signs
13. Capitulation to production pressure
14. Inadequate emergency preparedness
15. Deficient organizational staffing and structure
16. Poor leadership
17. Insufficient management oversight
18. Faulty teamwork
19. Degradation of operating limits

Once you and your team members become acquainted with them, you will be well-equipped to foresee and forestall industrial accidents. Let's examine each of the nineteen with attention to the questions, *How do they form?* and *Why do they create dangerous conditions?*

Defective Night Operations

An eery similarity among accidents is the disproportionate number that occur during late night hours. The *Exxon-Valdez* mishap was not an aberration. The *RMS Titanic* struck the fatal iceberg at 11:40 P.M., finally sinking at 2:18 A.M. on April 15, 1912. The Three-Mile Island reactor core meltdown began with events occurring near 4:00 A.M. on March 28, 1979. The fatal gaseous release at the Union Carbide pesticide plant in Bhopal resulted from actions starting around 9:30 P.M. on December 2, 1984, and escalating until about 1:30 A.M. Though the Space Shuttle *Challenger* accident occurred the morning of January 28, 1986, the critical decision to launch was made near 11:00 P.M. the night before. Finally, the explosion and meltdown of the reactor core at the Chernobyl Atomic Power Station occurred at 1:23 A.M. on April 26, 1986, after a buildup of conditions for hours.

Why are night operations so perilous? Midnight operations are not intrinsically harmful; but, the accompanying conditions can lead to added risk. Fatigue, low light, marginal experience levels, skeletal supervision, and complacency combine to degrade safety margins during night operations.

Fatigue is first among contributors to inadequate fitness for duty. It hinders awareness, slows response, and clouds the rational thought process. Since most people seem to be limited in their ability to adapt to nighttime hours, especially with a rotating shift schedule, fatigue takes its worst toll during late night shifts.

Fatigue is frequently accompanied by low light conditions, especially for outside tasks. When lighting conditions are poor, dangers that might be detected in adequate light go unnoticed. Though the *Titanic* sailed in calm, clear conditions, the lack of moonlight (and the lack of binoculars in the crow's nest) prevented detection of the iceberg until it was less than a mile away.

A third impediment to efficient performance in late night operations is the tendency to place the least experienced employees on the midnight shift because of low seniority. Yet, low experience level is not compatible with effective response to abnormal conditions. (Though shift rotation usually compensates for this problem, it exacerbates the issue of fatigue.)

Common Components of Accidents

Backshift operations also seem to foster complacency. Combined with fatigue, an attitude of "just making it until morning" may develop. Alertness wanes and people cease to think ahead. Without excellent, experienced leaders, such an attitude can quickly become prevalent on an operating or maintenance crew.

One key cause of complacency is an absence of experienced management oversight. Night operations often employ skeletal crews. Few senior leaders are present. With no one looking, the temptation to cut corners can grow. Further, without mature, experienced leaders on hand, potentially dangerous conditions may not even be recognized.

Finally, late night operations are dangerous because little help is initially available for combatting crises. Since the first actions taken are usually of greatest importance, delayed and incorrect initial responses often allow a situation to become uncontrollable.

Again, late night operation is not of itself an accident contributor, but late night *conditions* are. When failure occurs late at night, it usually results from deficient staffing, improper organization, non-conservative operation, marginal knowledge, or fatigue.

Unfitness for Duty

Fitness for duty is a term used to describe the physical and mental readiness of personnel to accomplish their assigned tasks. The best equipment and procedures cannot compensate for human controllers who are not fit for duty.

The NTSB investigation of the *Exxon-Valdez* grounding determined that both the master and third mate were probably physically unfit for duty at the time of the accident—the former because of alcohol consumption and the latter because of lack of rest.

Samples from the master taken ten hours after the accident revealed blood and urine alcohol levels of .06 and .09 percent, indicating a blood alcohol level of about .20 percent at the time of grounding. Federal regulations for deck officers on duty require blood alcohol levels to be below .04 percent, and further prohibit members of the navigation watch from consuming alcohol within four hours of duty commencement. The master testified that he had several drinks during that day in Valdez, his last about 7:00 P.M. Five people (including the harbor pilot, a security guard, and the Exxon representative) testified to smelling alcohol on the master's breath as he boarded the vessel. Though he stated that his elevated blood alcohol level resulted from drinking after the accident, recordings of his radio communications indicate otherwise.

Yet, one need not be intoxicated to be unfit for duty. Fatigue, mental distractions, adverse attitude, and inadequate training are other important factors in duty readiness.

Federal regulations governing rest requirements aboard vessels require all officers standing deck watch when a vessel departs port to have had 6 of the last 12 hours off. Regulations further state that other licensed individuals and seamen must not be required to work more than eight hours in one day. Yet, most of the crew had been awake since well before midnight on March 22 and had worked long hours to complete cargo loading operations. It appears that few of the *Exxon-Valdez* staff were in compliance with the law.

Fitness for duty is both a skill and a responsibility. Proper sleep, eating, and exercise habits are all necessary to achieve optimum individual and team condition. From that perspective, it should be taught as a skill. But, fitness for duty also has a component of professional desire. Team members must *want* to be fit for duty. That attitude can be imparted only through the example of the team leaders and through intolerance of unfitness on the job by both leaders and coworkers. If it is not perceived as a responsibility of position, it may be ignored by some.

Incomplete Communication

In July, 1945, the heavy cruiser, USS Indianapolis, sank with its crew of 1,196 after being torpedoed by a Japanese submarine. Eight hundred sailors survived the initial sinking; but, because of poor communication regarding scheduled arrival in the port of Leyte, a search and rescue mission was not launched until days later. The crew remained in the shark-infested waters for five days. Only 316 survived.

Communication failure is probably the most often identified accident contributor. Communication breakdowns catalyze accident development through unwarranted assumptions or misunderstandings. Conversely, disciplined communication habits frequently interdict the formation of accident conditions.

The *Exxon-Valdez* accident is no exception. In Prince William Sound, one of the most important methods of monitoring ice conditions was through ice reports from incoming and outgoing vessels. The VTC was required to pass ice information to tankers participating in the Vessel Traffic System. Unfortunately, vessels had no requirement to transmit ice reports to the VTC. The NTSB concluded:

> *Ice reports issued by the Vessel Traffic Center frequently are neither sufficiently timely nor sufficiently accurate to enable masters to ascertain before leaving Port Valdez the ice conditions that will be encountered in Valdez Arm. The policy adopted by the Coast Guard about 1985 to discontinue independent collection of ice information and statistics about vessel deviations from the traffic separation scheme probably contributed to the commanding officer and the operations officer not knowing that ice was causing vessels to depart from the traffic lanes and pass close to Bligh Reef.*

Many forms of communication are essential to effective industrial operations. Verbal communication can be divided into categories of face-to-face, telephone, public address system, and radio communications. Written communication includes log writing, recording machinery data, writing instructions to crews, and creating emergency change notices for equipment operation. Emergency communication usually takes the form of audible warnings, public address announcements, or computer generated messages. Other communication formats include status boards, message boards, procedures, and policy manuals.

Because communication is such a routine part of daily operations, team members or leaders may incorrectly assume that a message has been successfully communicated. Some of the most common errors include failure to identify sender or receiver, complication of a message by too much content, ambiguous message content, failure to provide feedback when the message is received, distortion of the message through a noisy or distractive environment, or even failure to send the message.

Successful communication depends not only upon the message, but upon the method by which the message is delivered (and received). The relationship of the sender and receiver, the environment in which the message is sent, and the viability of the feedback process all bear upon the success of the communication.

Inadequate Turnover

One form of communication is so important to the success of the industrial team that it requires special discussion. *Turnover*, or shift change as it is sometimes called, refers to the exchange of information between offgoing and oncoming work units.

The concept of turnover is familiar to anyone who has worked in a team environment. Track and field relay teams expend great effort in perfecting the baton pass. They have learned that the most likely source of failure in the relay is a poor handoff.

In an industrial environment, successful turnover is no less a coordinated effort. It depends as much on the offgoing unit as the one arriving. As such, it requires practice.

Turnover can occur between operating units, maintenance units, design teams, or construction teams. It is as relevant on a friendly sporting field or a hostile battlefield as it is in the industrial environment. But, regardless of setting, the elements of all turnovers are similar. Information critical to successfully continuing a process must be passed from one group to another in a fashion that minimizes or eliminates interruption of work.

Poor operational turnovers probably account for more accidents in industrial facilities employing multiple shifts than any other factor. On September 11, 1991, a Continental Express Embraer 120 crashed when the tail section separated from the aircraft. In Chapter 19, we will see that this crash was caused by 43 missing screws on the stabilizer leading edge—the result of failed turnovers.

Policies and practices for turnovers must be clearly delineated and rigorously enforced to avoid painful and costly errors. The "baton pass" must be taught, practiced, and monitored to maintain high standards. The decay of that process is a harbinger of impending failure.

Ignorance of Equipment Operating Characteristics

Another common contributor to accidents is inadequate knowledge of equipment operating characteristics. In the *Exxon-Valdez* accident, both the third mate and the helmsman, according to previous performance evaluations, were not experienced in ship handling. This may have led to improper steering actions.

The VTC radar operators' lack of knowledge of their equipment allowed the *Exxon-Valdez* to proceed unmonitored for a substantial time. The offgoing radarman had assumed that the radar unit was malfunctioning. He was unaware that the range scale selector for the slave radar unit he was monitoring did not change the sending unit radar pulse frequency. He was accustomed to monitoring the master unit which, when a different range scale was selected, would change radar pulse frequency as well as match tune the receiving unit. A failed circuit card in the master unit designed to overlay the TSS on the screen led the radar technicians to begin monitoring the slave unit instead. The master unit had not yet been repaired. The oncoming radar operator had failed to recognize the mismatched condition upon assuming responsibility for the station.

The causes of unfamiliarity with equipment are myriad. The first, of course, is inadequate training. Training can be deficient because of inadequate content, poor delivery, or failure to reinforce content. Proper training is expensive in time and financial resources. But improper training typically ends in crises far *more* expensive.

Related to deficient training is a failure to establish initial personnel selection criteria high enough. *Selection criteria* are the minimum knowledge, skill, experience, and character standards considered necessary for a prospective employee to enter an initial training program. These criteria represent the base line or starting point for training. A potential employee possessing only minimum criteria is likely to take longer to train and will probably require more supervision.

Even if selection criteria are appropriate and initial training is well done, operators stay proficient only in those tasks that they perform often. To keep operating skills and knowledge sharp, it is necessary to continue training. The purpose of continuing training is to review and reinforce old skills and knowledge as well as broaden the capabilities of operators to manage unusual operating conditions. Unfortunately, few industries conduct continuing training programs for their teams.

Finally, organizations often try to compensate for poor baseline skills and inadequate training with stricter and more detailed procedures. But, no amount of procedure detail can compensate for low selection criteria and poor training. Without clearly understanding the underlying design, construction, and principles of equipment function, operators can't consistently abide by safety constraints.

Non-Compliance with Procedural Requirements

In almost every industrial accident, analysis reveals that procedure violations played a significant role. Both the *Exxon-Valdez* crew and the Coast Guard ignored longstanding regulations governing navigation in Prince William Sound. The master of the *Exxon-Valdez* ignored Exxon policy regarding use and possession of alcohol on company property. Finally, federal regulations regarding blood alcohol levels and rest for the crew were blatantly disregarded.

As technologies are devised and refined, rules of operation develop, based on original design studies and experiences derived from the equipment's testing and restricted operations phases. The resulting procedures are usually comprehensive and, if properly followed, result in safe equipment operation. But, over time, compliance with procedure often deteriorates.

There are various "reasons" for such non-compliance, including: poorly written procedures, inadequate training, obsolete procedures, poor operator attitudes, and inadequate equipment maintenance (which forces operators to "work around" procedures to get their work done).

Most importantly, if the value and practice of procedure use are not exemplified and enforced by the organization's leaders, procedure usage and compliance will deteriorate. If leaders (actively or passively) communicate that procedures aren't very important, the procedures won't be used.

Failure to Monitor Equipment Status

Successful control of equipment requires correct alignment and monitoring. Technical procedures usually contain an *Initial Conditions* section to specify the state of equipment that must exist before initiating the procedure. Similarly, startup procedures for industrial complexes include system lineups, switch alignments, and instrument checks designed to ensure that correct baseline conditions exist to support operation. Then, once an operation begins, equipment must be monitored to ensure it functions as expected. If, at any time, the equipment fails to respond as anticipated, the operation must usually be stopped and the cause of the abnormality determined before proceeding.

Equipment monitoring is essential to industrial success. Failure to monitor equipment status played a critical role in the grounding of the *Exxon-Valdez*. The NTSB concluded:

> *Monitoring the Exxon-Valdez by radar as it transited Valdez Arm would have revealed to the Vessel Traffic Center watchstander that the vessel had changed course to 180°, had departed the vessel traffic separation scheme, and was headed for shoal water east of Bligh Reef.*

Members of the *Exxon-Valdez* bridge watch were also guilty of this deficiency. Neither the third mate nor the helmsman closely monitored the ship's heading after the initial 10 degree right rudder steering order was issued.

Whenever changes to equipment status are initiated, the instrumentation for monitoring the affected parameters must be observed until the changes are complete or at least properly progressing. Though testimony is inconclusive, evidence suggests that the *Exxon-Valdez* may actually still have been on autopilot when the third mate first ordered the vessel turned. Recordings show that the ship continued a 180° course for 6½ minutes beyond that point. If the vessel was still on autopilot, both the third mate and the helmsman were seriously deficient in the process of equipment monitoring.

Finally, failure to monitor equipment status may result from incomplete exchange of information between offgoing and oncoming shifts, inattentiveness to instrumentation through laziness, fatigue, not knowing how to monitor equipment, or inadequate tracking and documentation methods. Regardless of the reason, it is clear that if equipment status is not closely monitored, operating abnormalities will go undetected. Undiscovered malfunctions or forgotten status changes lead to unwelcome surprises as unanticipated conditions occur. Up-to-date knowledge of equipment status enables operators to control equipment during both routine and abnormal conditions.

Deficient Equipment Maintenance

Just as equipment must be operated within design constraints, it must also be *maintained* within design constraints. The entire process of preventive and corrective maintenance is vital to successful equipment operation. Without proper maintenance, equipment will eventually degrade below design specifications.

On the night of December 2, 1984, a release of methyl isocyanate (MIC) at Union Carbide's pesticide production facility outside the city of Bhopal, India, resulted in the deaths of as many as 8,000 people. Violation of operating limits combined with an improper valve lineup and piping flush, triggered an exothermic reaction in an underground storage tank containing 13,000 gallons of MIC. Later (Chapter 9) we will see that a number of other critical maintenance and equipment deficiencies contributed materially to this casualty.

Successful execution and control of maintenance in an industrial facility depends on the ability of operating and maintenance groups to jointly plan, schedule, and perform maintenance. Though some routine maintenance tasks are done daily by the operating staff, most preventive and corrective maintenance is performed by specially skilled maintenance technicians.

Thus, operations and maintenance groups together must ensure that maintenance needs are identified, conditions for maintenance are established, necessary inspections or repairs are performed, and the system is restored to its normal operating status. In a complex operating environment, the orchestration of maintenance activities with operational endeavors presents special problems of coordination and exceptional exposure to risk of failure.

One area of risk (and common accident contributor) is simply the failure to identify, report, and assess the impact of equipment deficiencies. Usually, equipment defects are found by operators in the process of regulating equipment; but, if an operator does not understand the equipment's normal functional characteristics, an operating deficiency or an *equipment limitation* may not be recognized. In the *Exxon-Valdez* accident, a failed circuit card in the VTC's master radar unit caused operators to monitor on the slave scope. (Circuit cards in both the master and slave units create a graphic overlay of the traffic lanes on the scopes—with a failed circuit card in the master unit, the slave unit was monitored.) The units differ, however, in that the range selector on the slave unit does not change the pulse frequency of the radar sending unit as does the master set. Lack of knowledge of this effect led the radar operators to assume that the radar was malfunctioning when, in fact, they were simply mismatching sending and receiving functions.

When equipment anomalies *are* discovered, immediate correction, though desirable, is not always possible. Without a method to record and tag the deficiency, it may be forgotten. Therefore, a formal identification and tagging system is necessary to ensure that deficiencies are recorded, analyzed, prioritized for repair, and tracked. Without such a system, deficiencies are forgotten and allowed to persist. Small deficiencies soon become major equipment breakdowns.

It is clear that the ability of an industrial organization to plan, schedule, coordinate, perform, and recover from preventive and corrective maintenance is vital to safe and efficient operations. The consequence of poorly coordinated and carelessly executed maintenance is often a degradation of the machinery's safety envelope and the formation of accident conditions.

Failure to Independently Verify Critical Tasks

Sometimes a task or event is so important, independent verification is necessary to assure its successful completion. There are many useful verification techniques. Second-checking vital valve positions, monitoring switch lineups, performing independent calculations, and verifying important instrument readings are but a few.

In the *Exxon-Valdez* mishap, rudder indication was one of those crucial parameters that should have been independently verified. Neither the third mate nor the helmsman could recall monitoring rudder angle indication. In reality, both should have monitored and verified that the rudder had responded and that the ship was turning.

In a control room, one vital function of the control room supervisor is to verify that instructions have been properly executed. Independent verification of critical parameter readings is an important method for accomplishing that task. Too often, however, operators complain that their leaders don't trust them, or leaders act as if their operators are invulnerable to making errors.

Mistakes are, however, both possible and probable. Admiral Hyman Rickover often quipped:

> [M]istakes must be taken into account....[I]t is important to recognize that mistakes will be made, because we are dealing with machines and they cannot be made perfect. The human body is God's finest creation and yet we get sick. If we cannot have perfect human beings then why should we expect, philosophically, that machines designed by human beings will be more perfect than their creators?

That's the reason for independent verification.

Sense of Invulnerability

Anyone who has experienced a serious automobile accident, death of a loved one, or other sobering event probably remembers an initial feeling of incredulity—a disbelief that the event is occurring. Humans seem to thrive on security. We tend to resist any event which forces us out of our perceived zone of protection. That tendency is so strong that, even as a startling event unfolds, the mind refuses to accept its reality.

At Chernobyl, a good operating record seemed to demonstrate that accidents only happened to other less skilled operators. As cavalier operating attitudes developed, operating limits and safety constraints were routinely violated. Finally, in a futile attempt to perform an electrical test, operators lost control of their equipment.

Days passed before operating personnel would even acknowledge that the reactor had been destroyed by an explosion due to maloperation. Following investigation of the event, the State Committee for the Use of Atomic Energy of the USSR concluded:

> *Violation of the established order in preparation for and performance of the tests, violation of the testing program itself and carelessness in control of the reactor maintenance attest to inadequate understanding on the part of the personnel of how to implement operating procedures in a nuclear reactor, and to their loss of a sense of danger.*

The sense of invulnerability that afflicted Chernobyl operators is a common accident contributor. Twenty-four successful launches of the space shuttle, despite mounting evidence of solid rocket motor joint performance below design specification, helped decision makers to ignore it as a mission-threatening problem. Nearly 9,000 safe tanker passages through Prince William Sound were evidence enough that navigation in the Sound was not dangerous.

The danger of a sense of invulnerability is a propensity to become sloppy and careless. Operators, maintenance technicians, and their leaders begin to take chances—to test the limits of luck. Such an attitude is the perfect medium in which to grow an accident.

Unfortunately, what most fail to realize is that the operating environment is always changing. During periods of operating success, subtle (almost imperceptible) changes in attitude, knowledge, and equipment capabilities will develop. In Prince William Sound, ice conditions in the traffic lanes progressively deteriorated over a ten year period. At Bhopal (originally believed to be a model facility) maintenance degraded incrementally and operating knowledge and skill retrogressed. The independent verification process which distinguished NASA through the Apollo Program disintegrated with the *Challenger* under the pressure of launch schedules.

Guarding against that sense of invulnerability is a continuous task. It falls to the organizational leaders to exemplify and foster a questioning and curious attitude in *all* members of the industrial team—an attitude that "it *can* and *will* happen here unless we watch what we're doing".

Acceptance of Abnormalities

This same complacency or sense of invulnerability manifests itself in another common but distressing tendency. In many accidents, equipment abnormalities and procedure violations become so common that they are accepted. *The abnormal is allowed to become normal.*

After the Bhopal accident, leaders and technicians alike testified to the *accepted* unreliability of instrumentation. They stopped believing critical parameter indicators that would otherwise identify the symptoms of impending disaster. Long before the *Challenger* accident, a faulty design had been identified but, because it hadn't caused a problem before, the decision makers chose to *accept* it. In Prince William Sound (before the *Exxon-Valdez* grounding), it became common practice for ships to leave the traffic lanes and steer around the ice. The practice was *well-established* and *accepted*, even though specifically prohibited by regulation (except in emergencies).

If equipment deficiencies aren't expeditiously identified and corrected, operators will compensate in some way for the condition—and, if such deficiencies are allowed to persist, they wil eventually be accepted. You have a serious problem if, when pointing out a deficiency, an operator responds: "But, it's always been that way." Whenever and wherever this malady is allowed to exist, accident components are being formed and are growing. Though acceptance of abnormalities may not manifest itself in an accident for years, it subtly and progressively erodes safety margins.

Ignorance of Warning Signs

Warning signs of impending danger are usually present long before actual failure. They are sometimes called *accident precursors*. Accident precursors are abnormal events which occur before an accident and indicate conditions which can promote failure. To the experienced observer, they are warning flags which occasion caution and conservative analysis of the situation.

The *Exxon-Valdez* accident was portended by near misses of Bligh Reef by other tankships. Just the day before the *Valdez* grounding, another tanker passed within one ship-length of striking the reef.

Failure to recognize and investigate warning signs is often an accident contributor because leaders and their team members are not taught *how* to recognize accident precursors. It is a skill that must be consciously developed.

The concept of precursors is neither new nor limited to accident interdiction. The process of predictive maintenance, for example, relies upon this same idea. Periodic analysis of pump or turbine lubricating oil can often identify an impending bearing failure. As small amounts of bearing material wear away, they are carried in the lubricating oil. Traces of bearing material in the lubricant are normal and a normal level of the wear products can be established. If a bearing begins to wear excessively, wear product amounts in the lubricant will be above expected levels. Analysis of the lubricant can thus give advance warning of impending failure, and the bearing can be replaced on a scheduled basis rather than a crisis basis.

Critical parameter anomalies are the precursors of process upset. The late Dr. W. Edwards Deming, renowned practitioner and educator of management principles and statistical quality control, advocated the same philosophy for industrial process control. He taught his students to identify the process critical parameters, establish their normal values, and then track them religiously. In so doing, abnormalities in the process can be identified and corrected before they become serious problems.

The same concept can be used to foil development of failure conditions in any industrial operation. If team members know what parameters are important, which conditions are normal and which are abnormal, they can intelligently intervene when things start to go wrong.

Capitulation to Production Pressure

Defensive driving courses warn of the danger at night of *overdriving* one's headlights. This phrase refers to the habit of driving at a rate of speed which exceeds one's ability to stop in time to prevent collision when an obstacle is seen in the road ahead.

Industrial teams often fall prey to a similar danger when the pressure for task completion outweighs serious consideration of consequences. *Production pressure*, as it is often termed, is a frequent accident contributor.

After the Chernobyl accident, the investigating committee wrote:

> *In the process of preparing for and conducting tests…, the personnel disengaged a number of reactor protection devices and violated the important conditions of the operating regulations in the section on safe performance of the operating procedures…. The basic motive in the behavior of the personnel was the attempt to complete the tests more quickly.*

The operators at Chernobyl, however, are not alone in succumbing to production pressure. In Prince William Sound, it had become common practice for tankers to leave the traffic lanes and travel near Bligh Reef to avoid ice. Navigating through the lanes took extra time and diligence. Originally, however, as ice calving from Columbia Glacier worsened, many shipping companies directed their tankers to transit the Valdez Arm during daylight hours only and at a speed of no more than 6 knots.

Often, people ignorantly take risks by not thinking ahead—by not considering what might happen if something goes wrong. (Anyone with driving experience has encountered the battle to stay awake when, deep inside, they know that they should pull over to rest.) In the *Exxon-Valdez* mishap, this tendency found expression in the master's actions as he navigated the *Valdez* out of the traffic lanes, headed toward shoal waters, placed the vessel on autopilot, initiated the loadup program to increase its speed, and turned over control to the inexperienced third mate. The speed of the tanker as it neared Bligh Reef precluded a timely recovery when it became clear that the ship was in hazardous waters. The third mate was clearly not equipped to handle this situation without assistance. Only through proper leadership, training, and experience is this tendency overcome.

Pressure to achieve is not inherently wrong. Problems arise when unnecessary urgency overshadows good hazards analysis and critical thinking.

Inadequate Emergency Preparedness

Emergency preparedness refers to the readiness of an individual or an organization to successfully combat and control abnormal conditions. Napoleon Bonaparte once wrote, "I never worry about what I will do if I win a battle, but I always know exactly what I will do if I lose one."

Achieving a high state of emergency preparedness in industrial operations is the result of anticipating probable aberrations that could cause unacceptable damage to personnel, equipment, or environment. In the *Exxon-Valdez* accident, the greatest emergency preparedness deficiency was the failure to formally reevaluate the ice hazards in Prince William Sound as Columbia Glacier continued its retreat. The NTSB concluded:

> *Ice in Valdez Arm is a significant hazard to navigation and requires closer monitoring and reporting.*

Once significant hazards have been identified, an organization must postulate mechanisms by which such events could occur, plan the means to combat each event, acquire the resources to execute the plan, and then train operating personnel to perform the necessary immediate and followup corrective actions to preclude or limit damage.

Failed emergency preparedness becomes an accident contributor for a host of reasons. Foremost is the neglect of a disturbing number of industrial organizations to do *any* emergency planning at all.

Even when the planning is complete, preparedness may fail in execution through insufficient allocation of resources or deficient training. The training of the *Titanic* officers and crew for the contingency that occurred was gravely inadequate. Only one lifeboat drill had been conducted. It was performed during the short 8-hour sea trial period twelve days prior to the accident. No lifeboat drills were conducted with the passengers.

Finally, emergency preparedness requires not only suitable initial training, but also ongoing classroom, seminar, and drill training to be effective. Leaders of many industrial organizations are unable to envision the value and the associated commitment of time and resources necessary for such a process. As a result, emergency preparedness training most often occurs after an emergency.

Deficient Organizational Staffing and Structure

During the opening years of the War Between the States, the armies of the Confederacy were plagued by the prohibition of line commanders to establish supporting logistical staffs. When Robert E. Lee assumed command of the Army of Northern Virginia, he established out of necessity a staff of officers to oversee the administrative, logistical, and informational needs of his army. Yet, not until late in the war was the concept of staff support legitimized by the Confederate Congress for their armed forces.

Deficient organizational staffing is a recurrent industrial accident contributor. It played a significant role in the grounding of the *Exxon-Valdez*. Following the accident, the NTSB concluded:

> *Exxon Shipping Company manning policies do not adequately consider the increase in workload caused by reduced manning. Exxon Shipping Company had manipulated shipboard reporting of crew overtime information that was to be submitted to the Coast Guard for its assessments of workloads on some tankships. The Coast Guard was unduly narrow in its perspective when it evaluated reduced manning requests for the Exxon-Valdez; it based manning reductions primarily on the assumption that shipboard hardware and equipment might reduce the workload at sea but did not consider the heavier workload associated with cargo operations in port and the frequency of such operations.*

Staffing required by the Coast Guard certificate of manning for the *Exxon-Valdez* was sufficient only if all hands were well and fit for duty. Barely enough staff was available to control routine operations. Emergencies could quickly place the crew in an untenable position. Also, such a small staff precluded thorough at-sea emergency training.

Industrial organizations must be careful to ensure that line organization leaders are staffed with specialty assistants who can take over the administrative functions that prevent the line leaders from being out in their facilities interacting with their workers. Patton once wrote:

> *Commanders must remember that the issuance of an order, or the devising of a plan, is only about five per cent of the responsibility of command. The other ninety-five per cent is to insure, by personal observation, or through the interposing of staff officers, that the order is carried out.*

Insufficient staffing, however, is not the only organizational flaw that leads to accident conditions. Organizational *structure* can also either enhance or detract from effective operations. For example, in organizations where operations, maintenance, training, and other supporting functions do not fall within the authority structure of a single leader, support organizations frequently fail to support. Line managers are faced with the task of *negotiating* for resources, a slow and ineffective process.

"Distance" between the top and bottom of an organization is another consideration. Too many or too few layers in the chain of authority can seriously affect communications. Great effort has been expended in American industry in the last few years to "flatten" organizational structures. The primary purpose in reducing the number of tiers in an organization is to limit the amount of distortion, filtering, and interruption of communication from the top to the bottom. As we have already noted, poor communication is a prominent contributor to inefficiency and accidents. Flattening an organization may, therefore, be a necessary and worthwhile process.

Too often, however, large organizations believe that the flattening process will compensate for the failure of leaders at all levels to get out of their offices and into the work place. Short lines of communication can't substitute for personal interaction between leaders and team members.

Trimming tiers out of organizations creates other problems as well. Too often after reducing the number of middle tier leaders, the work that they were performing is thoughtlessly reassigned to first-line leaders without the accompanying additional staff and training to manage the increased workload. The result is a dangerous overextension of supervisors who are often already overtaxed.

This is precisely what occurred in the VTC in the *Exxon-Valdez* accident. Staff had been cut from the original three required on duty to two. The OOD no longer was required to be in the control station. Both the operations officer and assistant operations officer became so overloaded with administrative duties that they were detached from VTC operations. The organization became *uncoupled*.

Another structural deficiency is the physical separation of leaders from their subordinates. Some organizations make the mistake of locating their leaders away from the work area. Leadership is closely tied to location. Just as in the *Exxon-Valdez* event, when the Officer of the Deck was physically removed from the Vessel Traffic Center, the organizational structure began to deteriorate. Unless there is a leader on scene who is charged with reinforcing the standards, coaching the team members, and overseeing the operation, the organization will degenerate.

In summary, industrial success is vitally dependent upon the ability to recruit and develop the human resources necessary for mission accomplishment, organize them in a structure which supports optimum performance, and then guide them to established goals with excellent leadership. If the organizational structure is inefficient, if insufficient personnel of the right talent and skills are available, if mission and supporting goals have not been clearly established, or if poor leadership is allowed to prevail, a foundation for failure has been laid.

Poor Leadership

Failed leadership is such a chronic and significant accident contributor that its effect must be examined in greater detail. The dramatic failure in leadership on board the *Exxon-Valdez* and in the VTC was a prominent contributing factor in that accident. The master of the *Exxon-Valdez* was not at the most important place of duty at the most critical time. His violation of federal law regarding both vessel navigation in confined waters and fitness for duty set a very poor example for the other officers and crew members.

The VTC leadership had also significantly deteriorated. The OOD was no longer required to be in the station at all times, a fact contributing to disintegration of the requirement to obtain permission prior to deviation from the TSS.

Every team eventually relies on a central leader who is able to inspire team members toward achieving high standards and coach them in the techniques necessary to achieve those standards. As in sports, industrial team performance depends substantially on a skilled leader. The values, attitudes, and often the skills of individuals within organizations are usually related to the standards, attitudes, and behaviors of their leaders. Incompetent leaders breed incompetent subordinates. Leaders who treat their subordinates with disrespect breed resentful, uncaring subordinates. Leaders who require high standards of their subordinates but fail to live them personally breed contemptuous, bitter subordinates. On the other hand, leaders who demonstrate technical and interpersonal skills tend to create subordinates with similar traits. Leaders who listen to their subordinates and treat them with professional dignity develop concerned, caring employees. And leaders who communicate and exemplify high performance standards breed subordinates with equally high standards.

Employees who receive improper or inadequate coaching are more likely to fail. Leadership, by its very nature, can't succeed without extensive personal interaction between leader and subordinates. As already noted, one obstacle to successful leadership is separation of leaders from their workers. Whether isolation is due to physical distance, encumbrance with administration and meetings, or simply avoiding the work place, a leader is unable to coach without being on the field of play.

More often, however, leadership failure is the result of negligence in training potential leaders in the proven principles, traits, and skills that have marked excellent leaders throughout history. Without practical training in what great leaders do and how effective leaders act, new leaders are adrift. They tend to behave in the same ways as their predecessors. If their role models have been good, they may succeed. If their role models have demonstrated crippled leadership traits and skills, they will probably fail. J. Paul Getty once wrote, "It doesn't make much difference how much other knowledge or experience an executive possesses; if he is unable to achieve results through people, he is worthless as an executive." It is clear that effective leaders must know how to influence subordinates through processes other than fear.

One final and disastrous leadership error warrants a note. *Leaders who do not understand the technology of their facility have difficulty understanding the problems presented by the team members, may be unable to establish credibility with those they lead, and are subject to poor technical decisions.* At Chernobyl, the technical incompetence of leaders dramatically exacerbated the development of accident conditions.

A similar failing developed in NASA. After investigating the *Challenger* disaster, the Rogers Commission made a number of recommendations to NASA. One stated:

> *The Commission observed that there appears to be a departure from the philosophy of the 1960s and 1970s relating to the use of astronauts in management positions. These individuals brought to their positions flight experience and a keen appreciation of operations and flight safety. NASA should encourage the transition of qualified astronauts into agency management positions.*

Though technical competence cannot substitute for the ability to lead, it is highly desirable for leaders to possess excellent technical skills. Certainly not all leadership positions require industrial leaders to be specifically qualified on the machinery which they oversee. But *some* level of technical competence is necessary. The decision regarding what level of competence is required must be based upon the potential consequences of failure.

In summary, one must never underrate the impact of effective leaders in catalyzing a team. A cadre of leaders who exemplify the values that they expect and who develop their subordinates through constant daily coaching build a strong, homogeneous work unit. Not only do such leaders improve their team's technical knowledge and skills, but they also impart to each member the characteristics of persistence, loyalty, integrity, accountability, and scrutiny.

Insufficient Management Oversight

One indispensable part of the coaching process is providing management oversight. Anyone who has ever coached a children's sports team understands the importance of management oversight. *Management oversight* refers to the process of observing, evaluating, and advising organizations and their members on how to improve performance. It is coaching in its purest form.

Unfortunately, many industrial organizations function as if they believe oversight to be unnecessary. In the *Exxon-Valdez* accident report, the NTSB concluded:

> *The Exxon Shipping Company did not adequately monitor the master for alcohol abuse after his alcohol rehabilitation program. Exxon Shipping Company did not have a sufficient program to identify, remove from service, if necessary, and provide treatment for employees who had chemical dependency problems.*

The master had a documented history of alcohol problems including two prior DUI convictions. He was also suffering the effects of a recent divorce. The Exxon Shipping Company was aware of the problem, but failed to exercise the control necessary to ensure that safety wasn't compromised.

Nor was management oversight failure isolated to Exxon. The NTSB stated:

> *The limited supervision of the Vessel Traffic Center probably contributed to the commanding officer's and operations officer's lack of awareness that tankships were departing from the traffic separation scheme to avoid ice and were passing close to Bligh Reef.*

During the investigation, the commanding officer of the Marine Safety Office at Valdez stated that he had no knowledge of tankers being allowed to leave the traffic lanes. The operations officer for the division also was unaware that tankers had regularly departed the lanes since 1986 to avoid ice.

Many excuses are offered for not performing the oversight function. One is that the leaders are too busy. It is true that leaders are encumbered with so many meetings and so much administrative paperwork that interactive coaching becomes difficult. But when it comes to a long-term choice between coaching or not coaching, there isn't a realistic choice.

Another is that the workers are competent enough that they don't need oversight. Yet, that argument would be laughable in the context of a professional sports team. If oversight is necessary for the extraordinarily talented members of the San Francisco '49ers, it certainly is needed for our own industrial teams.

Common Components of Accidents

A third excuse is that management oversight makes employees uncomfortable. When performed from an autocratic, punishment-oriented viewpoint, management oversight is intimidating. But when competent leaders interact frequently with the members of their teams, oversight becomes routine and expected. In fact, in an organization which effectively uses oversight, employees become concerned if their leaders aren't available for a few days.

The inference is clear. Without competent management oversight, there can be no improvement. Failed management oversight paves the road to aberrant operational performance.

Faulty Teamwork

A brief study of industrial accidents shows that most failures result in some way from a breakdown in teamwork of the operating team and its supporting elements. The *Exxon-Valdez* accident is as clear a failure in teamwork as any. The team structure not only failed on the bridge of the tanker, it also failed at higher levels in the Exxon Shipping Company, at several tiers in the Coast Guard's Vessel Traffic System, and in the pilotage support for Prince William Sound.

Teamwork is disrupted by many factors. As noted earlier, two of the most important teamwork disrupters are inadequate leadership and deficient communications. Others include inappropriate or poorly communicated goals, an improperly structured operating team, misdirected or inadequate training, unclear procedures or policies, and misunderstood roles and relationships.

In fact, failed teamwork usually results from a combination of other accident contributors. As the disrupters accumulate and the team disintegrates, operators begin to perform as individuals rather than as a coordinated team. Unfortunately, complex equipment operation may be unforgiving of anything less than coordinated team performance.

Disregard for Operating Limits

The *Exxon-Valdez* accident was characterized by a systematic deterioration of established safe operating boundaries. Tanker navigation in Prince William Sound was originally governed by a set of laws, regulations, procedures, and policies. The navigation system also incorporated charts, radar, sonar, vessel traffic control signals, radio communications, and pilotage services to ensure that vessels stayed within the established boundaries.

Yet, as so often occurs, the viability of the equipment and the sanctity of the governing limits sequentially disintegrated. Ice blockage of the traffic lanes worsened over a little more than a decade. A formal re-evaluation of hazards was indicated but did not occur; subsequently, the practice of leaving the lanes for convenience and speed became habit. Pilotage requirements were relaxed, allowing ships' masters to substitute for state harbor pilots with local area knowledge. Radio communication equipment, a critical part of navigational control in the sound, was inadequately maintained. The organizational structure of the VTC degraded as leadership in the center was removed. As a result, compliance with regulations and policies diminished. The requirement for the VTC to plot tanker traffic was administratively removed, masking the dangers of operating outside the vessel traffic lanes. Had the VTC tracked *near misses* (events in which no accident occurred but in which nearly all the elements necessary for occurrence were present), they would have seen that an accident was inevitable. As stated earlier, only a day before the *Exxon-Valdez* grounding, another tank ship came within one ship-length of striking the reef. Each deficiency breached one of the multiple design or operating barriers which ensured safe operation in Prince William Sound. As the barriers fell, the opportunities for disaster multiplied.

In the *Exxon-Valdez* accident, it is clear that those charged with the safe operation of equipment either did not value or did not understand the seriousness of the safety constraints that they violated. This tendency is a common contributor to accidents. After the Chernobyl accident, the investigating committee concluded:

> *The developers of the reactor installation did not envisage the creation of protective safety systems capable of preventing an accident in the presence of the set of premeditated diversions of reactor protection systems and violations of operating regulations which occurred, since they considered such a set of events impossible. An extremely improbable combination of procedure violations and operating conditions tolerated by personnel of the power unit thus was the original cause of the accident.*

Though archaic reactor design contributed, the proximate cause of the accident was the willful and deliberate violation of stringent safety boundaries by certified plant operators and their supervisors as they tried to execute an engineering test during a brief window of opportunity. They ignored important equipment performance characteristics and the consequences of violating facility safety limits.

Topic Summary and Questions to Consider

In this chapter, we identified nineteen common contributors to industrial failure:

1. Defective night operations
2. Unfitness for duty
3. Incomplete communication
4. Inadequate turnovers
5. Ignorance of equipment operating characteristics
6. Non-compliance with procedural requirements
7. Failure to monitor equipment status
8. Deficient equipment maintenance
9. Failure to independently verify critical tasks
10. Sense of invulnerability
11. Acceptance of abnormalities
12. Ignorance of warning signs
13. Capitulation to production pressure
14. Inadequate emergency preparedness
15. Deficient organizational staffing and structure
16. Poor leadership
17. Insufficient management oversight
18. Faulty teamwork
19. Degradation of operating limits

We used the *Exxon-Valdez* grounding as the primary case study because every one of these nineteen components contributed (in some degree) to that accident. Throughout this book, you will see that these same common components were contributing factors to every major disaster of the past century, including: the sinking of *RMS Titanic* in 1912; the loss of coolant accident and meltdown of the reactor core at Three-Mile Island Unit 2 in 1979; the 1984 gaseous release of methyl isocyanate (MIC) at Union Carbide's pesticide production facility in Bhopal, India; the loss of the Space Shuttle *Challenger* in 1986; and the reactor explosion and meltdown Chernobyl Atomic Power Station also in 1986. And there are hundreds more with names such as *Andrea Doria* and *Texas City, Texas*, that all tell a similar tale—"minor" deficiencies were allowed to root themselves in daily operations, grow, and finally bear bitter fruit as they united with other contributors to create accident scenarios.

But these are mere representations of problems that plague everyday industrial operations throughout the world. Though the technologies involved may differ, the mistakes do not. Violation of the fundamental principles and practices that played major roles in these disasters are the same violations that lead to tens of thousands of household, automobile, and agricultural accidents every year.

Ask Yourself Take a few moments to think back and briefly analyze the last accident—major, minor, or even near miss—at *your* facility. How many of these nineteen components contributed to that event? And, to what extent was each a factor?

Do you permit non-routine operations to be conducted after normal working hours? If so, you may be placing your team members in jeopardy. Make sure you take the necessary precautions to forestall failure.

Do you have a clearly written and understood policy on fitness for duty? Your team needs to know that fitness goes far beyond avoiding the illicit or unwise use of alcohol and drugs.

Do your personnel know how to initiate emergency procedures and obtain the proper technical assistance in an emergency? In many instances, disaster can be avoided if help is summoned quickly.

The number and severity of accidents *can* be reduced where well-trained, vigilant equipment controllers predominate. By learning these common contributors—through accident study—you will be far better prepared to identify and interdict potential disasters. That is the theme and purpose of this book.

Part II

A Strategy for Operating Success

○ Systematic Industrial Operations
○ Boundaries of Safe Operation
○ Principles of Operation
○ The Alert, Well-Trained Operator

Now that the need for a systematic approach to industrial operations has been established, Part II introduces the objectives and elements of a successful operating strategy.

Concepts of operating boundaries and operating principles are introduced and discussed in the following chapters.

Finally, we expand on the need for alert, well-trained operators as the most important element of the successful operating strategy.

Chapter 3

Systematic Industrial Operations

None of us are entirely immune to the common accident components discussed in the last chapter. The operating crews of Eastern 401 and the *Exxon-Valdez* were carefully selected, well-trained personnel.

What, then, does it take to safely control equipment and processes? Industrial leaders all over the world struggle with that question every day. Their task is to coordinate the efforts of thousands of employees to accomplish an operating mission effectively, safely, and with economic efficiency. It certainly is no mean task, but that objective is being attained regularly by the best of industrial leaders.

What is their secret? Is it a question of resources? Is it having the most talented employees in the workforce? Is it an issue of leadership? It is all of these and more. Yet, it is none of these alone. Safe and successful accomplishment of an industrial mission relies on talented, well-trained team players pursuing shared goals, directed by excellent leaders—leaders with a strategy for operating success.

This chapter proposes just such a strategy.

Objective of the Strategy

The purpose of an operating strategy must be to gather and employ human and material resources in a manner that leads to consistently safe and efficient operation. The strategy must be practical, cost effective (at least over the long term), and it must make sense to the operators who have to employ it. Otherwise, it cannot be taught and implemented.

Further, an operating strategy needs a clearly defined objective. For well-designed and well-constructed equipment and processes, the operators, the environment, and the general public will, under most circumstances, be protected if components and systems are intelligently controlled. Therefore, the objective of the strategy for operating success can be summarized in one imperative: **Control equipment and processes!**

Elements of the Strategy

In devising an operating strategy, industrial leaders have only three or four major variables with which to work: the operating equipment and facilities, the procedures and policies which govern equipment operation and maintenance, the skill and knowledge of operators and leaders, and the environment in which the equipment is used.

In one sense, an industrial operating strategy, though more complex, differs little from that necessary to safely drive a car. It must be predicated on alert, well-trained operators safely controlling well-designed and well-constructed equipment within prescribed design specifications during both normal and abnormal operating conditions.

Indeed, whether for driving or complex industrial operations, the strategy can be defined (as shown in Figure 3-1) within the framework of a few fundamental elements:
- An Underlying Operating Philosophy
- Reliable Equipment and Facilities
- Well-Defined Operating Boundaries

- Valid Procedures and Policies
- An Efficient Operating Structure
- Alert, Well-Trained Operators
- Superb Leaders
- A Team Approach

This chapter explores the role of each element in the strategy. As we progress, ask yourself, "Is it practical? Is it cost effective? And does it make sense?"

An Underlying Operating Philosophy

Successful team operations are founded on an underlying operating philosophy. In the movie **Hoosiers**, Coach Norman Dale (portrayed by Gene Hackman) began teaching the philosophy of basketball to his team during their first practice. He said, *"I've seen you guys can shoot; but there's more to the game than shooting—there's fundamentals and defense."*

Industrial teams are no less in need of a complete operating philosophy. Yet, unlike Coach Dale, many industrial leaders are satisfied simply to teach their subordinates how to "shoot"—that is, how to accomplish normal tasks under ideal conditions. Little effort is devoted to instilling the *theory* of equipment operation, teaching response to abnormal and hazardous conditions, and cultivating habits that prevent emergencies and accidents. These are the skills, however, that represent the "defense and fundamentals" of industrial operations.

The philosophy presented here is a systematic approach to industrial operations. It's a conservative operating philosophy that relies upon an operator's profound knowledge of the equipment and its operating characteristics—including detailed knowledge of machinery response during both normal and abnormal operating conditions. Such knowledge provides the basis for developing the skills to confidently and safely perform under all operating conditions. It is a philosophy not unlike that of defensive driving.

Figure 3-1: The elements of a successful operating strategy.

Industrial "Defensive Driving"

Most responsible drivers are familiar with the *defensive driving* concept. It is founded upon skilled and knowledgeable vehicle operators applying conservative operating principles in a changing (and often hostile) operating environment. The driver is expected not only to be capable of navigating the vehicle from point to point during ideal driving conditions, but also to control the vehicle safely during abnormal or emergency conditions.

In defensive driving, just obeying the rules isn't enough. Thousands of innocent drivers and their passengers are killed or seriously injured every year though meticulously heeding the traffic laws. Many could be saved by greater awareness of the environment and the actions of other drivers.

Since driving environments are dynamic and often uncontrollable, competent defensive drivers constantly analyze their surroundings for emergent (usually unexpected) hazards. They remain aware not only of themselves and their own movements, but also of the actions and movements of others around them. They don't blindly enter intersections on green lights even though it is their "right" under the law. They know how to handle their vehicles during abnormal conditions—whether on wet pavement, snow, or glare ice, they slow down and exercise special precautions. And, should an emergency arise or an accident occur, defensive drivers are prepared to respond with predetermined, well-thought-out plans and practiced actions.

This defensive driving thought process is also the root of disciplined industrial operations. *It is simply the mastery of humans over the equipment for which they are responsible.* It is an attitude, a state of awareness, wherein professional operators anticipate and recognize the formation of potentially dangerous conditions and take preemptive actions to forestall threatening situations. It is an understanding that *preventing* problems—or resolving them in their infant stages—is the key to operating success.

The Case for Disciplined Operations

The philosophy of disciplined (systematic) operations is neither new, nor confined to the highways. Joe Paterno, extraordinary Penn State football mentor, describes this concept in the arena of sports:

> *In a game of about 160 plays, only three or four are going to make the difference in the outcome. You never know from moment to moment when those three or four are going to happen. Every player has to bust his butt on every single play, so that when each of those three or four game-turning plays is suddenly happening, it happens to be turning your way. Winning or losing depends on that and on little else. Put another way, the team that wins is the team that makes the fewest bad plays. The best player is the man who makes the fewest individual mistakes. So we've got to keep spending time on fundamental movements and techniques. The minute you take for granted, say, your tackling and you spend less time on tackling drills, your tackling will fail you, and so will your blocking. You will lose a football game.*

Coach Paterno teaches that success in sports is rooted in consistent, correct execution of fundamentals—that solid team performance is built on doing the little things right each time. His advice assumes even greater significance when applied to moderate- or high-risk industry where a "loss" can be wholesale destruction of the operators, the environment, and the local populace.

In industry, disciplined operation doesn't imply a harsh or unforgiving operating environment. Rather, it means that a planned, step-by-step process integrates all team members to safely and efficiently accomplish a mission. Discipline provides consistency. The actions of the team are uniform and steady—the result of rigorous training. Team members are seldom surprised by anomalies in the operating environment; they anticipate and are mentally prepared for unusual conditions. Their thoughts and movements are sure and deliberate. They operate with confidence because they have practiced and been tested in operations under difficult circumstances. They know the characteristics and capabilities of their machines. They have learned that extended operating success results from performing the everyday, mundane operating tasks with precision each time. In other words, they know that success depends upon unrelenting execution of the fundamentals.

Such a level of sophistication in an industrial team is not achieved without cost. It comes from well-defined objectives, the best of tools and equipment, carefully chosen team members and leaders, a relentless commitment to practice, and excellent coaching. Further, it relies upon constant self-assessment—a never-ending review of individual and team performance by coaches *and* by the players themselves—in a continuing effort to improve. Team members and their leaders incessantly scrutinize their own performances, not only for deficiencies, but also to discover avenues for advancement. Constant improvement is the goal and vigilance is the price.

In a disciplined operating team, the team members know that the mortal enemy of vigilance is complacency. They know that the ingredients of past achievement are easily forgotten in the torrent of urgent daily requirements. They have learned that—just as there is little splendor in the routine performance of blocking and tackling drills—there isn't much excitement in habitually well-written logs, correctly recorded machinery data, precise communications, or faithfully executed procedures. So they guard against complacency, executing the fundamentals well day-in and day-out, knowing that the mastery of industrial operations is based upon this attitude.

Reasonable Risk-Minimization

The purpose of disciplined industrial operations is not to achieve *zero* risk—that's impossible. Every human activity has *some* associated risk. The simple act of driving an automobile exposes one to the risk of death or serious injury nearly every day. (Nearly one of every 125 drivers will be killed in an automobile accident.) Therefore, the goal of machinery control is not risk *elimination* but rather reasonable risk-*minimization*, which is both achievable and desirable.

In the disciplined approach, *level* of risk is the basis for determining the extent to which operators must be trained and the degree of operating conservatism required. For certain low-risk operations, rudimentary knowledge and skill levels may suffice. But, as operating risk increases, higher levels of competence are essential. Consider motorcycle operation as one example. Motorcycle riders have very little physical protection against injury; their chances of surviving a serious accident are poor. Accordingly, the theme of the National Motorcycle Safety Course is to foresee and avoid dangerous situations through continuous evaluation of the operating environment and application of conservative operating habits—far more conservative than for automobile operation. NMS course participants are taught the capabilities of their machines and to recognize and evade hazards that would escape the attention of other motorists.

Similarly, higher-risk operations demand greater awareness and skill. Equipment must be operated with knowledge, forethought, and awareness commensurate with the associated risk. Admiral Rickover wrote about applying higher standards for operations of greater risk and lesser experience:

> *From the very beginning of the naval nuclear propulsion program I recognized that there were a large number of engineering problems in putting a naval reactor into a submarine. Some problems were unique to submarine application, and some to the general problem of making a reactor plant work. I realized at the time that the use of nuclear power, as with any new sophisticated technology, would require the institution of novel requirements and standards. I realized that these requirements would necessarily be difficult to meet, and the standards would need to be more stringent than those which had been used in power plants up to that time. But when you are at the Frontiers of science you must be prepared to accept the discipline this requires in order to proceed.*

By applying disciplined operating standards, competent industrial operators are (like good defensive drivers) able to arrest emergencies and forestall accidents. By looking ahead, by anticipating the abnormal, they are better prepared for aberrant conditions. Through consistent, conservative operating habits, they are able to stop little problems before they become big ones. Potential accident contributors are not allowed to take root and blossom into mature, unmanageable problems. Thus, a disciplined operating philosophy is essential to achieving reasonable risk-minimization.

Reliable Equipment and Facilities

The second element in the strategy for operating success is providing operating personnel and their support teams with the best, most reliable equipment and facilities affordable with available resources. *Reliability*, in this context, refers to the ability of equipment and facilities to consistently fulfil their functions without unnecessary complexity and cost. This reliability is achieved through conservative design, precise construction, maintenance and modification, and intelligent operation.

If the design is too complex, construction substandard, or the machinery not maintained in prime operating condition, several problems arise. For example, if a machine's operating controls are unusually complicated, reliability will probably be reduced because of operator errors. If a component or system requires excess maintenance, its usability—and thus reliability—is similarly reduced.

Clearly, some machines are inherently more complex than others, and operators must, accordingly, receive training commensurate with that complexity and risk of operation. The issue here, however, is to select equipment of the best design and construction within resource constraints. Judgments as to the "reasonableness" of expenditures for equipment and facilities must consider the *additional* potential costs of operating mistakes and inefficiencies associated with lower quality resources.

Conservative Design

Operating success is jeopardized by poorly designed equipment. Unnecessarily complex equipment design leads to complicated system controls, tending to induce operating problems. When faced with unreasonably complex design, operators are apt to modify the equipment or its operating procedures on their own to improve ease of use—a *very* dangerous practice.

Rickover said this about the design process:

> *First, in any engineering endeavor…conservatism is necessary, so as to allow for possible unknown and unforeseen effects. This conservatism must be built into the design from the very beginning. If the basic design is not conservative, it quickly becomes impracticable to provide the needed conservatism. It then becomes necessary to add complexities to the system in an attempt to compensate for the inadequacies of the basic design. These complexities, in turn, serve to reduce conservatism and reliability.*

As far as practicable, equipment design should compensate for predictable human error. The largest share of human error can be eliminated through excellent training, responsible leadership, and teamwork. But, human error is inevitable. The equipment must be designed to "forgive" mistakes. Rickover:

> *I consider it essential that a reactor plant design be simple and straightforward and that it have a large degree of "forgiveness" built into it. With an inherently stable reactor, it is possible to design a simpler control system.*

Besides incorporating a degree of forgiveness, superior equipment design should also minimize maintenance and provide ease of repair. Anyone who has worked on an automobile can attest to this simple principle of design. If you have to remove a wheel and tire to change an oil filter, you probably won't change the oil very often. Rickover:

> *The propulsion plant design had to be readily maintainable so possible equipment failures at sea could be repaired. The fact that major maintenance operations would be infrequent and refueling possibly as seldom as once in a ship's lifetime, required that standards for materials and systems be very rigorous and that only premium products which had a proven pedigree could be considered for use.*

Superior Construction

The best design cannot succeed if it is not built (or bought) to specification. Obviously, then, design plans must incorporate terminology and conventions that qualified builders (or buyers) understand; and must leave no doubt regarding choice of materials, assembly techniques, and manufacturing tolerances.

Nor should manufacturing (purchasing) occur in isolation: Superior construction (especially for new design) is unlikely without consulting the designers and operating personnel who must eventually use the equipment. As the late Dr. W. Edwards Deming summarized:

> *Purchasing should be a team effort, and one of the most important people on the team should be the chosen supplier—if you have a choice—picked on the basis of his record of improvement…. The team should also include the product engineer and representatives of manufacturing, purchasing, sales, or whatever departments will be involved with the product. Other vital members of the team will be the people who have to use the equipment.*

As risk and complexity increase, designers and operators need to be assured that manufacturing progresses as specified in the design plans. Rickover:

> *Strict control of manufacture of all equipment, including extensive inspections by specially trained inspectors [is required] during the course of manufacture and on the finished equipment. This means that at many points during the manufacture an independent check is required, with signed certification that the step has been completed properly.*

Though such inspections slow the manufacturing process, they are designed to find mistakes or problems at stages of production when they can be corrected with the least financial upset and loss of time. The object is to do it right the first time rather than to fix it after building it incorrectly.

Precise Maintenance and Modification

Operating equipment will certainly require both preventive and corrective maintenance during its lifespan, and may require modification to address changes or improvements in operation. Therefore, if operating equipment is to remain reliable, precise equipment maintenance and modification must supplement conservative design and superior construction.

The purpose of *corrective* maintenance is to restore equipment to original operating specifications when parts or components fail. The purpose of *preventive* maintenance is to detect and forestall degradation so that equipment maintains its original operating specifications. Clearly, since design specifications are so important, repair parts and materials must be chosen and installed in a manner that maintains the integrity of the design.

Equipment *modification* adds additional challenges. Rather than simply restoring, modification *changes* the design specifications. As a result, modifications must be carefully engineered, reviewed, and tested to ensure that they don't introduce new hazards or negate other protective features.

Well-Defined Operating Boundaries

Before members of an operating team can successfully operate and maintain equipment, they must have a clear picture of how the operation interrelates with the designer's vision and the builder's execution of the design. Operators, trainers, maintainers, and engineers who don't understand the integrated functions and operating limits of their equipment are at a distinct disadvantage. They are subject to making conceptual errors with disastrous consequences.

If equipment design has been properly executed, the boundaries for safe operation and maintenance are clearly specified and incorporated into the operating procedures. Understanding those boundaries, then, becomes another critical part of the strategy for operating success.

Every operating facility establishes boundaries to guide and control equipment operation. Some use a systematic approach to develop the boundaries. Others, unfortunately, establish them haphazardly and subjectively. If operating risks are low, less stringent boundaries may suffice, but, if the risks are moderate or high, boundaries must be exacting.

Regardless of risk, the boundaries must be clear. When operating boundaries are vague, groundless, or inconsistent, they will not be valued or heeded. Conversely, if the boundaries have been wisely established after thorough research of the operating hazards, the rules will have meaning and relevance—and, will be observed.

The Nature of Boundaries

Nobody likes boundaries. They restrict action. Yet, most boundaries are established to prevent unfavorable circumstances experienced during previous operations.

Tom Landry (former Dallas Cowboys head coach) wrote:

> *You can't enjoy true freedom without limits. We often resent rules because they limit what we can do. Yet without rules that define a football game, you can't play the game, let alone enjoy it. The same thing is true in life.*

Boundaries define the field of play and the behaviors which are acceptable during the game. The same is true for industrial operations. Operating limits, regulations, guidelines, and policies, when properly established, guide operating team performance within established safe boundaries. They define the field of play. The greater the risk, the narrower is the margin for error and the stricter are the barrier rules.

When properly utilized, boundaries aid safe and efficient operation. Therefore, they must be based in reason rather than be arbitrary. If improperly established or poorly communicated, they often become an impediment rather than an aid.

Boundaries should be communicated not only from the standpoint of their letter, but also from the perspective of their spirit. Failure to teach the spirit or underlying reasons for boundaries eventually results in ignorant or malicious "compliance" with the text of the rules.

The Purpose of Operating Boundaries

In motor vehicle operation, a network of signs, signals, policies, laws, and conventions aids traffic coordination. They are based on the design capabilities of standard automobiles, the operating skills of ordinary drivers, the limitations of roadway design, and foreseeable driving conditions. Together, they describe the boundaries of safe motor vehicle operation for both normal and predictable abnormal conditions.

Industrial operations are bound by similar constraints intended to keep plant equipment functioning within design specifications. The composite of design features, operating regulations, and maintenance requirements which, if observed, ensure safe operation is sometimes termed the *safe operating envelope*.

As in motor vehicle operation, these regulations, policies, and guidelines are effective only when understood and obeyed by those charged with operating or overseeing the operation of equipment. Anyone who has driven where traffic laws are neither heeded nor enforced recognizes that defensive driving assumes most operators adhere to the concept. For defensive driving to succeed, operators must be trained in the rules and their underlying reasons, and then must be trained in the skills necessary to operate. Finally, the rules must be enforced.

Industrial "defensive driving" is no different. Operators, technicians, engineers, and administrators all need design and operating standards which delineate the boundaries of safe equipment operation in order to safely operate, modify, and maintain an operating facility.

The Safe Operating Envelope

One of the most important products of the facility design and development process is the safe operating envelope. The skilled operator finds a well-defined envelope indispensable for prudent and efficient equipment operation.

A **safe operating envelope** is a graphic or verbal representation (usually both) of the boundaries and margins of safety established for critical parameters of machinery design, construction, operation, and maintenance. These margins and boundaries are established to protect operators and equipment, the environment, and the general public from hazards characteristic of the technology.

Every safe operating envelope has three distinct regions: First, is the failure zone; second, the safety margins; and third, the normal operations zone. (See Figure 3-2.)

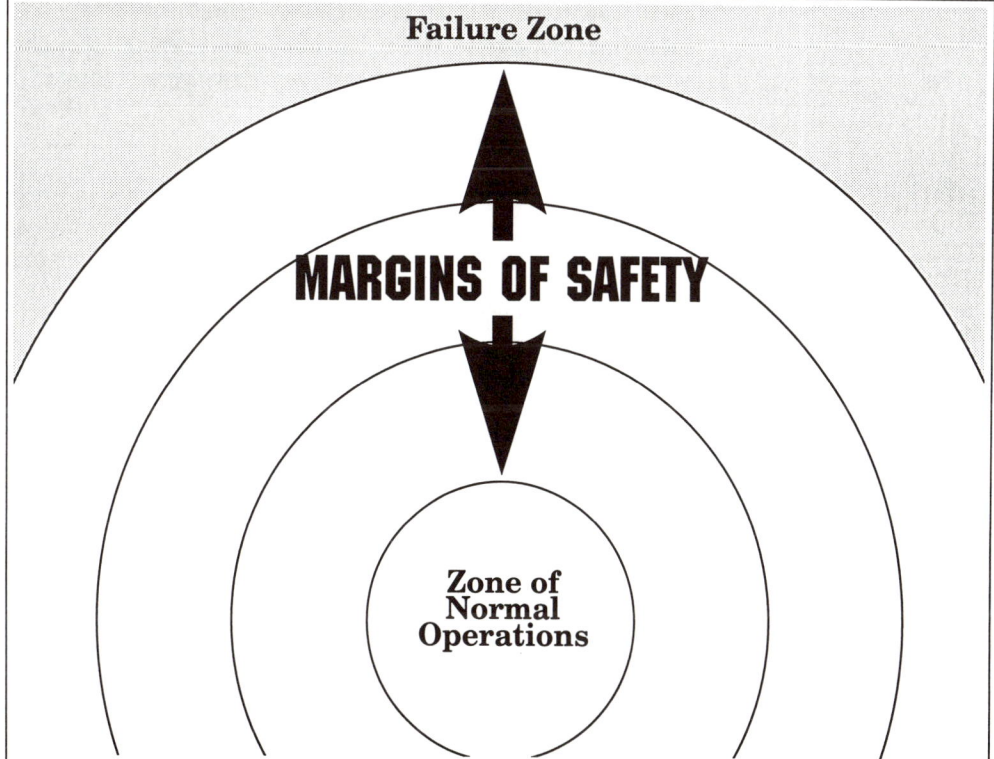

Figure 3-2: Margins of Safety

Systematic Industrial Operations

The **failure zone** is the region in which a condition for a given parameter is known to be unacceptably harmful to equipment, people, or the environment. In short, **failure** occurs when a material, component, or system ceases to meet its specified performance requirements.

The failure zone is established through **hazards analysis,** a determination of the dangers of operation and the consequences of failure. By determining the potential dangers of operation, **risk analysis** is performed whereby the risk of failure is analyzed. Once hazards are defined and risk characterized, **risk management** is employed to protect against (or mitigate) the consequences of failure.

Risk management establishes **margins of safety**, additional buffers to preclude equipment failure even in the presence of certain anomalous conditions. The depth of the safety margins are usually determined by two factors:

- Acceptable tolerances for critical parameters of the materials, components, systems, or operators.
- Ability of design features, automatic responses, and operator actions to counter the effects of excursions beyond specified tolerances.

After safety margins are applied, what remains is the **zone of normal operations**. When machinery is controlled within this region, there is little danger of failure.

A parameter is considered a **critical parameter** if excursion outside its specified tolerances will lead to failure of the component or system. Such excursions must be *probable* as well as *significant* for the parameter to be considered critical.

The safe operating envelope concept is germane to almost every kind of risky activity. Mountaineers establish safe operating envelopes for each climb and each pitch of the climb. Pilots know and observe the boundaries of an aircraft's safe operating envelope or they soon meet with disaster.

Similarly, when driving a car, we apply the safe operating envelope concept every day. Design features such as mirrors, anti-lock brakes, and independent front and rear brake systems provide safety margins to prevent accidents. Bumpers, seat belts, and air bags are features that mitigate damage to vehicle and occupants if collision (failure) occurs.

Since design and construction alone cannot ensure safe vehicle operation, driving rules and limits are imposed to promote safe operation. For example, safety of a car and its occupants may be threatened if the vehicle is operated at excessive speed. Therefore, speed limits are established based upon average equipment design, average driver capability, and normal road conditions. In other words, rate of speed for prevailing conditions has been determined to be a critical safety parameter in vehicle operation, and is, therefore, one boundary of the motor vehicle safe operating envelope.

Speed is obviously only one critical parameter associated with safe driving. Engine temperature, oil and tire pressure are some of the other critical parameters. Unless each is controlled within prescribed tolerances, failure may result. Most, then, require margins of safety.

For driving, safe following distance is a good example of safety margin. When the guidelines for safe following distance are observed, a margin to failure is established that allows recovery from potentially dangerous situations.

Whether for motor vehicle operations or chemical process operations, it is clear that every critical parameter has its own design, construction, or operating restrictions. The composite of boundaries and margins for all such parameters (along with design and construction) form the safe operating envelope.

Valid Policies and Procedures

An operating strategy will not succeed without clear operating *policies* and *procedures*. The term **procedure**, as used in this text, refers to an equipment operating or maintenance instruction, the purpose of which is to govern operation or maintenance of a component or system. For example, a camera's instruction manual includes step-by-step directions for installing/removing film, setting exposure, and cleaning lenses. Procedures provide the detail and sequence of actions necessary for operators to properly use the equipment.

Policy, on the other hand, has a more general meaning. The term **policy** means written guidance describing a plan or general principles for performance (as opposed to the step-by-step direction of procedures). Policies usually provide a conceptual framework for performing general activities, whereas procedures specify with exactness the steps necessary for operating equipment. For example, starting a power plant turbine would require a procedure. Conducting shift turnover between crews, however, would likely be governed by a policy (which might also contain specific steps to be accomplished during the turnover).

Legitimate policies and procedures form the interface between machines, processes, and their human controllers. In their absence, operators must "guess" at equipment characteristics, features and guidance necessary to run the equipment. To be legitimate, policies and procedures must be easily understood, available, and consistently functional. When well-written (and used), policies and procedures remove much of the "human factor" variability in operations.

No Substitute for Training

Industrial components and systems require unambiguous operating and maintenance procedures if they are to be operated within design specifications. Rickover:

> [We provide] extensive detailed operating procedures and manuals, prepared and approved by technical people knowledgeable of the plant design. These manuals are constantly updated as we learn from the operations of the many other reactors. What we learn on one plant is incorporated into all our plants.

As risk and equipment complexity increase, so must operating instruction detail and precision.

Procedures are sometimes considered encumbrances; but well-written procedures provide operators with proven guidelines for controlling their machinery. Since most facilities are too complex to rely on an operator's memory for exacting performance, procedure reviews before, during, and after each operation guard against forgotten steps.

Unfortunately, some believe that sufficiently detailed procedures can be safely implemented by nearly anyone—and thus use procedure detail to substitute for training. Regardless of detail, procedures can *never* substitute for a comprehensive understanding of hazards associated with equipment operation and the safety features incorporated to protect against those hazards. Good procedures are, rather, an essential adjunct to good training. They are the basis for conducting superior operator training.

Industries that desire their operators to consistently use procedures must educate them regarding the basis of the procedural requisites and the importance of procedure use. A dictum stating that procedures will be used without accompanying education/explanation is usually resisted. One result is *exact* compliance with the procedure; and, since few procedures specify *all* the thoughts and actions necessary to complete a task, little will be accomplished if such a "work-to-rules" attitude is adopted.

Above all, establishing a proper outlook toward procedure use begins with good procedures. If procedures are poorly written or if operators don't have access to a responsive system for correcting procedures, compliance will fail.

An Efficient Operating Structure

No operating strategy can succeed without a functional operating structure. Efficient group interaction, successful problem-solving and, more importantly, effective problem prevention are acutely dependent upon developing the right team structure to manage the organization's mission. Frank Lloyd Wright's prescription for structural style, "Form follows function", is as applicable in industry as it is in architecture. The organization (form) must be molded around the industrial mission (function); and, since the mission sometimes changes, so must the organizational structure change to meet new needs.

Mission

An articulate mission is the heart of organizational structure. It is fundamental to focusing the efforts of a team. A well-conceived and well-communicated mission galvanizes the efforts of teams and team members toward a preeminent goal.

The activities of an organization without a mission are likely to be diffused. Mission, however, converges the energy of team members. Light provides a fitting analogy. When comparing diffuse light and laser light, two of the greatest differences are *focus* and *coherence*. Laser light is coherent and focused whereas diffuse light is incoherent and dispersed in all directions. In an organization, mission focuses and synchronizes the team. If the members of an organization don't understand (and can't clearly state) their mission and how it relates to the overall company mission, the organization will lack both focus and coherence.

The mission of an industrial facility is the primary objective for which the facility is designed, constructed, and operated. In general, a facility mission has two parts:

1. To produce a product or service adhering to the specifications of the customer, and
2. To produce that product or service with limited adverse impact to the operators, the general public, and the environment as governed by the current policies, rules, and regulations.

For most industrial facilities, mission centers around one or more products. Earlier in our nation's history, the product was, for the most part, the *only* important part of the mission. Little regard was given to employees, environment, or the general public. However, we learned from hard experience that mission is incomplete unless it also considers and accounts for risks inherent to facility operation.

Therefore, as our understanding of technologies and their effects improves, facility missions are becoming "cradle to grave" propositions wherein the facility is responsible not only for the products that it currently produces, but also for the byproducts and the effects of operation for the lifetime of the facility and beyond.

A facility mission statement is a clear, concise description of objective. For example, the mission statement for a public utility power generating station might be:

> *The mission of the River City Power Generating Station is to safely and continuously produce electrical power for the Riverine Public Utilities District in accordance with current state and federal regulations at competitive market prices.*

Besides providing direction, a clear mission statement is fundamental to successful training and motivation of an operating team and its supporting elements. The mission defines the long-term operating goal and provides a reason for working. It puts the subordinate missions of each supporting team element into perspective.

Yet, many leaders fail not only to successfully state their mission, but also to indoctrinate employees regarding each department's role in that mission. Employee confusion and indifference often result.

In his classic work, **Further Up the Organization**, Robert Townsend, then CEO of Avis and one of America's most innovative leaders, had this to say about objectives:

> *One of the important functions of a leader is to make the organization concentrate on its objectives. In the case of Avis, it took us six months to define one objective—which turned out to be: "We want to become the fastest-growing company with the highest profit margin in the business of renting and leasing vehicles without drivers." That objective was simple enough so that we didn't have to write it down. We could put it in every speech and talk about it wherever we went. And it had some social significance, because up to that time Hertz had a crushingly large share of the market.*

Facility missions should be no less well-stated. Every employee should clearly understand the primary facility objective, how the mission of his or her organization supports that primary objective, and what part he or she plays in the overall role. Without such understanding, teamwork is difficult, if not impossible.

The Industrial Team

The safe execution of a facility mission takes root long before the first operator starts the first piece of equipment in a new plant. Mission accomplishment relies equally upon design, construction, and operations. (See Figure 3-3.)

The creation, development, and operation of an industrial facility is a complex team effort. It engages a conglomerate of teams whose work spans the life cycle of the facility from birth on the drawing table to retirement and disposal. Diverse teams from the disciplines of design, research and development, assembly and construction, staffing and training, testing, operations, maintenance, and technical support must act in concert over a period of years to guarantee the successful operation of the facility to meet its intended mission.

In its simplest form, the life cycle of a facility consists of only three phases: design phase, construction and startup phase, and operations phase. The operations phase depends not only upon the actions of machinery operators, but also includes maintenance, technical support, and administrative support activities.

Accident studies show that industrial failure is most often a failure within the operations phase of the facility life cycle—a failure in teamwork amongst those who must act together to operate, maintain, and modify the facility as it functions daily in its capacity to fulfil the mission.

We refer to the industrial team functioning during the operating phase of a facility as the *operating organization*. (See Figure 3-4.) This operating organization is a composite of the operators, maintainers, engineers, technicians, trainers, and administrators in specialized teams and whose actions are all vital to the daily execution of the facility mission. No single organization is able to operate without the other. The loss of any one ultimately results in failure or shutdown of the whole facility.

The Role of the Operating Team

Central to the operating organization is the team of machinery operators and leaders who physically control the operating equipment—the **operating team** that performs the functions to produce the goods or services that are the reason for the facility's existence. They are the core and the lifeblood of the operation. Of all organizations, they are usually closest to the execution of the facility mission.

For the most part, the role of safe operations devolves upon this operating team. This is the team that has daily contact with the components and integrated machinery systems that comprise the facility. Their actions will most likely determine the success or failure of the whole facility.

The role of the operating team is, in other words, to operate the facility components and systems within established bands of tolerance in accordance with approved procedures to accomplish the facility mission. To accomplish this role, the operating team is charged with having sufficient knowledge and familiarity with the equipment to control it safely during infrequent and abnormal conditions as well as during normal routine operations.

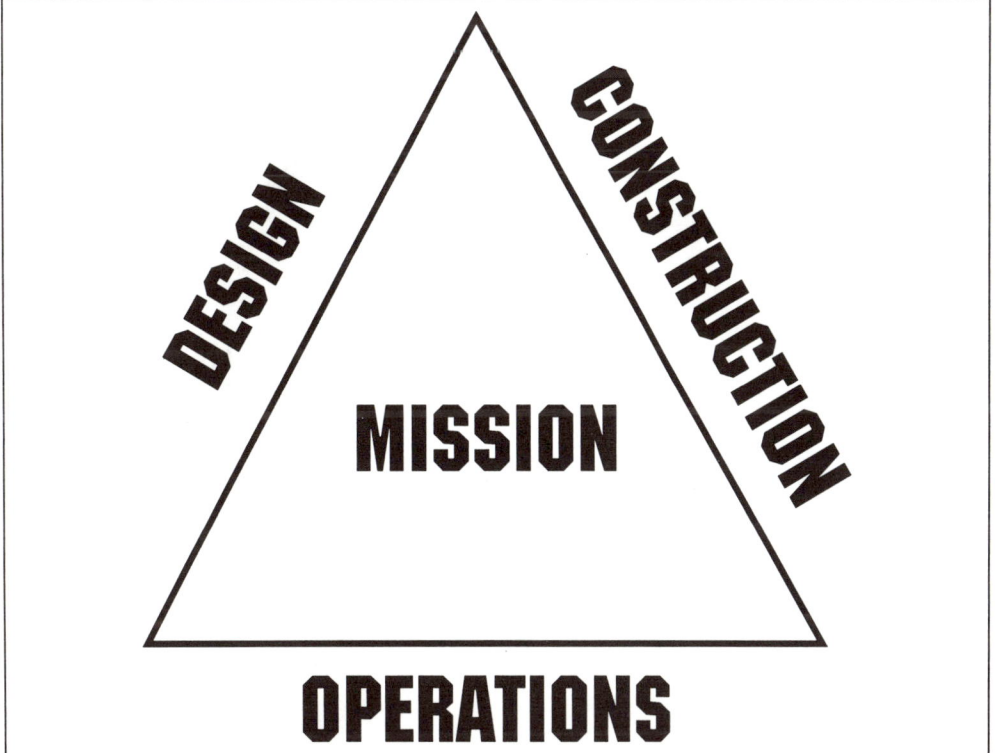

Figure 3-3: The Triangle of Safe Operation.

Systematic Industrial Operations

The Role of Supporting Teams

The operating team cannot successfully function without the cooperation of a number of *supporting teams*. Core **supporting teams** of most industrial facilities include such organizations as maintenance, training, engineering, quality assurance, safeguards and security, procurement, and human resources. Other specialty support teams may be necessary, depending on the nature of the operation and its associated hazards. For example, a facility engaged in producing, using, or recycling nuclear materials will probably have a staff of health physicists specialized in radiological controls.

Supporting teams are usually outgrowths of the operating team. In the early stages of most operations, support functions are performed by the operating team itself; but as the complexity of a facility grows, specialists evolve to perform tasks which would otherwise detract from the operating team's primary role.

In simplest terms, the role of supporting teams is to assist the operating team in producing the product or service central to the facility mission in a safe and efficient manner. At times, those supporting missions may *appear* to interfere with the central facility mission. The steps necessary to ensure safe and efficient facility operation sometimes hinder short-term performance; but, in the long run, they usually prove to be the cheapest and fastest path to mission accomplishment.

Sometimes supporting teams lose track of their primary function—*support* of the operating team. A role reversal often results, wherein the operating team must negotiate for the assistance of supporting teams.

This is a danger which requires constant vigilance. It is imperative that supporting team leaders and the central facility manager ensure that the appropriate support relationships are understood and practiced in order for the facility team to be successful.

It follows that support organization personnel must clearly understand both their role in accomplishing the central facility mission and their interrelationship with operations and other support teams. We wouldn't consider fielding a football team in which the quarterback and running backs had received training and practice while the guards and tackles had not. Nor should we consider fielding an industrial team in which team elements had never practiced together. An industrial facility's support teams cannot perform well as part of the overall team without training and practicing as team elements.

Figure 3-4: Operations Molecule.

Another danger to which industrial organizations often succumb is a belief that support teams need not be staffed with the same high quality personnel and leadership as the operations group. As a result, they are often understaffed or poorly led with the expectation that only the operating team personnel need to meet high standards of selection, training, certification, and performance.

In reality, *every* organization within the staffing structure of a facility plays a vital team role. Supporting organizations must, therefore, share the same intense desire to accomplish the facility mission and the same high standards of performance as does the operating team. General George S. Patton, Jr. recognized this need clearly when he told his commanders:

> *The desperate determination to succeed is just as vital to supply as it is to the firing line.*

As with any team, resolution of problems and conflicts requires cooperative, interactive leaders who educate and coach their subordinates toward proper conclusions. Successful leadership requires leaders who set high standards of performance for their teams and who, at the same time, teach them how to achieve those standards.

Alert, Well-Trained Operators

Accident-free performance requires superbly trained industrial "defensive drivers" who, through constant vigilance, intercede to prevent small problems from infiltrating their operating functions. Therefore, any reasonable operating strategy must select, develop, and maintain a cadre of superb operators.

Dr. Deming in his sixth point states:

> *Institute training. Too often, workers have learned their job from another worker who was never trained properly. They are forced to follow unintelligible instructions. They can't do their jobs because no one tells them how.*

Excellent operation requires alert, well-trained operators who thoroughly understand their machinery and its normal operating characteristics. They must be educated in the design basis of their equipment and know its operating limitations and the reasons for the constraints. They must know the underlying hazards of their technology and the avenues from which failure approaches. They must frequently practice the actions necessary to control their equipment for both routine operation, and also for abnormal situations and emergencies. They must follow written procedures and review them prior to performing tasks with attendant high risk or which are infrequently executed. And they must operate not only in accordance with prescribed operating rules, but also in accordance with conservative habits of operation. Truly excellent operators know that the most important method of preventing the formation of accident conditions is to deter the seemingly small conditions that lead ultimately to disaster.

Extensive Theoretical Knowledge

Theoretical knowledge is the basis for proper response to both routine and abnormal equipment conditions. Without that knowledge, operators may know *what* to do under routine conditions, but seldom *why*. And, when emergencies beyond the scope of superficial training arise, performance fails.

Admiral Hyman Rickover continually emphasized in-depth theory training for the officers and enlisted personnel charged with operating and maintaining U.S. nuclear-powered war vessels:

> *We train our people in theory because you can never postulate every accident that might happen.... [T]he only real safety you have is each operator having a theoretical and practical knowledge of the plant so he can react in any emergency.*

Just as defensive driving goes far beyond knowing how to steer a car, excellent industrial operations must reflect a thorough understanding of the capabilities and potential hazards of operating machinery, efficient routine operating regimens, and appropriate responses to emergent conditions.

Vital Operating Skills

Once the educational foundation has been laid, the skills of operation can be built upon it—skills which prepare operators to control equipment and processes safely and efficiently, the objective of the operating strategy.

Systematic Industrial Operations

For an industrial operation to succeed, every operating technician and leader *must* learn twelve vital operating skills—a systematic approach to equipment operation. Each operator must be able to:

- Conduct pre-task briefings,
- Understand and comply with procedures,
- Monitor important operating parameters,
- Independently verify critical task completion,
- Communicate vital information,
- Keep logs and record operating data,
- Recognize, prioritize, and respond to abnormalities,
- Combat emergencies and casualties,
- Oversee maintenance, modification, and testing activities,
- Isolate hazards or energy sources when required,
- Train others on-the-job, and
- Perform shift turnovers.

These fundamental skills of process and equipment control are so essential to the overall operating strategy that most of this text (Chapters 7 through 19) is devoted to detailed discussion of what they are, how they affect operations and how to implement them.

Attitude, Awareness, and Fitness

Good technical training is only part of the prescription for effective operators.

The ability to remain fit, alert, and aware is the other part. Acquiring operating knowledge and attaining vital operating skills are fundamental to individual and team success. Yet, some of the most knowledgeable and skilled people still fail because of poor attitude, awareness, or fitness. For one reason or another, they are not alert.

Attitude is an important component of alertness. Though intangible, attitude is a prominent factor in determining whether a task is well-performed or inadequately accomplished. *Attitude*, for our purposes, may be described as the mental outlook or approach toward task performance. When optimistic and constructive, attitude yields patience, persistence, and notions of improvement. When pessimistic and destructive, attitude engenders lack of focus, impatience, reluctance, and a desire simply to be done with the job.

Since concentration is critical to most operations, positive attitude is very desirable. Lack of concentration, on the other hand, is clearly related in accident studies to increased numbers and severity of mishaps. Fortunately, maintaining a positive attitude, even in the face of adversity, can be learned, usually through the example of excellent leaders and mature team members. Victor Frankyl, famous Austrian psychiatrist and author of the classic work, **Man's Search for Meaning**, stated it well:

> *The last of the human freedoms is to choose one's attitude in any given set of circumstances.*

One last word about attitude. Leaders must create an environment where operators feel both responsible and accountable for their actions. This feeling of ownership and pride in workmanship is indispensable for operating success.

Awareness is a close relative of attitude. In industrial operations, *situational awareness* implies cognizance of the operating environment, potential dangers, and attentiveness to the state of operating machinery.

Awareness is fostered by confidence in the machinery and one's own ability. Without that confidence, it is difficult for situational awareness to exist—when either is lacking, fear predominates. Fear causes a person to focus on the source of apprehension and the means to avoid it. One who lacks confidence in his or her operating skills spends much time worrying about what *might* go wrong rather than how to ensure that nothing *does* go wrong. Situational awareness suffers.

Besides confidence, awareness requires fitness. The term *fit-for-duty* implies that an operator is both mentally and physically prepared to handle the operating challenges that will be encountered. Fatigue, anger, worry, sickness, drug impairment, or mental distraction can all render an operator unfit for duty. They divert concentration away from the operating tasks and errors result.

Team Players

One special component of attitude is a willingness to learn and improve. Even the best leaders can't teach anything to players who are unwilling to learn. Coach Joe Paterno says it this way:

*I said in those days that to have a great team you've got to have **we** and **us** people. We won't have a great team unless you as an individual work at making this squad a team. At making so deep a total commitment to every guy who wants to win that he puts aside any thoughts of personal glory. Everything for the team. Don't look for what you're going to get, but for what you're going to contribute. Praise each other. Help each other. Be interested in each other. Learn what makes the other guy tick. Don't grandstand, because grandstanding only causes resentment. Don't sulk and pout and shoot your mouth off. All that does is alienate your teammates. Don't boast. Have confidence that you're the best—but don't talk about it.*

What, then, do good team players do?

Adopt a Teachable Attitude. A teachable attitude is one of the most important characteristics of a team player. A team member cannot learn without first committing to learn.

Accept the Authority of the Coach. Committed team players accept the authority of the coach even when they don't like the coach. Certainly there are circumstances in which a team member, out of principle, must refuse to participate on a team. But, personalities must, to the extent possible, be subjugated to the process of building a team.

Commit to Build the Team. Team players are loyal to the cause, the leader, and the team. They commit to the struggle to make a team and are willing to sacrifice for the sake of the cause and the team.

Practice Diligently. Good team members are not just "game players". They give their best effort during practice when there is no crowd to cheer as well as during the game.

Communicate Honestly. Successful team members communicate honestly with the coach and their teammates. They recognize the value and need of open communication in team-building and problem-solving.

Listen to Criticism. Team players learn to listen to criticism, evaluate their own performances, and adjust their behavior accordingly. They also develop the skill of rendering *constructive* criticism in a manner that does not obscure the message. They know how to provide constructive criticism to their bosses as well as their colleagues.

Superb Leaders

Beyond mission, nothing is more influential in pulling a team together than a talented coach. Whether leading in sports or in industry, great leaders learn to pilot a team to consistent success in the best and worst situations. Without competent leaders, an operating strategy is doomed.

Learning to lead is a lifelong experience. Fortunately, the lessons of leadership are neither mysterious nor obscure. The characteristics of leadership are written in the pages of history for any who are willing to study it. Great leaders are distinguished by shared traits, skills, and principles that may be analyzed and learned. What, then, must good leaders do?

Master the Game Concept. Successful coaches have a superior understanding of the game in which they are involved. They understand the purpose of the game, its rules, and the skills necessary for players to perform well. Industrial leaders are no different. They must know the technology of their business, the boundaries which define it, and the skills necessary for a successful industrial team.

Envision the Team Mission. The leader must be able to envision the team mission, the team's potential, and the intermediate goals necessary to accomplish the mission. If the leader has no vision of the future, the team has little chance of success. Great leaders also recognize that the best visions and the best ideas may come from the team members themselves. As a result, they encourage team members to stretch their own visions for the future.

Establish the Team Structure. Leaders understand that poorly structured teams are inefficient teams. Therefore, they clearly establish the structure, the lines of authority, and lines of communication for the team. Team leaders recognize that well-conceived and well-defined structure is an aid rather than a hindrance to open communications and participative management.

Communicate the Team Mission. Envisioning team potential and the mission of the team is insufficient for success. Great coaches know that they must recreate in the minds of the players the vision

Systematic Industrial Operations

that they themselves have. Leaders must communicate the team mission, the team potential, and the team goals clearly and unmistakably.

Inspire a Team Attitude. Successful leaders recognize the importance of a positive, enthusiastic, team attitude. They also understand that the leader is the source of that attitude. Unenthusiastic leaders create unenthusiastic subordinates. Leaders have a special responsibility to speak and act in encouraging ways even when they don't feel like doing so.

Define the Roles of the Players. Great leaders use their vision and experience to clearly define the roles of each team member. They wisely discern individual players' potential, and how each player can be best employed to assist the team. Leaders also encourage team members to expand their own roles and take on new responsibilities.

Set the Standards of Performance. Leaders set the standards of performance, not only through stating the standards, but also through *living* the standards. Team members need distinct role models that demonstrate the quality levels to which tasks are to be performed.

Create an Open Communication Environment. Good coaches understand the value of responsive, thinking team members, and therefore, create an environment that values honest, evaluative communication among subordinates. They encourage *constructive* criticism at all levels.

Develop a Training Program. Successful leaders are the result of successful subordinates. Great coaches recognize that success comes through training, practice, and player development. Therefore, at the heart of their programs is a structured program of teaching, training, and practice.

Demonstrate a Concerned, Caring Attitude. Demanding high standards of player performance is only half of the equation for success. The other half is the leader's demonstration of a concerned, caring attitude. People serve leaders, not causes. Team leaders must also serve players if truly high standards are to be achieved and maintained.

Garner Resources. True leaders know that a team must have the right resources to succeed. Good leaders do their best to acquire the resources necessary to achieve. But great leaders also know that the most important of all resources is the composite of players on the team. As a result, leaders invest much time and effort in developing the individual players.

Coach Individuals as Well as Teams. Great leaders recognize differences in people and work with them as individuals as well as teams. Coach Joe Paterno tells us that good coaches learn that everybody is different:

> *Eventually, through trial and error, I found out there are different ways to handle different people. So I began to be a coach.*

One great dichotomy of leadership is that successful leaders must, on one hand, be consistent in their treatment of the members of a team, and, on the other hand, coach each team member in a way that maximizes that member's potential. Not until Coach Paterno recognized that need, he says, did he *begin* to be a coach.

Lead by Example. Great organizations are not built by "Do as I say, not as I do" leaders. Good leaders model the values that they espouse. They understand the incalculable effect of leading by example on team-building and team performance.

Recognize and Respond to Change. The best leaders recognize new and changing circumstances and initiate team changes to meet new situations. They don't wait to get run over by the train. Yet, they do *not* bend where principle is involved, even in the face of adversity.

Teach Subordinates the Meaning of Winning. Great leaders are also great teachers. One of the most important lessons they teach their teams is the definition of winning. They know that winning is not winning every game but winning games consistently over a long period of time. They also know that winning is the result of every team member and the team itself living up to its potential.

Empower Subordinates. One of the hardest tasks of a leader is trusting subordinates to perform after they have been given a task. Patton said it was one of the most difficult leadership lessons he ever learned. There is a lot of talk about *empowerment* in many industrial organizations today. But empowerment is predicated upon trust and training. Trust is not evident in many organizations as we study the divisions between leaders and subordinates or management and labor. Trust is based upon honesty.

Finally, there can be no empowerment without training. An attempt to build empowered teams is doomed to failure unless it is based on a rigorous training program in which team members prove their ability to perform.

Provide Constructive Evaluation. The best coaches provide constant, constructive evaluation of individual and team performance. They encourage all members of the team, whether leaders or subordinates, to similarly evaluate individual and team performance.

Redirect, Retrain, and Retry the Team. Finally, good leaders redirect, retrain, and retry their teams until they reach their potential. They tune the team like an engine and continuously look for ways to improve.

A Team Approach

The final element in the strategy for operating success is teamwork. Though a simple word, *teamwork* is a complex concept. Many leaders seem plagued with a naive belief that a group of people gathered together to perform a task constitutes a *team*. But, a team is far more than just the right number of people with the right skills.

Admiral Rickover wrote:

> *Operating nuclear plants safely requires adherence to a total concept wherein all functional elements which support operations are recognized as important and each is constantly reinforced. Even after each support function—technical, training, quality assurance, radiological control, maintenance, etc.—is adequately staffed and trained, they must be effectively integrated if they are to support sound operating decisions.*

Integration is the difference. Bill Walsh, former head coach of the San Francisco '49ers and, more recently, the head coach of the Stanford University Cardinals football team has stated:

> *The real task in sports is to bring together groups of people to accomplish something.... Those teams that have been most successful are the ones that have demonstrated the greatest commitment to their people. They are the ones that have created the greatest sense of belonging. And they are the ones that have done the most in-house to develop their people.*

Coach Walsh tells us that it is not enough to develop excellent knowledge and skills. We must also create a sense of belonging—a sense that is only developed by coaches who take a personal interest in cultivating the team through the improvement of the team members.

Why Create a Team? There are at least three major reasons for creating a team. Teams can accomplish more than individuals acting alone, teams can usually solve problems better than individuals, and teams offer mutual encouragement to their members.

Greater Accomplishments. The ability of teams to accomplish more than individuals acting alone results from much more than sheer numerical strength. Teams of people, when properly integrated, provide strength through diversity of knowledge and skills, diversity of life experiences, and differing perspectives and viewpoints. The team *fabric* that is created, if woven by a skilled leader, can be stronger than if made of homogeneous threads.

Diversity lends strength by allowing members of the team to see problems and obstacles from different viewpoints. An obstacle which seems insurmountable from one team member's perspective, may have already been overcome by another. This leads to mutual encouragement and elevated confidence in the team's capabilities.

Mutual Encouragement. The members of teams are individually complex and different. Seldom are all team members simultaneously thinking and performing at their peak capabilities. Yet, sports and industrial history are both rife with inspiring stories of teams that pulled together under adverse circumstances to accomplish almost unbelievable feats.

Systematic Industrial Operations

Mutual encouragement is fundamental to that process. The desire and ability of team members to raise the expectations and spirits of their colleagues is a vital factor in accomplishing difficult tasks. It is clear that one of the leader's greatest tasks is to orchestrate the attitudes and efforts of the members of the team and to cultivate an environment of mutual support.

Better Problem-Solving Capabilities. Teams are often much better equipped through diversity and numbers to solve problems than individuals alone. Properly selected and properly led teams can efficiently study problems, develop alternatives, and propose recommendations.

Problem-solving teams need not be afflicted by the inefficiencies so commonly attributed to committees. The Toyota motor company has demonstrated the value of studying problems down to the lowest affected levels through quality circles. The success of Ford's Taurus program was based on diverse problem-solving groups. Problems can be solved rapidly when a team has the requisite technical knowledge, clear objectives, and a competent leader who encourages open communication.

Seeds of a Team

The seeds of successful team operation in the industrial environment are little different than those for the creation, development, and improvement of any team. Teams require:

Mission. A legitimate, well-stated mission, as previously stated, is essential for transmitting the team vision and setting the team direction. All subordinate team goals should support the team mission.

Failure to establish the team mission in the minds of all of the team members is a serious, but common problem, especially among industrial operating teams.

Talented Players. Talented players result from properly established player selection criteria followed by appropriate initial and continuing training. All three areas are sometimes slighted in industrial operations; but *continuing* training is probably the most often neglected.

It's important to have good players; but the best individual players aren't always the best team players. Team synergy can often overcome talent deficiencies. Good coaches know the value of developing *team consciousness* as well as developing individuals to their peak potential.

Tools and Equipment. Proper tools, equipment, and other material resources are essential to team success. Though good teams can often excel without top-of-the-line resources, there is a point at which talent cannot overcome resource deficiencies.

Successful team leaders assess the tools of competing teams and adopt those that they can use. Good coaches avoid the *not-invented-here* syndrome.

Rules of Play. As Coach Landry advises us, games cannot be played or even enjoyed if the rules are not well-established. Nothing is more frustrating or confusing than a set of rules which is continuously changing. Over time, the rules and standards certainly must change as a game matures, but the rule changes must enhance the game. Most often, they should evolve only as necessity dictates.

It is incumbent upon leaders to stay abreast of the rules and ensure that their players clearly understand them. That is a difficult task in the industrial arena, especially regarding the rules of hazardous waste and environmental protection. The best protection is to enlist the aid of experts to educate team members and then reinforce the knowledge through self-assessment.

Team Structure. Proper team structure organizes players in a way that uses their talents most effectively. Team structure also defines for players their relationships, the chain of authority within the team, and the lines of communication by which to coordinate their efforts with one another and with the team leader.

Inefficient structure is analogous to poor design. Regardless of execution by talented players, poor team structure will lead to errors.

Communications Network. A clear, open communications network is fundamental to fluid teamwork. Successful teams are bathed in a process of constant, open, evaluative communication. Players and coaches all feel confident to express their ideas and object to actions or plans which they believe are not in the team's best interest. If sports teams of just a few players rely so heavily upon communications, how much more should industrial teams with many players and a complex mission?

Conventions of Player Interaction. Every team must have a culture defined by written and unwritten conventions of player interaction. These conventions help to delineate player roles and interrelationships. In many ways, they define the personality of the team. They form an important part of the operational discipline which helps to glue the team together.

A team without such a personality or culture lacks identity. Players have difficulty subordinating personal goals to team goals in that circumstance.

Similarly, every industry has such a set of conventions and an industrial culture. The difficulty is in ensuring that the culture changes to reflect changes in risk.

Training and Practice. A directed program of training, practice, and player development is essential for any team to excel. We would not even consider fielding a serious sports team without such a program; yet, we throw together operators who have never operated or practiced together in industrial environments every day. When we consider the potential consequences of industrial failure, neglect of individual and team training is unreasonable.

Team training should include a structured program for administering controlled facility casualty and emergency drills. The drills should be conducted frequently enough to provide practice and opportunity to assess team performance.

Self-Assessment. A continuous self-assessment, analysis, and feedback process is necessary to improve individual players and the team as a whole. Healthy self-assessment involves both players and coaches. It thrives within the open environment of communications.

Self-assessment is poorly conducted by many teams because some players and coaches haven't learned that criticism can be positive. Coaches often use criticism for personal ego-building and players take offense. Such interactions result in petty arguments rather than team improvement.

For self-assessment to be effective coaches must teach players by personal example to subordinate ego. The sacrifice of short-term personal goals can be overcome by the value and personal satisfaction in building a successful team.

Talented Leaders. In many ways, the success of a team relies upon talented, competent, caring leadership. Teams with the best players often fail when poorly coached. The leader gives meaning to the mission, direction by example, and motivation through coaching. All other team elements can be in place, but without this catalyst, the team will degrade.

The question is often asked regarding where to find such leaders. The answer is that good leaders are cultivated by serving under other good leaders. They pick up their styles, their habits, their deportment, and their values. On the other hand, poor leaders learn their trade in the same way. Every leader, then, has the job of cultivating subordinates into leaders for the future.

Cultivating the Team

The commitment to build a team should not be made lightly. It is expensive in time for both team members and leaders, in financial resources, and in emotional dedication. It cannot be done in a sterile, uninvolved manner. It demands team leaders and team members willing to subordinate their own personalities and desires to create the team environment. It does *not* mean that team members sacrifice their principles. The best teams are those whose members work closely together while maintaining a sense of personal responsibility and accountability for the outcome.

Team building is accomplished most effectively by solving problems together. Nothing draws a group of people together more completely than facing and resolving difficult issues. As already discussed, team building requires a worthy, unifying cause; talented, committed people; demanding, caring leaders; a directed program of training, practice, and player development; and time, dedication, and persistence.

Continuous training is critical. Just as for sports, training in industry should be a never-ending process. No team ever gets good enough that it doesn't need practice. The skills that served so well yesterday are dulled without practice. Constant evaluation and improvement are hallmarks of a vibrant team.

Time must be allocated for team practice. Many industrial operations don't budget time to train individuals beyond their initial training. After the Three-Mile Island accident, it became clear to the commercial nuclear power industry that training had to improve dramatically. To accommodate better training, some companies have increased the number of operating shifts (crews) beyond the traditional four. Intelligent modification of the shift cycle can provide valuable time for operating organizations to train as teams rather than in a piecemeal fashion.

It is clear that good coaches cost a premium. The planning, execution, team coaching, and individual counselling necessary to effect good training only occur when a team has good coaches. The practice of choosing trainers (and leaders) from the ranks of those who can't get along with others or who have shown that they cannot perform adequately is exceptionally short-sighted and destructive to an organization. The *best* members of operating teams who also show a penchant for teaching and demonstrate a positive team attitude are the best candidates for training (and leadership) positions. The team structure must be built in a way that desirable candidates have incentives to serve in coaching positions. The loss of shift differentials, overtime pay, or other monetary benefits must be offset at least with prospects of better future work opportunities. Without incentives, candidates with the requisite skills and knowledge will often not be inclined to become coaches.

Time, Dedication, and Persistence

The process of cultivating a team is time consuming and frustrating. Sometimes it seems that progress is more often lost than advanced. But experienced leaders recognize that most lasting changes in organizations take time, dedication, and persistence to plan and implement. General Creighton Abrams once admonished his staff that elephants can only be eaten one bite at a time. President Eisenhower similarly expressed the sentiment:

> *The older I get the more wisdom I find in the ancient rule of taking first things first—a process that often reduces the most complex human problems to manageable proportion.*

Leaders must expect setbacks and frustrations. If you know that they will occur, they aren't so hard to accept when they happen.

Instilling Operating Discipline

Developing operating discipline is not a one-time task—it is a constant, never-ending job. Since it must be pervasive in the thinking in order to be consistently exhibited in the behavior, it is often spoken of as a *culture*.

Even in the best of teams, complacency and failed attention to detail can subtly creep into the operating environment. Perhaps the most important trait of truly great teams is their commitment and vigilance to avoiding complacent operation. They have witnessed the mechanism of team deterioration and its devastating effects at other times in their operating careers. As a result, the best teams have developed well-rehearsed, defined operating habits and roles that they perform relentlessly, day-in and day-out, even when it may seem unimportant to do so. They are committed to continuous performance improvement, both as individuals and as a team. And they have learned that no improvement will occur without incessant review by themselves, by their peers, and by their leaders. Finally, they have learned that solving complex problems is often a team effort, relying heavily on evaluative communication.

Teaching operating discipline is akin to teaching morality to teenagers. Unless it has been a lifelong process, it's like teaching a new way of life that seems burdened with overbearing, trifling rules.

Operators who haven't been educated in the reasons and the methods of the systematic approach to operations often resist the process because the rules sometimes seem unreasonable. A common response is, "We've done it like this for the last twenty-five years and it hasn't gotten us in trouble yet."

The difficulty—as seen in the first two chapters—is that simple problems are insidious. If allowed to exist, they grow and join with other problems to complicate the situation. The result is a complex problem.

Trying to teach the systematic approach through a series of "industrial morality" lectures won't work. It must be a logical, methodical process of introducing the student to the overall concept and then demonstrating the practical relevance of the concept to the student's facility.

The training objective is for each operator to understand and embrace the central purpose of the guidelines of the systematic approach. Rules and policies alone might otherwise appear to be an onerous compendium of laws.

The only effective way to teach the practices and conventions of disciplined operations is to educate operators in the root philosophy and then demonstrate how failure to observe those practices and conventions has led to disaster or how adherence has averted disaster. The conventions and practices must then be reinforced daily by leaders who supervise as coaches through observation and constructive feedback, taking time to explain and redirect their subordinates when they observe deficient behavior and to reinforce excellent performance through recognition.

The Team Feeling

When an operating unit has been properly schooled in the systematic approach, a team feeling begins to grow. Almost everyone has been on a winning team at some time. The team feeling is one of camaraderie at overcoming difficulties and a feeling of pride in the team itself. The French called it *esprit de corps*—the spirit of the corps.

The team feeling is created by many elements. Among them are:

Well-Defined Goals. A common objective pulls teams together. When team goals are well-defined, team members can pull together and the team feeling is enhanced.

Challenging Work. Teams need challenging work and high performance standards. Low standards lead to complacent attitudes and careless performance.

A Concerned, Caring Leader. There is little that creates the team feeling more than a leader who not only sets high standards of performance, but also demonstrates commitment to the team and to the players through individual coaching and counselling. A team must have good players but it is unlikely to win consistently without a good coach.

A Leader Who Serves as a Role Model. Leaders who live up to the standards that they set are invaluable in creating the team feeling. Little is more destructive to the morale of teams than leaders who demand more of the team than they themselves are willing to give. A leader who serves as a role model is a part of the team. A leader who doesn't will be effective only in proportion to the amount of coercive power that they are able to wield.

Success on the Field of Play. Success—realizing that all the hard work has paid off—is an important factor in developing the team feeling. When a team learns how to win and what it feels like to win, winning becomes easier. Players begin to realize that the team can accomplish more working together in a concerted effort than any one member and more than all the players working independently. Players who have learned that the essence of teamwork is helping one another and relying on one another never forget the value of a team.

Employing the Team Concept

Teamwork emerges again and again as an important element in operational success. In the decade from 1979 to 1989, the aerospace, chemical, nuclear, and transportation industries all received grim reminders of the dangers of complex equipment operation. Investigations of the Three-Mile Island accident, the Bhopal gaseous release, the *Challenger* disaster, Chernobyl, and the *Exxon-Valdez* grounding all identified critical breakdowns in execution of the principles and practices of conservative design, disciplined operating and maintenance routines, and organizational communications. In short, all six resulted from failures in industrial teamwork.

Tom Landry, head coach of the extraordinarily successful Dallas Cowboys for twenty-nine years, has studied teamwork perhaps as much as any person in America. In his autobiography, he states:

> *The very best football players have to depend more on their teammates. All eleven men on a team have specific roles on every play. Unless each successfully does his part, the play won't work. It's a coordinated effort. It's not enough for ten out of eleven defensive men to perform their assignments perfectly. Ninety percent performance can mean one hundred percent failure.... What this means is that in football no one individual can be more important than the team. Sometimes this is a hard lesson for the most talented athletes to learn; it's why some very gifted players bounce from team to team. It's also why in football the greatest superstars may not be the ones with the most physical talent; they're the ones who learn to adapt and fit their talents into the team concept. Prima donnas don't often make great football players and it's even more seldom that they make for a successful team.*

Teamwork is no less important in the operation of a complex industrial facility. In fact, teamwork is probably *more* important since the number of variables which must be controlled to have a successful outcome are far greater. Only through exceptional teamwork can complex problems be solved. More importantly, it is through teamwork that the formation of complex problems is prevented.

Topic Summary and Questions to Consider

In presenting this strategy for operating success, we identified eight fundamental elements of the disciplined approach to industrial operations. One might be tempted at this point to ask, "Which one is most important?" When asked the same question, Admiral Rickover responded:

> *Over the years, many people have asked me how I run the Naval Reactors Program, so that they might find some benefit for their own work. I am always chagrined at the tendency of people to expect that I have a simple, easy gimmick that makes my program function. They are disappointed when they find out there is none. Any successful program functions as an integrated whole of many factors. Trying to select one aspect as the key one will not work. Each element depends on all the other elements.... [A]ll of these elements must mesh for the system to work. You cannot separate out and use the pieces which you like, and discard those which are "too hard".*

It's no one thing. It's everything together.

Ask Yourself Do you have a strategy for operating success? To answer that question, you must first consider:

- Are your plant and process operating boundaries clearly defined and based on safety studies?
- Can your team members identify the hazards associated with their work as well as the design and operating barriers incorporated to protect against those hazards?
- Do clear policies and procedures (understood and used by your personnel) control operations and maintenance at your plant?
- Do your leaders understand the equipment and processes at least as well as your operators?
- Does an underlying "defensive driving" operating philosophy permeate your team?

If you can't answer all of these questions affirmatively and with certainty, your operating strategy may be missing (or deficient in) one or more of these eight fundamental elements for success. Subsequent chapters will help you to better evaluate, understand, and improve your facility operating strategy.

Chapter 4

Boundaries of Safe Operation

Hazards are inherent in the use of tools. Whether a tool is simple or complex, we assume risk to gain advantage. Even as uncomplicated a task as handling paper presents the danger of a painfully cut finger. Risk is unavoidable.

However, observant people soon learn that risk is also manageable. Managing risk is a familiar experience for everyone—something each of us has been learning and doing since we were children. When you were learning to ride a bicycle, you probably were a little afraid. You knew that you might fall down, possibly scraping an elbow or skinning a knee. There was risk, but the usefulness, prestige, and fun of riding was compensation enough to take the chance. Besides, with some assistance (and perhaps a set of training wheels) you knew instinctively that the risk could be minimized.

Some people believe that the essence of all risk management is to avoid the activity which presents the risk. If the issue in doubt is smoking, avoidance is probably wise, because smoking is a matter of luxury. But if the issue is driving or generating electricity (activities which most consider necessary), cost-effective risk minimization achieved through suitable design, construction, operation, and maintenance is preferable to abstinence.

Chapter 3 introduced well-defined operating boundaries as a fundamental element of the operating strategy. Because these boundaries are so essential to operating success—by defining the conditions of failure qualitatively and quantitatively—we have devoted this entire chapter to promote a deeper understanding of the composite elements, derivation, and purpose of safe operating boundaries.

Safety Analysis

Whether the subject is bicycle riding, chemical manufacturing, or nuclear power, the process of identifying operating hazards, determining the risks associated with each hazard, and erecting protective barriers and boundaries of design, construction, operation, and maintenance is termed **safety analysis**.

Safety analysis is a part of the invention, production, and use of every implement. Though it may not be formally performed for simple tools, it is, nonetheless, never far from the thoughts of responsible designers and users. For an industrial complex (especially in the context of today's operating environment), formal safety analysis is a necessity.

The purpose of safety analysis is, through reasonable means, to determine and implement protective features for the operating personnel, the environment, and the general public—features devised to shield them from the hazards of operation. Protective features include conservative (and forgiving) component and system design (including engineered safeguards to provide physical barriers), superior construction (to ensure that the design is appropriately implemented), and operating and maintenance restrictions which, *if observed*, ensure safe operation for normal and expected abnormal conditions. Note, in particular, the emphasis attached to compliance with operating restrictions. At the Chernobyl Atomic Power Station, as we will see in Chapter 5, operators disabled important protective features and ignored safe operating limits imposed by design. No safety analysis, regardless of how well performed, can compensate for ignorant or malicious circumvention of protective features. Without proper operation and maintenance, the safety analysis is invalid.

Identifying the Hazards

Without thorough identification of hazards, effective protection can never be developed. Therefore, the first step in safety analysis is hazard identification.

A **hazard** is a potential danger, usually of personnel injury or equipment damage, incurred as a result of engaging in or being near an activity. Hazards are identified through the process of **hazards analysis**, referred to in some quarters as a *hazards and operability study*.

Regardless of title, the objective of hazard study is to determine what dangers could result from an activity, process, or technology. The question that hazards analysis poses is: "As a result of this activity, what could happen that would adversely affect the operators, the environment, or the general public?"

For example, a hazards analysis of mountaineering would reveal that the sport is characterized by the dangers of falling, having something fall on you, or being exposed to adverse climatic conditions. If the subject is motor vehicle operation, hazards include damage to the vehicle and its occupants due to collision or overturning. In chemical process operations, exposure to harmful gaseous or liquid compounds are among the hazards. Nuclear operations are uniquely characterized by radiological hazards.

In Prince William Sound, the hazards analysis identified two overriding dangers—collision (with other ships or icebergs) and grounding on a reef or shoal. Hence, protection imposed by the safety analysis for the sound was designed to prevent accidents resulting from grounding and collision.

Determining the Risk

Protecting against *all* hazards under *all* circumstances is neither practical nor cost-effective. Though a hazard may be possible, it may pose little risk—so improbable as to be unworthy of consideration, or of low consequence if it occurs.

Therefore, safety analysis dictates that, following hazards analysis, *risk analysis* must be performed. **Risk analysis** is the process of determining whether a hazard has a significant probability of occurring, and whether the consequences of the hazard are worth protecting against. In fact the question that risk analysis poses is: "What is the probability and consequence of this hazard being realized?"

Risk is the product of probability and consequences. (Risk = Probability x Consequences.) If the probability of an event occurring is low and the consequences (should the event occur) are also low, the risk associated with the hazard probably does not warrant a large outlay of capital for protection. In other words, if the event occurs, you can afford the outcome. Conversely, if the probability of occurrence is high and the consequences of failure are also high, protection is required. The situation exhibits the characteristics of high risk.

If you pay insurance for a teenage driver, you are familiar with this concept. Young males are more expensive to insure since actuarial statistics show that, on average, they are more likely to be involved in automobile accidents. Since the consequences of automobile accidents (especially those involving teenagers) are often great, so are insurance premiums for these *high-risk* clients.

High risk indicates that either the consequences of failure or the probability of occurrence is high while the other is in the range of moderate to high. Figure 4-1 illustrates the concept of risk as a product of consequence severity and probability—thus establishing risk management priorities.

In industrial operations, just as in the insurance industry, designers attempt to quantify risk to assist in making protection decisions. Decisions regarding whether to protect against high risk situations are usually not difficult. But, deciding whether a situation is moderate or low risk, whether to protect against moderate risk events, and how much to spend for the protection are more difficult questions. A good thumb rule is: "Always protect when the situation is high or moderate risk."

Establishing Protection

Once hazards have been identified and associated risks determined, barriers of protection can be established.

The process of erecting protective barriers for risk-laden hazards is called **risk management**. The operative question posed in risk management is: "If this is a hazard for which protection is required, what can I do about it? And what *should* I do about it?"

Options abound for managing risk in industrial operations. Classes of protection include:
- Improvements in design and construction,
- Restrictions on equipment maintenance and modification,
- Limitations for equipment operation, and
- Administrative prerequisites, usually for the operating personnel.

Design and Construction Characteristics

Conservatism in design and construction are the first (and usually the most important) barriers of protection for safe equipment operation. In Prince William Sound, design requirements included, beyond the fundamental design of the port and the tankships, communications equipment with a required reliability of 99.9%, radar on board tankships, and a VTC radar system capable of tracking vessels beyond Bligh Reef. Further, an oil spill response barge was required to be loaded, on-station, and capable of responding to the scene of a spill within 2 hours of notification. Unfortunately, communication reliability had degraded to less than 80%, the VTC radar master unit had an operating deficiency, and the spill response barge was not loaded.

Inferior design makes future operation and maintenance of equipment far less reliable. The Chernobyl RBMK-1000 reactor design was inherently unstable, complicating the operation and control process and enhancing the risk of a reactor accident. The Tacoma Narrows Bridge which collapsed on November 7, 1940, was flawed in design—doomed to fail from the outset. The West Atrium walkways in the Kansas City, Missouri, Hyatt Regency Hotel, though soundly designed, were built using an unevaluated modification to the drawings. Their box-beam supports failed on July 17, 1981, killing 114 and injuring 200 others.

Safety analysis can reveal needs for design improvement. Preliminary hazards and operability analysis is predicated upon established (or, in the case of new technology, postulated) equipment design. In the process of conducting preliminary safety analysis, it may become clear that improvements in design and construction would enhance safe operation. If the proposed improvements are cost-effective, design or construction modifications may ensue and the safety analysis be changed to incorporate the upgraded features.

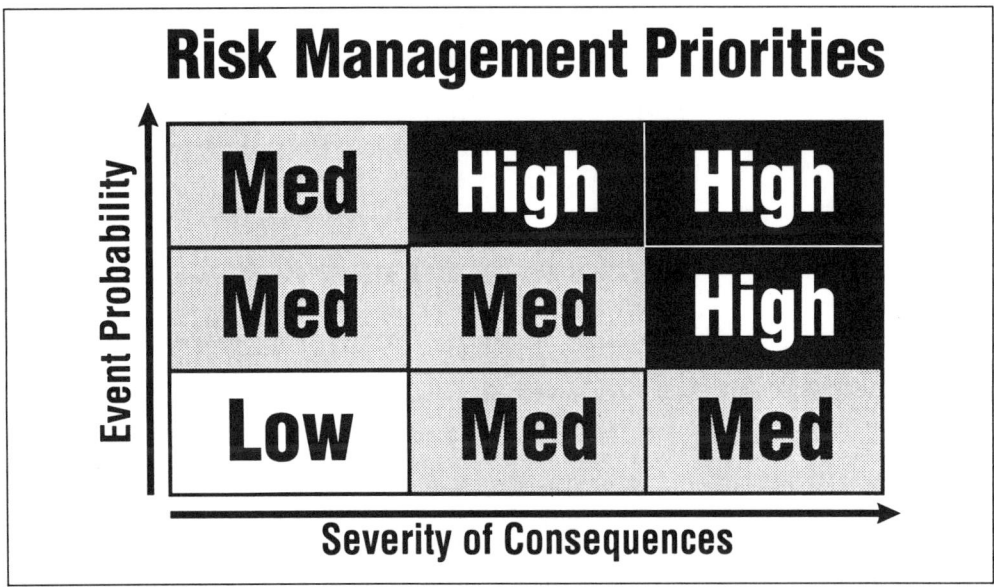

Figure 4-1: Risk Management Priorities based on event propbability and consequence severity.

If the original design is sound, but safety margin for certain activities or actions is minimal, *engineered safeguards* may be incorporated. **Engineered safeguards** are design features that physically protect against unsafe acts or conditions. A blade guard on a table saw is an excellent example of an engineered safeguard. Other examples include guardrails on highway overpasses, seatbelts in automobiles, or twin operating buttons on sheet metal shears. If the consequences of failure are great and the margin for error small, physical restraint is appropriate.

For conditions in which operators are unlikely to act quickly enough to forestall failure, automatic protective features may be necessary to counteract hazards. An automobile air bag is a good example of automatic protection. To effectively protect a driver, the air bag must activate within a few thousandths of a second after impact. If controlled by a button on the dash, it would do little good. Therefore, an impact-inertia switch in the front bumper is used to trigger air bag deployment. The setting at which an automatic protective feature activates is known as a **limiting control setting**.

Maintenance and Modification Restrictions

During the life of components and systems, repair or modification is usually required. If improperly performed, either can void the original safety analysis.

As we learned in Chapter 3, maintenance is the process of restoring equipment to current design specifications. Since the safety analysis for an operating complex is based upon precise equipment design, deviations from valid maintenance requirements and practices may invalidate the conclusions of the safety analysis. For example, if repair parts or materials don't meet design specification, the equipment in which they are used may not function as designed. Accordingly, strict control of the maintenance process is critical to safe operations.

Modification, by definition, is a change to original design specifications. When contemplating modification to any part, component, or system affecting safety, the safety analysis must be revisited and, if necessary, redone to accommodate the change. That's why manufacturers' warranties are invalidated when unauthorized modifications are performed.

Since improperly performed maintenance and modification can adversely affect safe operation, restrictions are imposed on their planning, performance, control, and review to ensure that no new hazards are introduced and that protection for all originally analyzed hazards is still intact. The principles of control are best summarized in two simple rules: "Maintain within design specifications" and "If modifications are contemplated, evaluate how they will affect the safety analysis."

Operating Limits

Just as poor maintenance or unreviewed modification can invalidate a safety analysis, so can inept operation. If equipment is not controlled within design specifications, it may not perform as designed. Therefore, safety analysis imposes *operating limits*. **Operating limits** are limits for control of critical operating parameters which, if observed, ensure safe operation. They are normally specified by procedure and rely upon the intelligent action of operators to be effective.

Machines are usually controlled by monitoring important operating indications such as temperature, pressure, speed, distance, fluid level, concentration, and purity. For example, in automobile operation, vital operating parameters include vehicle speed, safe following distance, engine oil sump level, oil pressure, coolant temperature, fuel quantity, fuel purity, and tire pressure to name but a few. Depending upon the circumstances, each parameter may be critical to the safe operation of the vehicle. Hence, as we learned in Chapter 3, they are known as *critical parameters*.

Most parameters critical to automobile operation are sensed and indicated on the dashboard. Others, such as tire pressure may require periodic checks and adjustments. Regardless, operators are responsible to monitor the parameters at reasonable intervals and maintain them within prescribed operating tolerances. If improperly maintained, vehicle safety may be jeopardized.

Prince William Sound navigation requirements provide several examples of operating limits. To protect against grounding and collision, the safety analysis imposed vessel navigation limits:

- Tankships were to remain within the boundaries of the Traffic Separation Scheme (TSS).
- Tankships were not to leave the lanes of the TSS without permission of the Officer of the Deck.
- Only one ship at a time was allowed in the Valdez Narrows and, speed was limited to 6 knots.

- If given permission to deviate from the TSS, at no time was the ship to encroach into the red sector of Busby Island Light.
- VTC personnel were to continuously monitor (and plot) the position of vessels in the traffic system while in radar range to ensure that they remained within the traffic lanes.

Unfortunately, many (if not most) of these limits were violated, contributing to the grounding of the *Exxon-Valdez*.

Administrative Prerequisites

We have seen that safety analysis may impose forms of protection through design and construction characteristics, maintenance and modification restrictions, and operating limits. One other type of protection is called an *administrative prerequisite*.

An **administrative prerequisite** is a requirement, usually placed upon those who control equipment, specifying the qualification, certification, and condition of operators for operating equipment. For example, before you are legally allowed to operate a vehicle on public thoroughfares, you must have a valid driver's license and your blood-alcohol level must be below the limit prescribed by the state. You must also be wearing proper corrective lenses if your license requires them. Similarly, airline pilots must be certified and proficient on their aircraft and they also must meet specified prerequisites of rest and blood-alcohol level. Each of these requirements is an administrative prerequisite.

Administrative prerequisites for tankship navigation in Prince William Sound included:
- Officers and seaman aboard tankships had to hold current licenses for the positions they occupied.
- The Coast Guard OOD position in the VTC was to be filled by a certified Deck Watch Officer.
- Every person performing the duties of OOD was to be certified on the VTC's radar units.
- While operating within Prince William Sound, tankships were (at all times) to be navigated by a licensed pilot with the assistance of a certified Deck Watch Officer and a helmsman.
- Tankships were required to meet the minimum manning requirements as specified in their Coast Guard-approved certificate of manning.
- The blood-alcohol level of any officer standing a navigation watch was required to be below .04%.
- No officer scheduled to stand a bridge watch was allowed to drink alcoholic beverages within 4 hours of the scheduled time of watch.
- Before standing a navigation watch, officers were required to be off-duty for 6 of the previous 12 hours.
- No member of the seaman's union could be forced to work more than 8 hours in succession.

In retrospect, it is startling how many of these administrative prerequisites had been abandoned.

Cost vs. Benefit

As protective features and requirements are evaluated, cost of protection versus the security that they provide must be weighed. Protection analysts continuously ask: "Can I do the same thing cheaper and more reliably another way?" An improvement in design may, in the long run, be far cheaper than unwieldy operating restrictions. Conversely, if design is fundamentally sound, better training and leadership may be the only options for improved protection.

In our study of the *Exxon-Valdez* accident, we could be tempted to indict the decision which didn't require double hulls on tankships entering the sound. But double hulls on the *Exxon-Valdez*, though they may have reduced the spill by half, would have added $31 million to the cost of a $125 million vessel and would have reduced cargo capacity 25 percent. Multiplied by all of the tankships navigating in the sound, the cost would have been extremely high. One might argue that double hulls would still have been cheaper than the cost of the accident; but, remember to ask, "Is there an easier way to accomplish the same objective?" In this instance, a better way would be to enforce procedures that were already in place. Had the procedures been followed, this accident would not have occurred. Wise protection uses the most cost-effective means that is both reliable and effective.

Categories of Operating Limits

Operating limits typically are categorized as: (1) *safe operating limits*, (2) *normal operating limits*, and (3) *abnormal operating limits*. (See Figure 4-2.) Safe operating limits are usually the least restrictive of the three categories since they allow more "room" to operate. Abnormal operating limits are normally the most restrictive since they dictate additional conservatism based on abnormal conditions.

Safe Operating Limits

A **safe operating limit** (sometimes termed safety limit, technical specification, or operational safety requirement) is the maximum limit beyond which a parameter may not be permitted to proceed if avoidance of failure is to be guaranteed. For example, in Prince William Sound, the dividing line between the red and white sectors of Busby Island Light represented a safe operating limit for navigation. (See Figure 4-3.) Violation of a safe operating limit doesn't automatically imply damage or injury, but *avoidance* of damage or injury is not guaranteed beyond the limit.

Normal Operating Limits

A **normal operating limit** is a routine limit for controlling an operating parameter during normal conditions. Excursion beyond a normal operating limit seldom threatens damage or injury but warns of an undesirable direction or trend. Within the confines of Prince William Sound, the boundaries of the Traffic Separation Scheme lanes represented normal operating limits. They were to be obeyed under normal circumstances and crossed only with special permission.

Abnormal Operating Limits

Operating conditions are not always normal. In motor vehicle operation, speed limits are normal operating limits. They specify maximum speeds for normal conditions. If, however, the roadway is

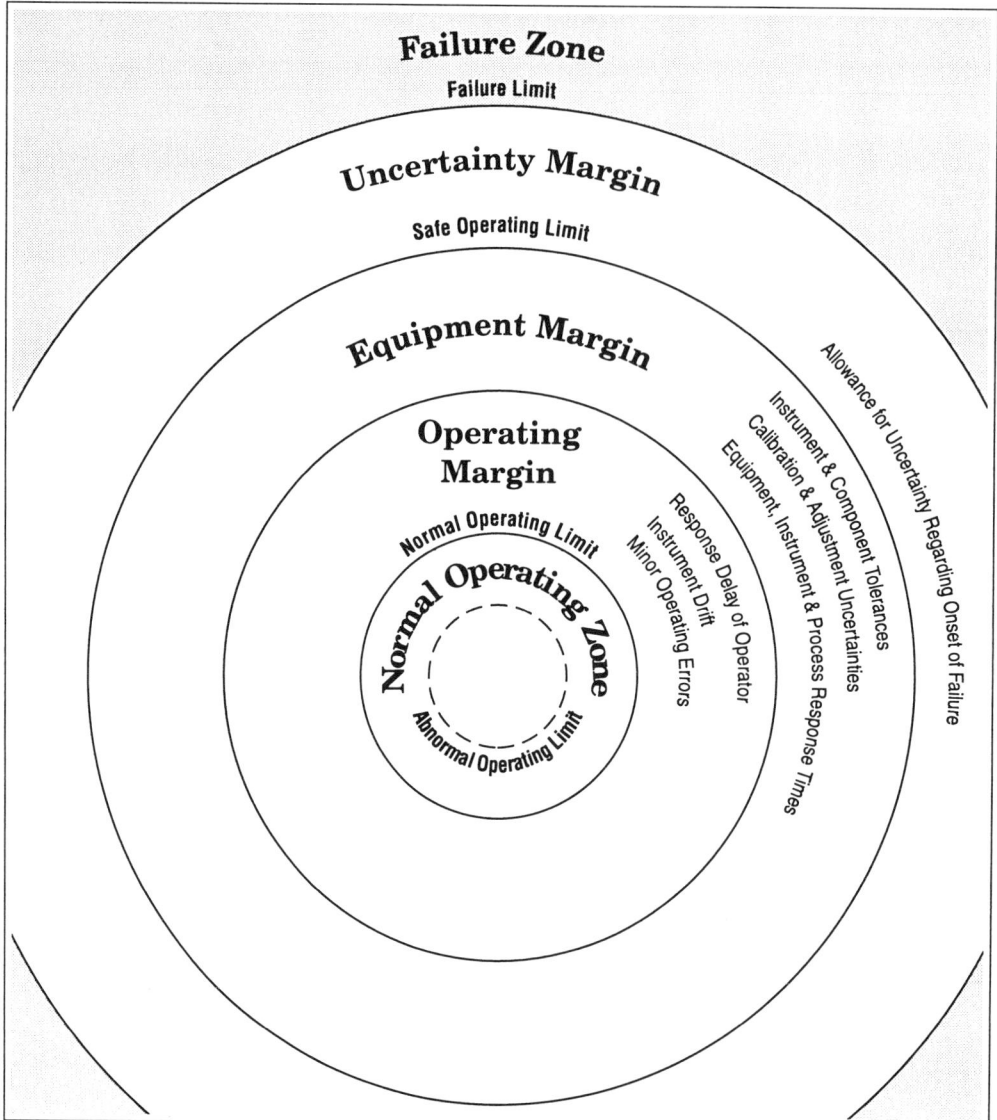

Figure 4-2: A Model Operating Envelope.

covered with ice or snow, state laws require drivers to slow down. No slower speed is specified by law because the requirements of each situation vary. But operators are held responsible to *self-impose* lower limits based upon good operating judgment.

A temporary limit which is usually more restrictive than the normal operating limit and is imposed to compensate for an abnormal operating condition is called an **abnormal operating limit**. The condition causing it to be imposed is called a *limiting condition of operation*. In Prince William Sound, as ice calving off Columbia Glacier worsened in the early 1980's, the Exxon Shipping Company (and several other transport companies) self-imposed a limitation on travel. This abnormal operating limit stated that when ice conditions in the shipping lanes were severe, tankship operation through that portion of the sound was limited to daylight travel only and a speed of no more than 6 knots. The limit was later abandoned as ships navigated through the mile-wide stretch of ice-free water just west of Bligh Reef.

Developing Operating Limits

As we learned in Chapter 3, properly established operating limits guide operating team performance within established safe boundaries.

For leaders and their subordinates to appreciate operating limits, it is important that they understand how limits are derived. Development of limits follows a few rudimentary steps:
- Define failure for the parameter under consideration,
- Establish a failure limit against which safety margin can be added to prevent failure,
- Add safety margin to account for uncertainty regarding the onset of failure,
- Determine the equipment and operating factors which affect failure,
- Add safety margin for equipment factors,
- Add safety margin for human and operating factors, and
- Add safety margin, if necessary, for abnormal environmental factors.

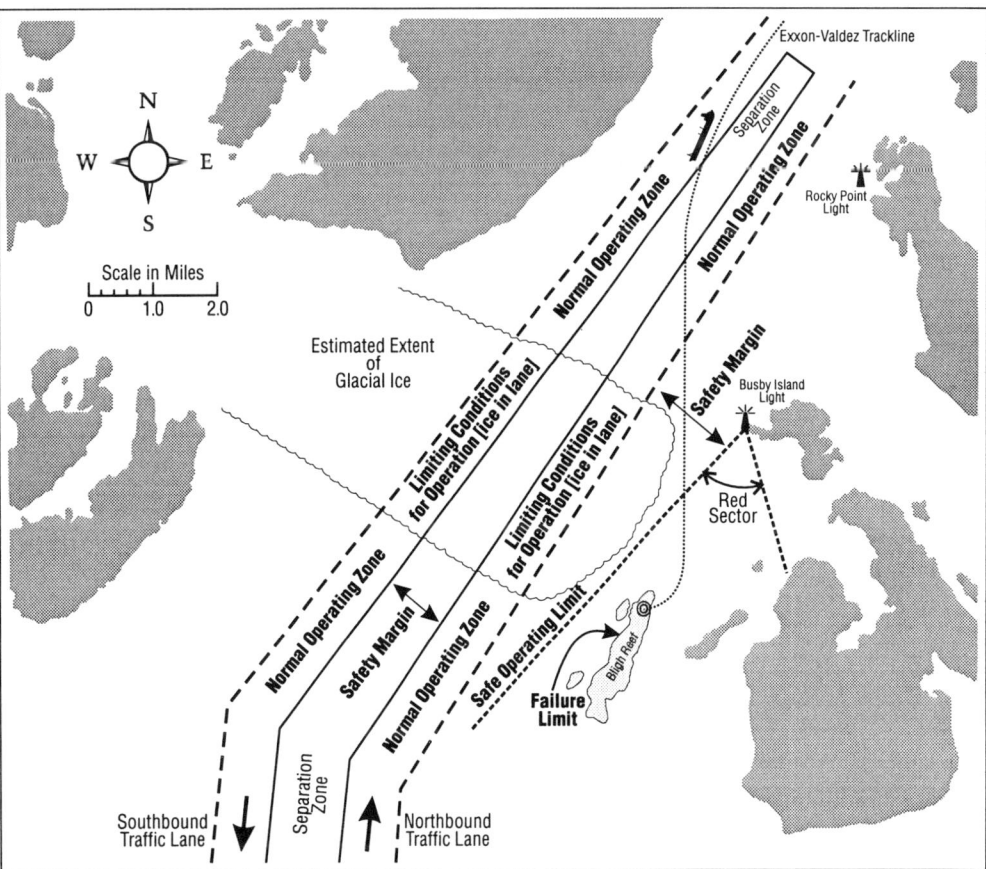

Figure 4-3: Prince William Sound, a "real-world" operating envelope example. (See also Fig. 2-2.)

Boundaries of Safe Operation

Defining Failure

In machinery operations, **failure** is when a material, component, or system ceases to meet its specified performance requirements. Usually, failure results from the deterioration or destruction of a part, component, or system beyond a specified limit. In motor vehicle operation, for example, a broken piston rod would clearly constitute failure since the part can no longer perform its intended function.

Not all deterioration, however, constitutes failure. Parts, components, and systems are designed with bands of tolerance for production and operation. The expected lifespan of an automobile engine, for example, is predicated on normal engine wear—the slow deterioration of components under reasonable usage. So long as the components remain within the tolerances specified by the designer, the engine, with proper care, will probably serve a full design life.

Practically speaking, some deterioration is expected and accepted. Clearly, the definition of *failure* for a part, component, or system must be linked to an amount of deterioration that interferes with the normal mission, well-being, and life of equipment, its operators, and the environment in which it functions. In the case of the *Exxon-Valdez*, failure occurred when the ship struck Bligh Reef.

Failure Limit

For failure to be usefully defined, a failure threshold must be established. For simplicity, we shall term this threshold a *failure limit* (sometimes called a design limit).

Literally every critical parameter has a *failure limit*. Simply stated, a **failure limit** is the maximum limit of a physical parameter beyond which failure of the material, component, or system is likely.

A failure limit provides a basis against which design decisions can be made to preclude failure. (See Figure 4-2.) In the *Exxon-Valdez* mishap, the failure limit was that distance from the reef at which the hull of the tankship would contact the reef based upon ship's draft and tide conditions.

The area beyond the failure limit is termed the *failure zone*. As we learned in Chapter 3, the **failure zone** is a region in which a situation or condition for a given parameter of a material, a component, or an integrated system of components has been shown by previous experience, experimentation, or simulation to be unacceptably harmful to the equipment, the operators, the environment, or the general public.

Uncertainty Margin

Designers usually know when failure *has* occurred. In fact, a failure limit is simply a line drawn to exclude all previously known, modeled, or predicted points of failure. But, for many technologies, designers are unable to precisely determine when failure *will* occur for a particular parameter. Sometimes this is because the failure phenomenon under consideration develops very slowly. Other times it is because the onset of failure may be affected by unknown or changing factors. Finally, uncertainty may result from an inability to measure well enough to detect the exact onset.

When there is uncertainty regarding the onset of failure, designers incorporate safety margin to guarantee that, under the worst case circumstances, the failure zone will not be reached if the equipment is controlled as intended. Therefore, **uncertainty margin** (also called design margin) is that margin added by the designer to account for uncertainties regarding the onset of failure for a critical parameter. In Prince William Sound, the distance between Bligh Reef and the dividing line between the red and white sectors of Busby Island Light is a good illustration of uncertainty margin.

Safety Limit

To preclude surpassing a failure limit and to avoid the uncertainties of the onset of failure, the next step in developing limits is to impose a safe operating limit.

A safe operating limit is the lower boundary of uncertainty margin. Recall that it is a limit for the parameter under consideration which (if not exceeded) ensures that failure will not occur. Had the crew of the *Exxon-Valdez* not encroached into the red sector established by Busby Island Light, the ship would not have struck the reef.

Factors Affecting Failure

In order to determine a normal operating limit, two other areas of safety margin must be calculated. They are termed *equipment margin* and *operating margin*. Each is composed of a series of tolerances for factors which affect the critical parameter under consideration.

Figure 4-4 (Vehicle Safe Following Distance) provides a simple example of this concept. Motor vehicle safe following distance can be affected by many factors. The design and condition of the vehicles involved in the event, the skill and condition of the drivers, and the driving environment all influence whether a rear-end collision (failure) occurs when the leading vehicle slows or stops.

To establish a reasonable safety margin for the safe following distance critical parameter, each of those factors must be considered. If a factor is determined to have a significant effect on whether a collision will or will not occur, the operating limit must incorporate a conservative element of margin to account for that factor.

A partial list of machinery and operating variables that affect safe following distance is provided below. To simplify the analogy, we assume that the vehicle is of a given weight and design traveling at a fixed rate of speed on concrete pavement.

Equipment Margin Items:
- Condition of the front disk brake rotors and pads
- Condition of the rear brake drums and shoes
- Condition of the front and rear tires
- Mechanical coupling of the linkages which operate the front and rear braking mechanisms
- Hydraulic coupling of the hydraulic braking systems (affected by amount of residual air in the system)
- Accuracy of the speedometer (verified to be within design tolerance at the time of installation)

Operating Margin Items:
- Normal operator reaction time from viewing the brake lights of the vehicle ahead to applying brakes of the vehicle being driven (verified as within legal limits during the driver's most recent examination)
- Speedometer instrument drift
- Degradations in visual acuity and depth perception of the driver since the last driver's examination
- Minor operating errors (such as having one's attention momentarily distracted in advance of the event resulting in a delayed response)

If we were considering speed limit through the Valdez Narrows rather than safe following distance, factors of concern would include how long it takes to reverse engines upon discovering an imminent hazard, how long the engines require to slow the vessel, turning radius of a typical tankship, rudder response time, and a host of other ingredients.

Equipment Margin Equipment margin is safety margin incorporated into the operating limit to account for inexact and delayed component response. All components (including instruments used to sense and indi-

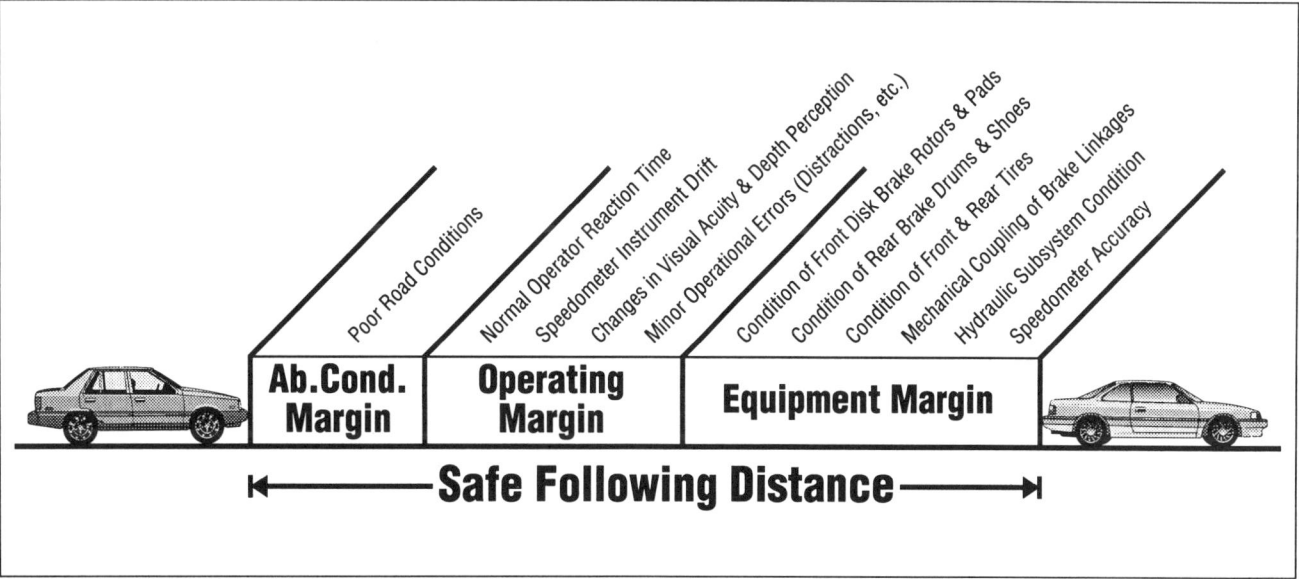

Figure 4-4: Vehicle Safe Following Distance.

Boundaries of Safe Operation

cate the status of critical parameters) contain an inherent amount of error due to manufacturing tolerance deviations and unavoidable response delay times.

Previously, we learned that important operating values called critical parameters are sensed and instrumented so that operators can monitor and control them. In fact, instruments which provide important safety readings are required to be *calibrated*—aligned to a known standard—to ensure that they read accurately and within the tolerances specified by design. But, no matter how good, instruments do not provide perfectly accurate readings. They have inherent errors including manufacturing tolerances, range non-linearities, and response delays.

To account for inherent instrument (and component) error and delay, a segment of equipment margin must be added to compensate for acceptable component and instrument inaccuracies. In the case of our safe following distance example, an increment is added for safety-related factors such as tire tread depth and condition of the brakes.

Operating Margin

Operating margin is a safety margin factor incorporated into normal operating limit calculations to account for reasonable and expected changes that occur during (or as a result of) operation. It includes such things as instrument drift, average human response delays, and minor operating errors.

People have response delay times just as instruments. When the brake lights of a car you are following light, you can't respond instantaneously. You have to allow for response delay in establishing safe following distance. The same is true for human response to equipment anomalies. Therefore, response delay must be a component in operating margin.

Also, during normal operations, instruments drift out of their calibration tolerances. To monitor and correct drift, most critical machinery parameters are checked in accordance with a periodic schedule. The purpose of the checks is to identify abnormalities or unhealthy trends in time for corrective action. Such a schedule is a part of preventive maintenance. If a parameter-sensing unit or an indicating instrument is found to be performing outside of the tolerances prescribed by design specifications, that unit or instrument must be adjusted.

Minor Operating Errors

One operating margin element deserving special discussion is the concept of *minor operating error*. As we have seen too many times already, humans make mistakes. Operating margin must, therefore, incorporate tolerance for minor errors which occur during equipment operation.

For our purposes, a **minor operating error** is defined as an error which constitutes (or leads to) a minor mistake or delay of action beyond normal reaction time (usually prescribed as a time limit). For the safe following distance operating limit, a minor operating error might be failure to notice the brake lights of the leading vehicle for a full second (which is 88 feet at 60 mph). Adjusting the radio and not watching the road (an error which may lead to an unsafe act) is another.

Depending on the parameter and the methods available to recognize changes, a reasonable delay in observing and responding to an abnormal machinery symptom might be a few seconds, a few minutes, or a few hours. Certainly the complexity of the machinery involved and the seriousness of the situation would have to be considered in the definition.

Few machines, however, can offset the effects of a *major* operating error. For example, in developing the operating limit for safe following distance, no margin of safety is added for an intoxicated driver. The philosophy of safe vehicle operation assumes that an operator's normal skills won't be impaired beyond specified limits. The same is true for industrial operation.

Normal Operating Limit

With all three regions of safety margin accounted for—uncertainty margin, equipment margin, and operating margin—a normal operating limit can be established. In the safe following distance analogy, the summation of all margin increments provides a measurement of minimum safe following distance for optimum road conditions.

The operating region bounded by normal operating limits is termed the **zone of normal operation**. For safe following distance, the zone of normal operation would include all distances up to the minimum for a given rate of speed as long as pavement conditions and visibility are optimum.

Sometimes, it seems, normal operating limits are administratively established more conservatively than necessary for a specific operator or piece of equipment. Such limits usually are imposed through the judgment of knowledgeable authorities based on the characteristics of a representative group of operators or machines. A speed limit is just such an example. It is administratively established by a state or county highway commission based on assumptions about typical vehicle performance, typical operator performance, and best-case road conditions. Though *you* and *your* vehicle might travel safely at higher speeds, the maximums (or minimums) are established to facilitate safe, efficient traffic flow *under normal circumstances*.

Abnormal Condition Margin

One last consideration is needed to complete our discussion of operating limits. For less than optimal conditions, it is necessary to add an extra margin of safety—an **abnormal condition margin**.

As we learned earlier, environmental abnormalities (or other unusual conditions) may require a self-imposed limit more restrictive than the normal operating limit. In Figure 4-4, additional safety margin has been added to account for poor road conditions. If the road surface is ice-covered, extra stopping time will be required to avoid failure.

The decision to add safety margin for abnormal conditions often depends upon an operator's ability to recognize abnormalities in the operating environment. If the pavement is dry, vision is not obscured, and the vehicle has no significant performance deficiencies, a minimum safe following distance (or time) may be observed. But, if road conditions are not ideal, the driver must make a judgment whether to reduce speed based upon the environment. Usually, if conditions are degraded, additional restrictive margin—margin beyond normal operating limits—is warranted.

As previously stated, a condition which requires a limit more restrictive than the normal operating limit is a **limiting condition of operation**. It may be based either upon the condition of the equipment, the condition of the operating environment, or the condition of the operator.

The Safe Operating Envelope

The foregoing safe following distance illustration portrayed only one operating limit for prudent vehicle operation. In reality, there are dozens of critical parameters and associated operating limits. It is clear that, for any operating component, system, or plant, nearly every critical parameter will have its own design safeguards, construction requirements, and operating limits. As such, it is equally clear that developing the safe operating boundaries for an industrial facility is a very complex task.

Definition of the Envelope

As discussed earlier in this chapter, the objective of safety analysis is to protect the operators, environment, and general public from the hazards of operation. This objective is achieved through a composite of design features, construction characteristics, operating limitations, maintenance restrictions, and administrative prerequisites.

One of the most important outgrowths of safety analysis is the *safe operating envelope*. Previously we learned that a **safe operating envelope** is a graphic or verbal representation (usually both) of the boundaries and safety margins established for critical parameters of machinery operation. These margins and boundaries are formulated to protect the operators and their equipment, the environment, and the general public from the harmful effects of the hazards (determined through hazards analysis) characteristic of the technology.

A properly developed safe operating envelope is one of the most helpful and important products of an industrial facility's design and development process. Skilled operators find a well-defined envelope indispensable for prudent and efficient equipment operation.

Origin of the Envelope

The safe operating envelope concept is germane to almost every kind of equipment operation. Skiers talk of *pushing the envelope*. Mountaineers establish safe operating envelopes for each climb and each pitch of the climb. Firing ranges have established safe operating envelope boundaries and rules with accompanying margins of safety. Racing-car drivers learn to operate their vehicles safely at the edge of their envelope.

As technologies have grown more complex and as the risk to operators, public, and environment have increased; military and industrial organizations have established formal methods for identifying hazards, assessing accident risks, and developing devices, controls, and procedures to preclude their occurrence. Leaders in this process include the branches of our armed forces, particularly the U.S. Air Force and Naval Nuclear Power Program; NASA; and commercial nuclear power generating utilities and their consortium, the Institute of Nuclear Power Operations. More recently, two other groups—chemical manufacturing/processing industries and Department of Energy facilities dealing with radioactive waste—have made great progress in safe operating envelope development.

In general, those technologies with the greatest risk have had the greatest incentive to systematically develop safe operating envelopes. High risk operations demand a thorough, systematic approach to the development of the boundaries. Low risk operations may function well with less stringent and less formally established boundaries. But, regardless of the level of risk, the basic process of designing the safe operating envelope is the same.

Operating Maps Since most complex equipment is characterized by numerous parameters which are critical to safe equipment operation, a representative safe operating envelope would have to be multi-dimensional. Therefore, a single, comprehensive, visual representation of an overall safe operating envelope for a facility (or even a complex piece of equipment) is not practical.

For that reason, safe operating envelopes often employ two-dimensional representations of a few of the most important parameters together with written procedures to depict equipment operating limits. A two-dimensional graphic representation of important operating boundaries for a piece of equipment is often referred to as an **operating map**.

Operating maps are provided to assist operators in maintaining equipment parameters within safe, specified bands of operation. Since the human mind tends to think in pictures, operating maps provide clear, visual representations. They serve in much the same way as painted traffic lane markers and road signs on public thoroughfares.

The navigational charts for Prince William Sound are an excellent example of an operating map. The charts clearly show normal operating limits, safety limits, and other visible protective features.

A combination of operating maps, procedures, and instructions governing the control of critical

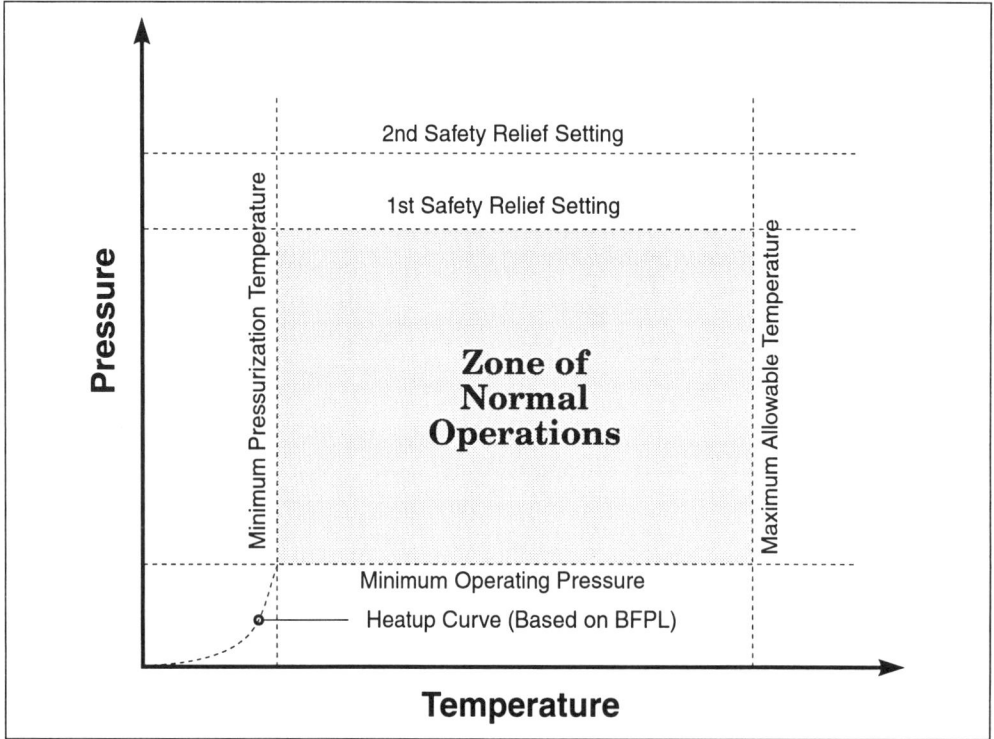

Figure 4-5: Steam Boiler Operating Map.

operating parameters creates the composite safe operating envelope for an industrial operation. This composite delineates the most important safety boundaries by which facility operators must operate and maintain in order to stay within design limits.

Typical Operating Map

A pressure vessel is a good example of a component governed by an operating map. Figure 4-5 depicts the relationship between boiler temperature and steam pressure. Though many other parameters such as feed-water level and chemical concentrations affect boiler operation, temperature and pressure are of greatest concern. They are related and are easily compared.

Most complex systems have a number of different operating maps for the many different types of equipment in the facility. For example, a pressurized water reactor will have one operating map for the primary coolant boundaries and another for the steam generators. Electrical equipment, pumps, and condensers may also employ operating maps to illustrate operating limitations.

Boundaries for other than the critical parameters are usually defined by written procedures or operating instructions. Together, the visual representations and verbal descriptions of safety boundaries and margins create for the operator a safe operating envelope.

Changes in the Envelope

Safe operating envelopes often change as a result of changing environments and circumstances. Just as the normal operating zone for a vehicle becomes more limiting as road conditions deteriorate, so will safety margins and boundaries for equipment operation under varying conditions.

In Prince William Sound, retreat of Columbia Glacier confounded the original safety analysis. The hazards upon which the safety analysis was originally predicated slowly changed over a ten-year period beginning about 1979. Protection, however, was not modified to address changing hazards.

Industrial operations are no different. New hazards (not originally envisioned) can develop subtly. To offset the effects of a changing operating environment, industrial teams and their leaders must be taught to recognize hazardous conditions not accounted for in the original operating envelope.

One common source of change is unavoidable degradation of materials used in equipment construction. For example, the pressure containment vessel for a pressurized water reactor begins its life rated for maximum pressures at certain minimum temperatures. But, exposure to neutrons from the fission process slowly embrittles the containment vessel—it loses some of its ductility. Thus, later in its rated life, it can't be pressurized above specified limits without first being heated to higher temperatures than originally required if the phenomenon of brittle fracture is to be avoided.

As a result of changes in operating environment, in equipment, and sometimes in capabilities of operators, periodic formal re-evaluation of safe operating envelopes is both necessary and prudent.

Topic Summary and Questions to Consider

Understanding the design limitations and the operating boundaries of equipment in an industrial environment is a critical part of the strategy for operating success.

Members of an operating team can successfully operate and maintain equipment only if they understand how equipment design interrelates with operator control and the operating environment. Without thorough knowledge of safe operating boundaries and limits, industrial team members—whether operators, maintenance technicians, or engineers—are predisposed to failure.

Ask Yourself

Are the design, construction, and operation of your plant or process based on formal safety studies? Do you have and periodically review those studies? Are they current? Are safety studies revised to reflect modifications to safety-related equipment?

Do your team members know and abide by established operating limits? Do they know the basis for each of those limits? Do they know when (and how) to invoke abnormal operating limits?

The well-being of your team, equipment, and facilities depends on your answers to these critical questions. Without thorough up-to-date analyses, operating limits are arbitrary at best and cannot guarantee safety. If operating limits and their bases aren't thoroughly understood, they won't be heeded. And if operating limits aren't heeded and enforced, safety is seriously compromised.

Chapter 5

Principles of Operation

The safe operating envelope, whether in procedural or graphic form, serves to transmit important operating constraints which must be observed if equipment is to consistently function as designed. It is the team members' road map for staying within the boundaries of safe operation. It also serves the operator as a guide to gauge the nearness of critical equipment parameters to danger or failure.

Most critical parameters are within the operator's control. A few are not. But, one thing is certain: rules alone will *not* guarantee safety. Safe operation is vitally dependent upon operators *choosing* to control machinery within established boundaries based on current "driving" conditions.

Building on our discussion of the safe operating envelope concept, this chapter presents a cardinal rule of operation and two fundamental principles for observing safe operating boundaries.

Chernobyl Atomic Power Station Accident, April 1986

Just three months after the explosion of the space shuttle *Challenger*, reports from Scandinavian countries of elevated radiological contaminants in the environment hinted of a radiological disaster in the Soviet Union. The world would soon learn of a catastrophic reactor accident, the magnitude of which had not before been experienced in the history of nuclear-powered electrical generation.

During the early morning hours of April 26, 1986, two explosions destroyed the reactor and part of the Unit 4 reactor building at the Chernobyl Atomic Power Station near Kiev, USSR. The explosions were caused by a severe, uncontrolled power excursion which resulted in massive thermal damage of Unit 4's reactor core. An unparalleled spread of radioactive contaminants blanketed the area for hundreds of miles around the site and a radioactive cloud spread eastward over much of Scandinavia. Chernobyl would be known as the world's worst reactor accident.

RBMK-1000 Description

The RBMK-1000 reactor units (Figure 5-1) are large, boiling water reactors with graphite-moderated uranium cores. Unit 4 was one of four on-line RBMK units at Chernobyl. A fifth unit was under construction. All were rated at 1,000 megawatts of electrical power and 3,200 megawatts of thermal power. All were capable of plutonium production for military weapons through the conversion of Uranium (U^{238}) fuel to Plutonium (Pu^{239}). Unit 4, however, had been run only for energy production since it began operation in December, 1983.

RBMKs operate at an average coolant temperature and pressure of approximately 530°F and 986 psig. The 14 meter diameter, 7 meter high cylindrical reactor core uses 1,700 tons of graphite blocks which are honeycombed vertically with 2,000 thin-wall pressure tubes designed to accept uranium fuel rod clusters, control rods, and instrumentation. Approximately 1,700 of the pressure tubes accept fuel clusters, consisting of 18 zirconium-clad, cylindrical fuel rods, all supported by a central support tube. Each fuel rod casing is hollow and is loaded with pellets of uranium-oxide fuel.

Power Generation

To cool the fuel rods during operation and generate the steam used to power the RBMK's two turbine-generators, water is pumped by main circulating pumps into headers which distribute the cooling water into the pressure tubes. (See Figure 5-2.) Water flows around individual rods in the fuel clusters, absorbing the heat of fission. As it travels up the pressure tubes and around the fuel rods, the coolant is changed to a steam/water mixture and is dried by bubbling through moisture separators when entering the steam drums. From there, the steam is routed to steam turbines to power two 500 megawatt electrical generators which supply power to the electrical grid. Makeup water needed to replace the water mass converted to steam is pumped into the steam drums via two main feed pumps at a temperature of approximately 165°F.

The Heat Source

Fission from the fuel generates the heat in the core necessary to turn water into steam for use in the turbines. The fission reaction in the RBMK core is obtained using a mildly enriched U^{238} fuel. Most naturally occurring uranium is in the form of U^{238}. But, U^{238} atoms will not normally fission when struck by a neutron. U^{235} atoms, however, will fission if the impinging neutrons are travelling slowly enough. The number of U^{235} atoms in the fuel must, therefore, be plentiful enough for sufficient fissions to occur to allow a self-sustaining reaction. The process of raising the number of U^{235} atoms from their naturally occurring level of 0.7 percent is termed *enrichment*. RBMK core fuel is enriched only to 2 percent. As a result, the core must be very large.

Reactor Core Characteristics

Moderation (slowing) of neutrons requires a relatively light substance such as graphite or deuterium (heavy water). Light water may also be used but only if the level of enrichment is substantially higher, a very expensive process. If enough fission neutrons are slowed through moderation and strike U^{235} atoms, a sustainable fission reaction is possible. In the RBMK units, pure graphite is used as the moderator since it is far cheaper than heavy water to produce.

The neutron balance in a reactor core is the key to a controlled, sustained reaction. If the ratio of neutrons generated to neutrons used in fission or absorbed by other materials is less than one, the reaction is termed *subcritical* and will not be sustained at necessary levels. If the ratio is greater than one, the reaction is said to be *supercritical* and power level will grow in relation to how many additional neutrons are produced above the value for an exactly self-sustaining reaction.

If the reaction relies only upon neutrons generated directly from fission (prompt neutrons), the reaction is said to be *prompt critical*. A prompt criticality will cause the number of fissions in the core (power level) to rise uncontrollably in a split second. It is, therefore, necessary to sustain the reaction using *delayed* neutrons from the decay of other fission products emitted about a minute after the original fission.

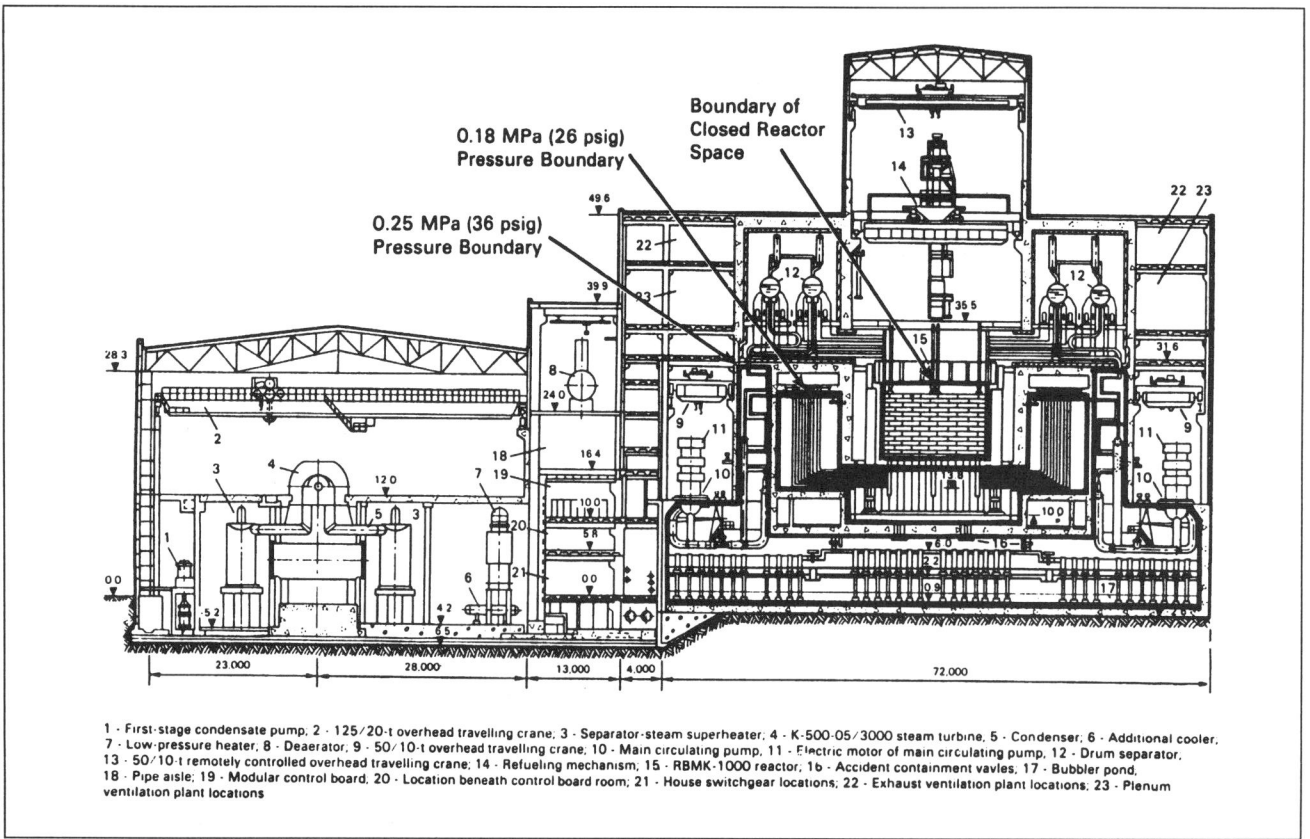

Figure 5-1: Cross-sectional view of reactor building, Charnobyl Units 3 and 4. [From NUREG 1250, Rev. 1]

Controlling the Fission Process

Control of the fission reaction is accomplished, in part, by neutron-absorbing *control rods* which are inserted or withdrawn from the core to balance the abundance or absence of other poisons. In the RBMK unit, 211 control rods could be withdrawn or inserted into the core through driving motors and inserted in case of emergency through the force of gravity. The latter function is termed a *scram* and is used in case of emergencies to rapidly shut down the fission reaction in the core. Unlike Western nation designs in which a scram results in the control rods free-falling directly into their core channels, a scram of rods in the RBMK units results in the rods dropping into their channels while attached to cables which unwind from a spool. Though still a gravity scram, the process takes approximately four times longer than the same process in a Western nation design.

Safety regulations for Unit 4 required that control rods be managed such that, at any time, an equivalent of a minimum of 30 full control rods worth of negative reactivity be inserted into the core to offset the excess positive reactivity of the core load. (In other words, it was unnecessary for 30 specific control rods to be fully inserted at a time. But the 211 control rods had to be managed in such a way that the combined value of all partially inserted rods was equivalent to 30 fully inserted rods.) This precaution was designed to allow sufficient time and margin to stop an unanticipated power excursion before it became uncontrollable. Some of the rods were automatically controlled and some manually controlled. A computer program known as the *Skala* kept the operators informed of the equivalent number of rods inserted.

RBMK-1000 Peculiarities

The Soviet RBMK-1000 design differs dramatically from U.S. and other Western nation designs which are predominantly light water reactors (LWRs). LWRs use water not only for cooling the fuel rods, but also for moderating neutrons to facilitate fission. In a LWR, if water is lost from the fuel channels during power operation, a rapid reduction in the number of fissions (power) occurs in the core and the reactor core shuts itself down. Such cores are *inherently stable*. In the RBMK design, however, cooling water pumped through the fuel rods is *not* the moderator. The graphite blocks slow the neutrons and water is used only to remove heat generated by fission in the fuel. But water also serves as a *poison* by absorbing neutrons. Should water in the cooling channels be lost during power operation in an RBMK core, neutrons otherwise absorbed by water will be moderated in the graphite and will cause *more* fissions rather than less. As a result, the core power level will rise rather than decrease. This behavior characteristic is termed a *positive void coefficient of reactivity*.

From the standpoint of safety and stability of a reactor core and its supporting systems, positive void coefficient is a very undesirable feature. The Soviets believed, however, that active control measures and operator vigilance were sufficient to offset this inherent instability. Also, in the RBMK unit, the positive void coefficient was offset (when operating above 20% power) by the reactor's *negative temperature coefficient*, the characteristic of the reactor core to thermalize (slow) fewer neutrons as temperature in the fuel and moderator rise. With less neutron slowing, fewer fissions occur and less power is generated. As a result, power operation below 20% was prohibited.

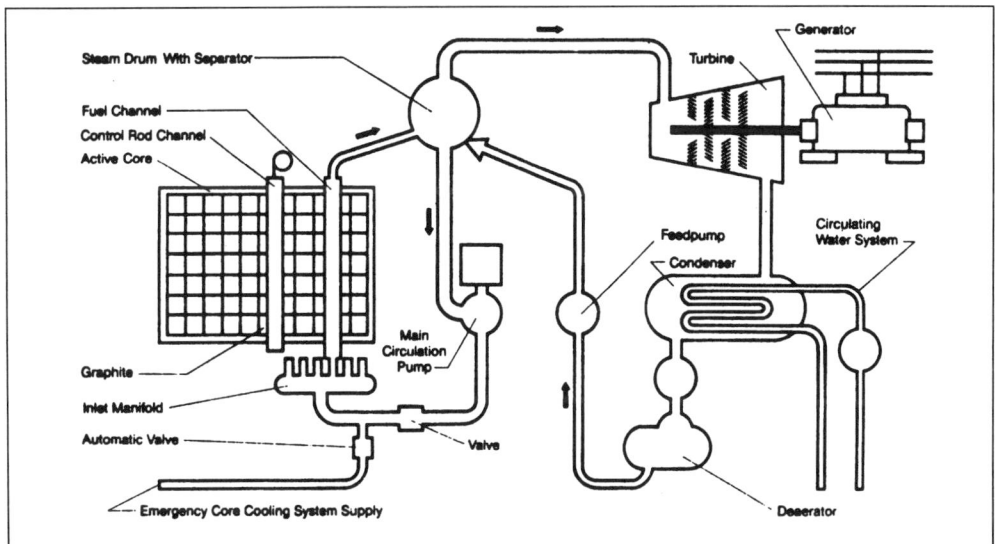

Figure 5-2: Schematic diagram of the RBMK-1000. [From NUREG 1250, Rev. 1]

Principles of Operation

Another major difference of the RBMK is the absence of a containment building. Containment buildings are protective structures designed to contain radiological contaminants even in the case of a complete core meltdown. Because of the RBMK's design to operate and refuel at the same time, the area over the top of the core must be prohibitively large. The Soviets chose (for convenience and cost) to forego a containment building.

An Engineering Test

On the afternoon of April 25, Unit 4 had been scheduled for an electrical engineering test. The test was designed to determine whether the unit's steam-powered turbine-generators could continue to generate sufficient electrical energy to power emergency loads after the turbine steam supplies were secured. If turbine coastdown energy could be used, main circulating pumps, main feed pumps, emergency cooling pumps, and other emergency electrical loads could be self-powered long enough for the diesel generators to come on line. Startup and preparation of diesel generators to carry these emergency loads required approximately three minutes.

Arrangements had been made to perform this test just prior to a scheduled reactor maintenance shutdown. The test procedure had been prepared by electrical engineers who were not qualified as plant operators and had little training in nuclear power reactors. Further, many of the unit's managers had backgrounds in conventional electrical power generation rather than nuclear power source generation. As a result, the test procedure had not received an appropriate operational review or nuclear safety review.

Disabled Safety Features

Test procedure required that the test be initiated from a power level of between 700 and 1,000 megawatts (thermal) following a reactor trip (rapid shutdown of the reactor core through insertion of control rods). Plant management, however, chose to disable portions of the reactor trip function so that the experiment could be rerun immediately if necessary. Following a reactor trip, the buildup of the neutron poison Xenon 135 (Xe^{135}) from Iodine 135 (I^{135}) decay may preclude rapid reactor restart, especially in a core such as the RBMK which is not loaded with as much fuel. By disabling the shutdown function, the reactor could be kept operating. If tripped, xenon buildup would have precluded a rapid restart and a second experiment, if required, could not have been run for days. Additionally, the test called for disabling the emergency core cooling system based upon a belief that initiation of emergency core cooling would result in a harmful thermal shock to the reactor system.

Testing Delay

At approximately 1:00 A.M. on April 25, power reduction was initiated in anticipation of starting the test at about 2:00 P.M. the same day. Twelve hours later, reactor power had been reduced to 50 percent of rating and one of Unit 4's 500 megawatt turbine-generators was secured. The second turbine-generator continued to operate and supply power for four of Unit 4's eight main circulating pumps and two main feed pumps. At 2:00 P.M. the emergency core cooling system was disabled in preparation for the test. Shortly thereafter, the electrical load dispatcher contacted Unit 4 and requested that they maintain power and continue to operate the single remaining turbine-generator to supply electrical power to the grid. Unit 4 complied and power reduction was not again resumed until approximately 11:10 P.M. that night. For nine hours the reactor operated at power with the emergency core cooling system disabled, a dangerous violation of the plant's operating limits.

Control Rod Limit Violation

As power reduction resumed, power was reduced too rapidly due to an operator's control rod programming mistake. Power fell to 30 megawatts (thermal), far below the procedural limit for power operation. Xenon poison buildup which began with the rapid power reduction created a situation in which the operating control rods could not be withdrawn quickly enough to restore power.

According to the test procedure, testing could not proceed at this power level. A shift change occurred at midnight and as the morning of April 26 began, the deputy chief of reactor operations, on scene for the test, began a tyrannical rage at the operators, coercing them to restore power and continue the test. But the only method then available to raise power was to begin withdrawing control rods in excess of the 30 control rod equivalent safety limit. The operators, under duress, did so.

Flow Restriction Violation

Then, at approximately 1:00 A.M., in preparation for the test, two more main circulating pumps were started to ensure coolant flow through the reactor core during the test. Eight main circulating pumps were now supplying coolant flow through the core. Four pumps were running on an outside power source and four were running on the turbine-generator powered by reactor steam. The concept was that, when the existing turbine-generator's steam supply was secured for the test, the operators would attempt to keep the turbine-generator-powered main circulating pumps running while the turbine coasted down. The other four main circulating pumps would continue to run on external power, supplying coolant flow through the core.

By operating all eight main circulating pumps at once, a maximum flow restriction was violated. The result was a reduction in the amount of steam in the reactor coolant channels and a subsequent reduction in fissions in the core due to the effect of the core's positive void coefficient. In order to offset this new reduction in power, operators elected to withdraw rods farther beyond the 30 rod safety limit to the point that only 7 or 8 rods worth of safety margin remained.

Prompt Criticality

With power barely stable at approximately 200 megawatts (thermal), the test was initiated at 1:23:04 A.M. by securing steam supply to the running turbine generator. As the turbine began its coastdown, flow through the core reduced as the turbine-generator-powered pumps slowed. The flow reduction allowed heat to increase in the reactor coolant and steam voiding (boiling) occurred in the core. The positive void coefficient asserted itself and a reactor power rise began. Operators saw the rise in power and activated motor-driven control rod insertion to counter the rise. The motor-driven rod insertion was insufficient to turn the power rise, so the operators initiated a rod scram. Unfortunately, because of rod design, the first portion of the control rods to be inserted were their graphite ends. As a moderator, the graphite ends actually *raised* the fission rate so rapidly that a prompt criticality occurred before the neutron absorbing portion of the rods could take effect. The rods traveled only 2.5 meters of their 7 meter length before heat distortions in the core obstructed their paths. Power rose in a split second to as much as 100 times the maximum rated value.

Steam and Hydrogen Explosions

Within 40 seconds of initiating the test, a steam excursion of explosive proportions occurred in the reactor core, disrupting the inner core configuration and bouncing hundreds of the 770 pound head blocks above the core channels from their seats. Graphite blocks and fuel rods fragmented. A second explosion, probably a hydrogen explosion, occurred shortly after the first. As superheated steam contacted the zirconium fuel rod cladding, a metal-water reaction probably caused a release of hydrogen. The hydrogen, now exposed to air, mixed with oxygen and sparked an explosion which lifted the 1,000 ton reactor head off the core. Graphite and fuel were hurled out of the destroyed reactor building and the ensuing graphite fire spread an estimated 50 million curies of radiological contaminants over the Soviet Union, Scandinavia, and Europe.

Consequences

The heroic effort of on-scene personnel to battle this catastrophe is almost legendary. Two operators died as a result of the explosion. Twenty-nine died as a result of combatting the disaster. Estimates of radiation exposure to the total population range from 20–50 million man-rem. But the final casualty toll will not be known for years. Scientists believe that approximately 9,000 lives will ultimately be forfeited as a result of the Chernobyl accident.

The Primary Cause

The Chernobyl disaster is a monument to the failure of humans to understand the design and operating constraints which govern safe and efficient equipment operation. Investigation of the event by the USSR State Committee for the Use of Atomic Energy concluded:

> *In the process of preparing for and conducting tests of a turbine generator in a coastdown mode with a load of system auxiliaries of the unit, the personnel disengaged a number of reactor protection devices and violated the important conditions of the operating regulations in the section on safe performance of the operating procedures.... The basic motive in the behavior of the personnel was the attempt to complete the tests more quickly. Violation of the established order in preparation for and performance of the tests, violation of the testing program itself and carelessness in control of the reactor maintenance attest to inadequate understanding on the part of the personnel of how to implement operating procedures in a nuclear reactor, and to their loss of a sense of danger. The developers of the reactor installation did not envisage the creation of protective safety systems capable of preventing an accident in the presence of the set of premeditated diversions of reactor protection systems and violations of operating regulations which occurred, since they considered such a set of events impossible. An extremely improbable combination of procedure violations and operating conditions tolerated by personnel of the power unit thus was the original cause of the accident.*

Principles of Operation

Though the archaic RBMK design contributed dramatically to this event, the direct cause of this accident was willful and deliberate violation of stringent operating limits by certified plant operators and their supervisors as they tried to conduct an electrical engineering test during a brief window of opportunity. In so doing, they ignored important performance characteristics of their equipment and the consequences of violating crucial equipment operating boundaries.

Principles of Operation

Competent equipment design always incorporates conservatism and safeguards to protect against hazards. Yet, no design can compensate for operating personnel who choose to mismanage machinery beyond its design constraints.

Chernobyl is a clear illustration of that concept. Had operating personnel simply followed procedure and controlled their equipment within design constraints, the accident would have been prevented.

Cardinal Rule of Operation In Chapter 3, we learned that the objective of the strategy for operating success is to intelligently control equipment. One rule summarizes better than all others this responsibility. Here it is termed the *cardinal rule of operation*.

Simply stated, the cardinal rule is:

> **Humans must remain in control of their machinery at all times. Any time the machinery operates without the knowledge, understanding, and assent of its human controllers, the machinery is out of control.**

At Chernobyl, the operators lost control. They didn't understand their equipment and nuclear technology well enough to sense impending failure. The officers of the *Exxon-Valdez* and the VTC operating personnel performed little better.

Both at Chernobyl and in Prince William Sound, computers played a role in equipment control. The cardinal rule does not preclude using computers to control equipment. But it *does* require operators to monitor the critical parameters associated with functions that computers control. Humans are *always* responsible.

From the cardinal rule are derived two fundamental principles. They address the need to stay within operating boundaries and prescribe general action when boundaries are exceeded.

Principle 1 **Operate and maintain equipment within the boundaries of the safe operating envelope.**

As we learned in Chapter 4, industrial operations are bound by operating constraints and maintenance limits fashioned to keep plant equipment functioning within design specifications. In the *Exxon-Valdez* accident, the boundaries deteriorated over a period of years as they were routinely violated by shipping company and Coast Guard employees alike. In the Chernobyl accident, strict operating limits were breached with impunity, apparently with the blessing of leaders whose primary responsibilities included enforcing the limits.

Staying within the boundaries requires operators to both *know* and *respect* them. At Chernobyl the limits were certainly not respected, raising great doubt as to whether they were understood.

For boundaries to have value and meaning, operators must understand their bases. In Chapter 18, we will review the elements necessary for a training program to succeed. Yet, the best of training is negated by poor leadership such as that exercised at Chernobyl.

Principle 2 **If an anomaly carries you outside the boundaries of the envelope, place the equipment in a safe and stable condition and notify appropriate authorities.**

The second principle recognizes that limits will sometimes be exceeded. Unforeseen conditions, lapses in attention, equipment malfunction—all may lead to excursions beyond normal (and even safe) operating limits. (This, in part, is why the safe operating envelope incorporates so much conservatism.)

Violations of operating envelope boundaries vary in seriousness. Exceeding a normal operating limit is not as dangerous as violating a safety limit. Each facility must establish policies for actions required in the event of departure from the normal operating zone. At a minimum, deviations from the zone of normal operations should be logged and reported.

If a *safety* limit is violated, though, an engineering evaluation is usually necessary to determine whether (or the extent to which) equipment was damaged. Recall that a safe operating limit is the lower boundary of the area of uncertainty regarding failure onset. Therefore, safety limit violations raise the question of whether or not the failure limit has been reached. If damage occurred, equipment operation must cease until repaired, or additional operating constraints must be imposed.

In the *Exxon-Valdez* accident, a safety limit was violated when the tanker crossed into the red sector of Busby Island Light. Then, a failure limit was breached when the tankship ran aground. The master's initial actions were proper. He ordered the chief mate to perform stress and stability calculations (an engineering evaluation) to determine sea-worthiness of the hull and its structural members. After performing the calculations, the chief mate phoned the master and told him that the ship was "marginal", meaning the ship's structural members were holding but were on the edge of their stress limits. He advised the master not to move the tanker. The master acknowledged, but still tried to force the tanker off the reef for a period of nearly one full hour.

The master's actions clearly violated the second principle of operation. When excursion beyond a safety limit occurs, equipment normally should be immediately shut down and stabilized.

Usually, "shut down and stabilized" means returning parameters back to the zone of normal operations. Situations may arise, however, where conditions are so unstable that sudden changes cannot—and should not—be made. For example, if a steam boiler's temperature falls below its nil ductility limit, rapid temperature or pressure changes may cause catastrophic brittle fracture failure. Careful, slow pressure and temperature changes are required to stabilize critical parameters within acceptable limits, prior to depressurizing and cooling the vessel.

Understandably, equipment operators and their leaders must be thoroughly familiar with the characteristics and capabilities of their machinery and unusual conditions which they may encounter. Then, when anomalies occur, they are prepared to manage their equipment safely.

Limiting Condition Corollary

We observed in Chapter 4 that normal operating limits are designed for normal operating conditions. Operating abnormalities, however, require operators to selectively impose more restrictive operating limits (based on good judgment) to ensure safe operation.

Recognizing this concept, a corollary is offered:

> **When circumstances are not normal, it is incumbent upon equipment operators to self-impose more restrictive limits to ensure that equipment is operated in such a manner as to avoid failure.**

For Prince William Sound, we learned that some shipping companies did impose abnormal operating limits (daylight travel and 6 knots maximum speed) when ice conditions across the lanes were severe. Though the ice conditions causing the imposed abnormal operating limit never changed, the limit was abrogated when tanker captains began to routinely steam around the ice, an action which brought them precariously close to Bligh Reef.

Topic Summary and Questions to Consider

From this brief study of operating principles, it is clear that successful machinery operation relies, not only upon proper design and construction, but, most importantly, upon competent operation and maintenance. Safe operating boundaries are developed as guidelines to assist operating personnel in controlling their equipment within design specifications. They are, however, of little use to operators who do not understand them or purposefully choose to ignore them.

In designing Chernobyl's RBMK-1000 nuclear generating unit, the hazard of principal concern was an uncontrolled release of fission products to the environment. Plant systems, protective features, and operating controls were devised to provide margins of safety against such a release. Yet, by disabling automatic shutdown features, withdrawing control rods beyond the 30 rod equivalent safety limit, violating the minimum power level for operation guidelines, and disobeying the minimum power restriction for test initiation, these operators and their leaders were able to breach the failure zone for a host of this reactor plant's components and systems. One critical breach was the reactor's pressure boundary. The pressure boundary of the closed reactor space was fashioned to contain the rupture of only one of the 2,000 pressure tubes. However, most, if not all, ruptured during Unit 4's severe power transient and subsequent steam explosion, resulting in the catastrophic release of intensely radioactive fission products to the environment.

At Chernobyl, many critical parameters associated with reactor control were ignored and the margins of safety for this operating system were obliterated. The actions of the operators and their leaders imply that either they did not understand the significance of their operating limits, or they felt that they could compensate for the reduced margin of protection by their own knowledge and skills. In either case, they were proven inadequate to the task. Thousands have suffered as a result. This was an operating "team" that did not understand the principles of safe operation.

Ask Yourself How does *your* team compare? Does *your* team understand and apply these principles of safe operation? If not—or if excursions outside established operating limits for your equipment and processes are frequent—it's just a matter of time before you have a serious accident.

Chapter 6

The Alert, Well-Trained Operator

Following the Three-Mile Island accident, President Carter asked Admiral Hyman G. Rickover (his former mentor) to study the accident and circumstances surrounding it, and to outline underlying weaknesses which led to the accident. In December of 1979, Rickover reported deficient training as one of the principal accident contributors. Then, in discussing the success of his own creation, the highly successful Naval Nuclear Power Program, he told Congress:

> ...[T]he basic thing we depend on is training of people.... If you have to depend on people then they must know what they are doing. That means training not only once but constantly.

Alert, well-trained operators are perhaps the most important element of the (Chapter 3) operating strategy. Operators are the final defense against failure and are ultimately responsible for safe and successful operations (remember the Cardinal Rule). Excellent design and construction must be supplemented with superior operating and maintenance instructions and then supported with the best-selected, best-trained people possible. This chapter, therefore, focuses on the selection and training that begins the process of developing highly skilled industrial team members.

Big Bayou Canot Bridge Accident, September 1993

While lost in a heavy fog in the early morning hours of September 22, 1993, the towboat *Mauvilla*, pushing a cargo of six barges, struck the Big Bayou Canot Railroad Bridge, dislocating a section of rail on a CSXT rail line. Eight minutes later at 2:53 A.M., the *Sunset Limited*, an Amtrak passenger train, derailed while crossing the bridge, plunging all three locomotives, the baggage and dormitory

Figure 6-1: Big Bayou Canot Bridge accident site. [From NTSB Report]

The Alert, Well-Trained Operator

cars, and two of six passenger cars into the Big Bayou Canot tributary of the Mobile River. The locomotives, baggage cars, and dormitory cars were quickly engulfed in flame when fuel tanks aboard the locomotives ruptured. Despite heroic rescue efforts mounted by Amtrak passengers and crew, the Coast Guard, the crew of *Mauvilla*, and other vessels in the area, forty-two passengers and five crewmembers of *Sunset Limited* were killed and 103 passengers injured.

Summarizing the conclusions of its accident investigation, the National Transportation Safety Board (NTSB) stated:

> …*the probable causes of Amtrak train 2's derailment were the displacement of the Big Bayou Canot railroad bridge when it was struck by the MAUVILLA and tow as a result of the MAUVILLA's pilot becoming lost and disoriented in the dense fog because of (1) the pilot's lack of radar navigation competency; (2) Warrior & Gulf Navigation Company's failure to ensure that its pilot was competent to use radar to navigate his tow during periods of reduced visibility; and (3) the U.S. Coast Guard's failure to establish higher standards for inland towing vessel operator licensing.*

The Accident A few minutes before 1:00 A.M. on Wednesday, September 22, 1993, the *Mauvilla*, a towboat owned and operated by Warrior & Gulf Navigation Company (W&GN), departed the National Marine Fleet northward on the Mobile River pushing an array of six barges upriver, three abreast. *Mauvilla* had just completed a trip downriver to the National Marine Fleet docks (at mile 5), and had picked up a new tow cargo of barges. (Locations on the Mobile River are designated by mile markers along the river in a northerly direction, starting at mile 0 in the city of Mobile.)

Staffed by a crew of four—a captain, a pilot, and two deckhands—*Mauvilla* acquired a replacement pilot on the trip southward as a substitute for one of her own licensed tow operators who had become sick. (Tows of *Mauvilla*'s class are required to have on board two Coast Guard licensed tow operators. The senior is the "captain" and the junior is the "pilot".) At the time of departure, the replacement pilot, a thirteen-year employee of W&GN, was serving his midnight to 6:00 A.M. navigation watch. The captain and pilot, according to W&GN operating procedure, exchanged duty as navigation watch every six hours, the captain normally standing watches from 6:00 A.M. to noon and again from 6:00 P.M. until midnight. The two deck hands alternated on a similar schedule.

As *Mauvilla* departed the National Marine Fleet, the weather was clear with visibility at 4 miles and air temperature of 72°F. Fog had not been forecast by the National Weather Service, but was known to form quickly along the river. *Mauvilla* pulled out behind another tow, the *Thomas B. McCabe*, and followed her upriver to Twelve Mile Island with *McCabe*'s stern lights in view.

As *Mauvilla* approached Catfish Bayou near mile 8.6 on the port passage around Twelve Mile Island (Figure 6-2), visibility began to degrade as the towboat entered an area of fog. *Mauvilla*'s pilot slowed the towboat and radioed ahead to *McCabe*'s tow operator, learning that visibility near the north end of the island had closed to zero. By the time *Mauvilla* reached mileboard 9.6 at approximately 2:30 A.M., the pilot could no longer see the front of his tow string, some 200 feet ahead.

While *Mauvilla* searched its route along the head of Twelve Mile Island, Amtrak Train Number 2, designated the *Sunset Limited*, on an eastward journey from Los Angeles to Miami, arrived in the city of Mobile. Approximately three hours earlier, the *Sunset Limited* had departed New Orleans after a brief delay for minor repairs. *Sunset Limited*'s crew of 18 embarked with 202 passengers in a northeasterly direction on the CSXT rail line, 33 minutes behind schedule. The route crossed several rail bridges spanning numerous tributaries, including the Big Bayou Canot tributary, which emptied into the Mobile River from its western drainages. About four miles north of Big Bayou Canot, the Amtrak liner would cross the Mobile River over Fourteen Mile Bridge, wending its way eastward.

Having lost sight of the *Thomas B. McCabe* and with the fog closing, *Mauvilla*'s pilot began looking for a place to tie up on the starboard bank, an accepted and common practice among tow operators during periods of limited visibility. Trees or other fixed objects along the river bank serve as common tie-off points. In an attempt to locate an anchor point, the pilot ordered the on-duty deckhand to take a station on the bow of the lead barge in the right-hand column of barges while the tow maneuvered slowly along the starboard bank. While the pilot probed the bank with his searchlight, the deckhand attempted to secure a tie-off. After several failed attempts, the pilot, worried about the safety of the deckhand who was out of view, ordered him back to the tow boat and continued to search the bank.

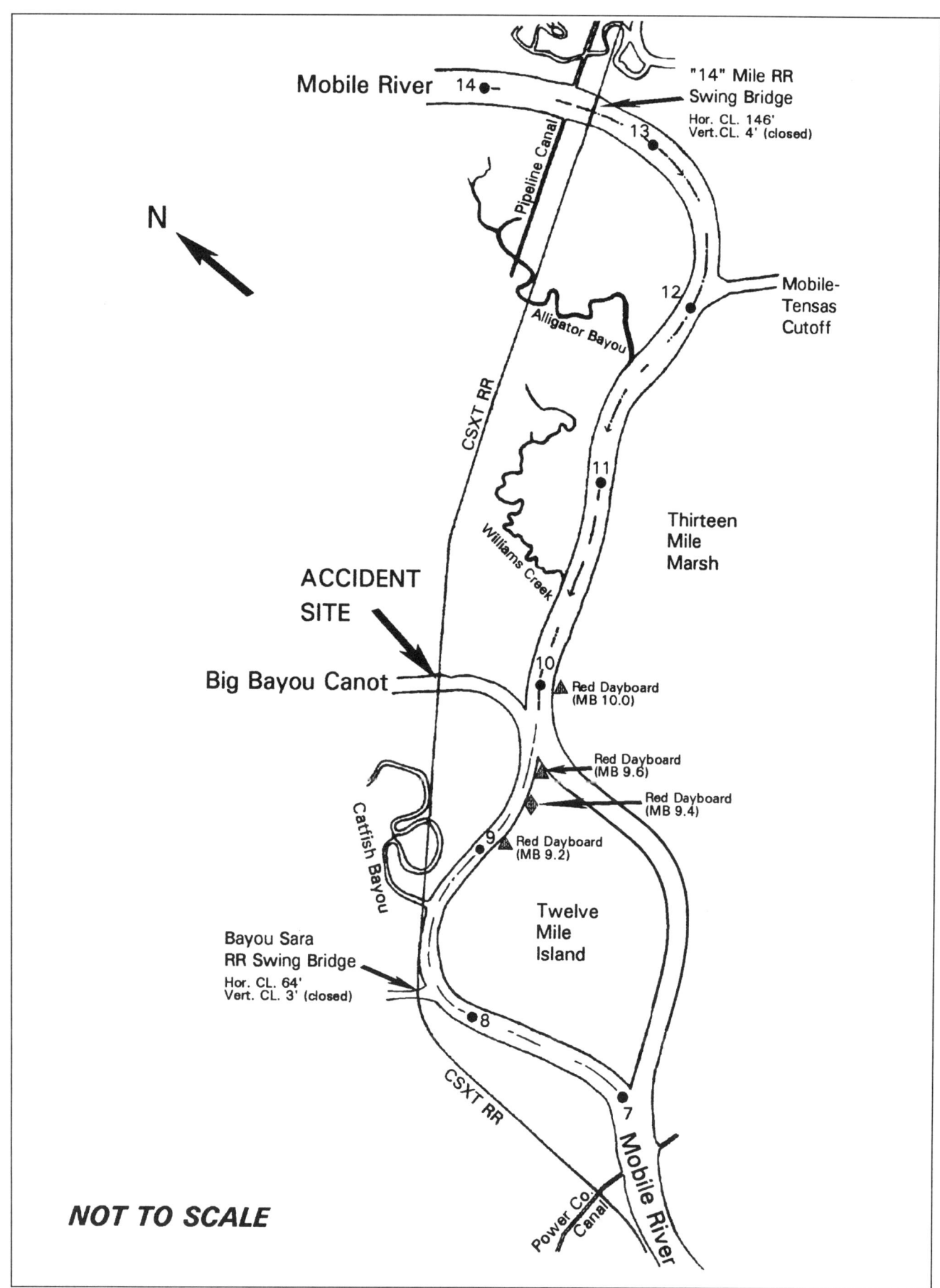

Figure 6-2: Mobile River chart. [From NTSB Report]

By now, the *Mauvilla* was locked in fog. Rather than pushing into the bank and holding the barges in place with engine force, the pilot pulled away from the shore and proceeded slowly upriver. Easing ahead at a speed of 2–3 knots, he noted the tow's swing meter (gyroscopic course deviation indicator) shift to the port. Consulting the tow's radar and observing what he believed to be the banks of the Mobile River to the north and northeast, the pilot followed what he thought was the river's main channel.

Confused about his location, the pilot concluded that he was near mile 12 upriver of Thirteen Mile Marsh, entering the sweeping turn of the Mobile River as it crossed under Fourteen Mile Bridge. The *Mauvilla*, like all other vessels of the uninspected tow vessel (UTV) class, was not required to have river charts or a compass. Operators of UTVs usually navigate by "dead reckoning", relying on experience and familiarity with the river and surrounding landscape. Though not mandated, most UTVs use radar to assist during conditions of poor visibility. Yet, operators were not required to demonstrate proficiency with radar to maintain UTV operator certification.

Robbed of visual cues and pre-occupied with the danger of navigating in dense fog, *Mauvilla*'s pilot had become disoriented. He had not been monitoring the radar closely and had lost track of his vessel's position while attempting to find a docking point. As he headed his barges into Big Bayou Canot, he thought he was in the Mobile River main channel.

Still looking for a place to put in, the pilot observed on radar a fixed structure in his path. Believing it to be a tow with barges, swung perpendicularly across his course, he called the *McCabe* to determine whether she had seen any other barges stopped along the river bank. Though the *McCabe* responded negatively, *Mauvilla*'s pilot still believed that the object must be another tow. But, rather than another vessel, *Mauvilla* was closing on the Big Bayou Canot railroad bridge at a speed of 1–2 knots.

As he later testified, the pilot's intent was to move in close to the other tow and secure the *Mauvilla* to it—another accepted river barge practice. Attempting to contact the other "tow" by radio, he got no reply. He also tried to identify the object using his searchlight. Throughout the approach, the deckhand, who previously served as lookout, stayed in the galley below decks rather than forward on the barges.

As *Mauvilla* neared the object, the pilot reversed engines and the tow "bumped" the unidentified obstacle. (See Figure 6-3.) The deckhand below, feeling the bump, noted the time as 2:45 A.M. The pilot thought that they had run aground. But *Mauvilla*'s port and starboard lead barges, according to the NTSB accident report, "*...had struck the south and center piers of the bridge's through-girder span...[while] [t]he forward center barge, which protruded about 5 feet ahead of the port and starboard barges, struck the east girder between these two piers, displacing the south end of the girder span 38 inches to the west*". The east girder of the bridge had been displaced by the force of impact into the path of rail traffic.

Mauvilla and her crew, now alerted to the crisis, were soon occupied with another problem. The left hand column of two barges had broken away at the time of collision and were drifting slowly back downriver. The pilot backed the tow into the river bank in an attempt to retrieve the barges but was pinned by the current and his cargo.

While *Mauvilla* attempted to extricate herself, the *Sunset Limited* raced toward the Big Bayou Canot railroad bridge from the southwest at a speed of 72 miles per hour, 2 miles per hour over the maximum speed for the track. As the train approached Bayou Sara, the track signal indicated that the road ahead was clear. (The signal system operates based upon track continuity. Since the rail over Big Bayou Canot had been displaced but not broken, the signal did not show a fault in the line.)

But at 2:53 A.M., as the *Sunset Limited* crossed the Big Bayou Canot bridge, the lead locomotive struck the displaced girder half-way across the span and derailed, plummeting into Big Bayou. Two other locomotives and four cars quickly followed. Four other cars remained on the track, protected by the steel truss bridge span surrounding them. (See Figure 6-1.) The NTSB report states:

> *Following the derailment, the three locomotive units came to rest on the east side of the bayou. Part of the lead unit, 819, was buried in about 46 feet of mud, and the part protruding above the embankment burned. The second unit, 262, also burned. The fuel tank of the third unit, 312, separated from it, and all equipment along the bottom of the unit below the frame was sheared off.... Baggage car 1139 and dorm-coach 39908, also on the east side of the bayou, were gutted by fire, and parts of both cars*

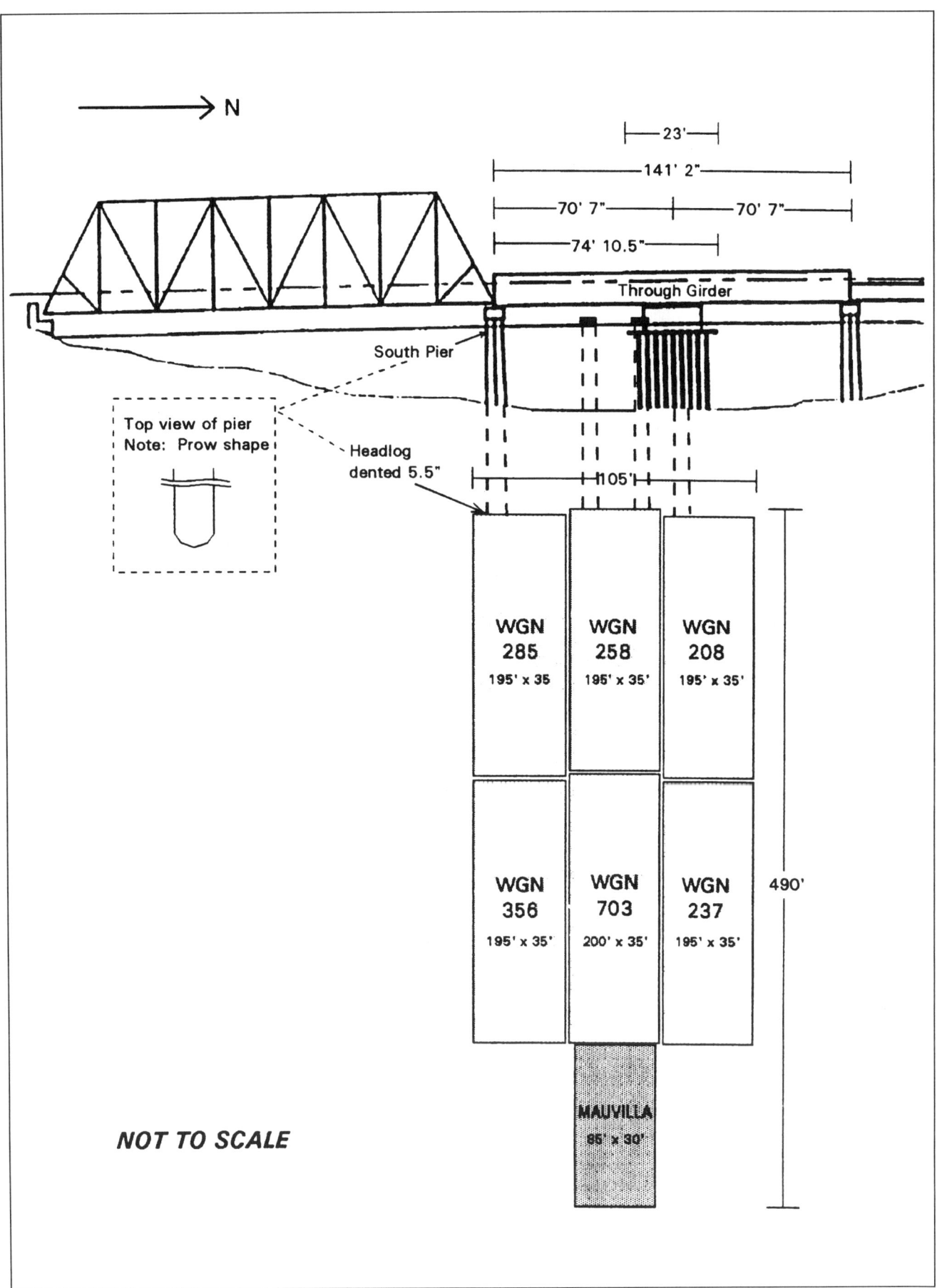

Figure 6-3: MAUVILLA tow configuration and Big Bayou Canot Bridge. [From NTSB Report]

The Alert, Well-Trained Operator

sustained major structural damage. About half of coach 34083, which rested against the bridge after the accident, was submerged, and coach 34068 was almost totally submerged.

At the time of the accident, the conductor and the assistant conductor were in the diner, next to the last car in the train. Both testified that there were no indications of trouble until they were thrown forward at the time of derailment. The conductor immediately tried to contact the lead locomotive with his portable radio but received no reply.

The *Mauvilla* pilot testified that he heard a "swishing" noise and saw a fire at the time of the derailment but was still not aware that he had struck a bridge or that a train had derailed. The on-duty deckhand testified to hearing "a hiss like a roar but not a boom or nothing like that".

Approximately 3 minutes after the derailment, the assistant conductor radioed a "mayday" on the railroad's frequency. The distress call was heard by both CSXT train 579 and Sibert Yard personnel in Mobile. Yard officials quickly notified the Jacksonville dispatcher, the Mobile Police Department, and the U.S. Coast Guard (after a delay because of an incorrectly printed phone number).

Because of confusion about the accident location, dispatch of emergency personnel and equipment was delayed until 3:20 A.M. Meanwhile, the surviving *Sunset Limited* crew and passengers organized a courageous rescue effort. They were soon assisted by the *Mauvilla*, other river craft, and Coast Guard helicopters which were launched at 4:04 A.M.

The Problems

About an hour after the derailment, the pilot of *Mauvilla* stated during radio communication with another tow vessel:

I made a wrong turn. I guess I can tell you. You know when you come out of the upper end of Twelve Mile Island you got a left; a river go[es] back to your left.

The pilot discovered that, while attempting to control *Mauvilla* in dense fog, his navigation skills had failed. The "tow" to which he had hoped to anchor was, in fact, the Big Bayou Canot railroad bridge, and the fire which he had observed was not, as he had originally believed, an unrelated event.

A thirteen year veteran of Warrior & Gulf Navigation Company operations, the pilot had been hired as a deckhand on April 23, 1980. He attended "sea school" in 1988 in preparation for a Coast Guard licensing exam, part of the certification process to become an operator of an uninspected tow vessel (OUTV). Following several exam failures, he finally passed the written certification test in October, 1990. The next month, Warrior & Gulf Navigation Company appointed him as an operator trainee, a position in which he underwent additional training prior to classification as a pilot for W&GN.

The pilot was promoted by W&GN to the rating of second class pilot in October, 1991, and to first class pilot in January, 1993. Then, in accordance with their practice, Warrior & Gulf placed him in a one-year on-the-job training (OJT) program under the instruction of a senior licensed tow operator prior to allowing him to perform independently as an OUTV. The pilot testified that the program included training in tow operation, riverine navigation, and radar operation.

W&GN's OJT program was not formally structured and documented. Rather, it was informally conducted by senior tow operators. Regarding the quality and content of training, the NTSB reported:

The general manager stated that he did not know whether operator trainees have the opportunity to listen to lectures, review written material, or demonstrate proficiency in the position based on a written test. He said that the W&GN towboat operators who conduct the training do not themselves receive formal instruction in their role as trainers. Nor does the company give written guidance to its assistant fleet captain, whose responsibilities include oversight of towboat operator training and evaluation.

Formal radar certification was not required. (Recall that radar is not required aboard UTVs.) Yet, most UTVs employed radar for assistance in riverine operations. The NTSB concluded:

The pilot probably received adequate OJT in towboat and barge maneuvering and was quite likely qualified to operate vessels under most conditions. Nonetheless, he did not use his radar properly on the night of the accident and certainly was not using it to determine his position on the river. The Safety Board concludes that W&GN did not adequately train the pilot to navigate by radar. If the pilot had received formal radar training, he might have known how to use the radar when visibility began to deteriorate. Considering that W&GN had equipped all its towboats with radar before September 22,

> *1993, the argument for radar training is compelling. The Safety Board believes that a structured radar training program enhances an operator's ability not only to determine his position but also to navigate his towboat to a safe mooring location and that this training should be required of all operators of radar-equipped towboats.*

Of 24 conclusions drawn by the NTSB following the Big Bayou Canot accident, 7 were related to the alertness, training, and skill of *Mauvilla*'s pilot and to the training process that led him to failure. Though a good pilot with a respectable record, he was ill-equipped and ill-trained for the situation in which he found himself during the early morning hours of September 22, 1993.

Tenets of Training

Accident studies demonstrate repeatedly that the human factor is a critical (if not the *most* important) element in safe and efficient equipment operation. Advising Congress after TMI-2, Admiral Rickover stated:

> *I consider the training of officers and men to be at least as important as any other element of the Navy Nuclear Power Program.*

Rickover then outlined the tenets of selection and training of personnel upon which his program was based. They included:

1. Careful screening of operator candidates, including special interviews by Admiral Rickover himself, for officer candidates;
2. Rigorous initial training including stringent classroom instruction, exacting on-the-job training under the tutelage of qualified OJT instructors on actual plant equipment, and frequent written, oral, and performance examinations during every phase of training;
3. Clearly delineated qualification and requalification requirements for all operators and their leaders;
4. Continual reinforcement of principles and procedures through recurrent and proficiency training designed to improve the knowledge and skills of each operator;
5. Weekly emergency preparedness practice through the execution of demanding casualty drills prepared by operating personnel and approved by plant leaders; and
6. Continual performance review of individuals and the entire operating team through on-the-job proficiency evaluation, inspection, and formal operational readiness testing.

Rickover's principles represent nothing less than a rigorous, practical approach to managing human factor variables in the equation of operating success. His pragmatic strategy for training can serve any industry well, whether in manufacturing, process operations, laboratory sampling and testing, or power generation. Through proper personnel selection, demanding initial training, comprehensive certification testing, continuous reinforcement of operating principles, frequent emergency preparedness exercises, and habitual evaluation of both individual and team performance, any team can achieve consistent operating success.

Determining Job Requirements

Before operator candidate selection and training can proceed, the jobs and their associated risks must be defined. Without a clear delineation of job requirements, the baseline educational and skill levels for potential candidates cannot be effectively determined.

Determining job requirements need not be a complex task, especially for jobs that are not new. Through "roundtable discussions" with experienced operators and their leaders, the vast majority of circumstances and responsibilities to which an operator may need to respond can be identified. For further insight, skilled observers can follow experienced operating personnel through their daily routines, questioning and recording to determine job requirements. Also, questionnaires administered to experienced operators may be used to catalog requisite tasks and skills. This process is known as *job and task analysis*.

Job requirements must also be assessed in terms of the risk that they present if improperly performed. Risk assessment is a decisive factor in determining the depth of training and evaluation required. Rickover once said:

> *…the more complex it [the equipment] is the far better trained the people have to be or they are going to get in trouble.*

If improper task performance will expose the operators themselves, the general public, or the environment to significant hazards, clearly the candidates chosen for the jobs must have above-average capability and must receive extensive instruction.

Selection of Candidates

Once the job requirements and the risk level are established, selection criteria for operator candidates can be determined. Successful training of operators depends heavily upon selecting the right people. Selection standards, or **selection criteria** as they are often known, are the minimum attributes and characteristics deemed necessary to successfully complete a qualification program and perform the duties and responsibilities required by the position.

If operator selection criteria are established at too low a level, an inordinate amount of time will be required to develop the skill and knowledge base necessary for intelligent and efficient performance. Therefore, selection standards should be carefully studied and chosen before initial training.

Rickover recognized this and stated:

> *When the nuclear propulsion program started, more than thirty years ago, I realized it was necessary to have excellence in operating personnel. In view of the possible serious consequences of a reactor accident I considered it of utmost importance that the operation of nuclear powered ships be entrusted only to those whose mental abilities, qualities of judgment and degree of training were commensurate with the public responsibility involved. The personnel selection and training procedures for the Naval Nuclear Propulsion Program were developed with these considerations in mind.*

When considering characteristics desirable for candidate operators and their leaders, the most common selection attributes—education and previous experience—may not be the only pertinent considerations. Some jobs may require special criteria for: strength and size, visual acuity, motor skill capabilities, or learning aptitudes. Other qualities that should be considered include character, integrity, and leadership acumen. ANSI standards are an excellent aid for developing selection criteria.

Developing the Training Program

After student selection criteria have been established, a training program must be mapped which will sequentially lead to the qualification of operating personnel. Creation of the training program includes development of a training plan, establishment of qualification standards, design of supporting instructional materials, selection and qualification of instructors, and generation of the records necessary to document completion of the training requirements. Additionally, the program must include a process to integrate and evaluate training within the confines of operations and maintenance.

Training Plan A critical part of training program development is the *training plan*. The **training plan** is a blueprint for the training program. It is a policy document which describes the training philosophy and the paths and methods by which candidate operators will be trained and evaluated to meet the job requirements previously established. Without a training plan, operator training is apt to be inconsistent and poorly controlled.

The plan should clearly describe the phases of training through which a candidate must progress to achieve a particular qualification. For example, most skill training programs include, in one degree or another, a combination of classroom training, on-the-job training, and proficiency training. For complex or risky jobs, each phase should include formal evaluation including written examinations, oral examinations, and skill demonstration evaluations.

Qualification Standards The training plan should require that qualification standards be developed for each operator rating. **Qualification standards** are written statements of the knowledge and skill requirements that an operator candidate must achieve in order to complete qualification. Thorough qualification standards separate the job into individual tasks that an operator must be able to perform. For each task, the desired knowledge or skill is described, a method for demonstrating the achievement of the knowledge or skill is specified, and a means of authenticating achievement of that knowledge or skill (usually an instructor signature) is provided.

Written qualification standards help ensure that each operator candidate receives consistent training. They also provide a format for documenting the training process.

Instructor Selection

No training program can be successful without good instructors. Therefore, the training plan must provide for the selection and training of instructor candidates who, upon qualifying as instructors, have the requisite skills to accomplish the training program objectives.

Clearly, the first prerequisite for an instructor candidate is exceptional knowledge and skill in the subject to be taught. An instructor who doesn't understand what he or she is teaching has little credibility and may create a dangerous operating environment. Accordingly, most instructor training programs establish, as an essential entry requirement, high skill and knowledge levels verified either by past performance evaluation or through examination.

The next consideration is aptitude for teaching. Regardless of knowledge and skill level, some people just don't have the aptitude to teach. Nothing is more frustrating for a student than to be instructed by a teacher who knows how to perform a task but can't explain it. Accordingly, it is very important to determine (prior to selection, if possible) whether an instructor candidate is likely to become a good teacher.

It's usually not difficult to ascertain a candidate's aptitude to teach, because those who have the propensity and desire do it naturally every day as they interact with others. Discussions with their leaders and colleagues quickly disclose that aptitude.

Another important criterion for selecting instructor candidates is character. Knowledge and skill levels are important, but so are the attributes of integrity, respect, and the absence of arrogance. Students will not respect nor learn from an instructor of low character. Further, instructors with questionable integrity may be predisposed to authenticate student qualification requirements when the candidate has not met the specified standards. A favoritism-based, "good old boy" training process usually results in inconsistent training and incompletely prepared operators. As with teaching aptitude, character can be determined through observation of performance, candidate interviews, and discussions with past leaders.

Finally, the intangible of attitude must be determined. You can't afford to entrust your precious, new employees to someone with a sour outlook, since the instructor who displays a bad attitude also transmits a bad attitude. Perhaps the best maxim for selecting instructor candidates is summarized in this one imperative: *Never clone a jerk!*

Instructor Qualification

Having selected high-potential instructor candidates, the training program must then equip the candidates with excellent instructional skills. Whether for classroom instruction, small group seminars, or on-the-job training, necessary instructor skills range far beyond the ability to stand in front of a group and talk. As with so many disciplines, there is more to teaching and training than meets the eye.

Depending upon the type of instruction, mastery of the teaching process may include the ability to research subjects and develop lessons, write lesson guides, develop examinations, present classroom lessons, facilitate small group seminars, and verify student skill or knowledge through observation and questioning during OJT. Certainly, any instructor must be able to hold the attention of students, explain abstract concepts in terms which different types of students can understand, demonstrate performance of tasks, and make judgments regarding student mastery of each knowledge or skill requirement.

Many instructor skills are not easily or naturally acquired. Without an instructor training program, instructors typically perform as they have seen others perform. Unless their role models have all been very good, they will have acquired some false impressions and some bad habits. Instructor training must, therefore, be more than a 2-hour class in "Here's how you do it".

Rickover provided this advice for training classroom instructors:

> *The initial training of a new instructor takes about three months. During this initial training the new instructor is first required to take the subject he will teach. He will give practice lectures and become familiar with related Nuclear Power School subjects. The new instructor must pass oral boards on the*

technical content of the course, and present a certification lecture for the division director, the department head, and the commanding officer. He must also pass an oral certification board by the division director, the department head, and the commanding officer. After qualification, the training continues so that the instructor will remain current and knowledgeable. An annual written examination is administered to all instructors to determine any weak areas. The instructor's classroom presentation is audited at least twice during each period he teaches a subject. The commanding officer, the executive officer, and the department heads are required to audit one instructor each week. Also Bettis Technical Consultants randomly monitor the instructors. Evaluation reports are filled out by the auditors and discussed with the instructor. These reports are forwarded up the chain of command and filed in the instructor training folder after any necessary corrective action has been taken.

Your instructor training may not need to be this rigorous; but, the principles are similar. Remember, as risk increases, so must attention to detail. If you can't afford failure, you have to take the actions necessary to avert it.

One final note. The instructor pool must not be allowed to stagnate. It is best to require the instructors to stay proficient in the tasks that they teach. As a result, their credibility remains high and they don't get out of touch. Therefore, instructor qualification must also include periodic requalification in the technology that they teach.

Lesson Materials

Beyond furnishing excellent instructors, the training plan must provide for developing materials to support lesson presentation. As qualification standards provide a consistent basis for determining what a student must know or be able to do, well-developed lesson materials—tools such as lesson plans, visual aids, and examination banks—afford a dependable and repeatable guide for teaching.

Developing lesson materials is often a complex, time-consuming task. But, well-prepared materials facilitate successful instruction. Lesson materials are best developed by those who will do the teaching. Through the process of lesson research and development, most instructors acquire a sense of ownership of the lesson and a deeper understanding of the subject.

Lesson material creation is not a one-time endeavor. Materials must be constantly reviewed and updated as equipment, knowledge, circumstances, technology, and job requirements change.

Clearly, there is an administrative burden which accompanies lesson materials. Files of lesson plans, visual aids, and examinations must be maintained if the process is to be regulated and monitored. Computer technology, thankfully, can greatly reduce this hardship.

Training Records and Documentation

For operating personnel in jobs which require proof of qualification or certification, archives of training records and associated documentation are necessary. Part of the training plan should, therefore, prescribe how records are to be made and preserved.

The greatest burden (and often the weakest area) in training record management is keeping records current. To overcome that problem, the record-keeping process must be "user-friendly" and provide for easy review and revision as well as security from unauthorized access.

Training Schedules

With materials, instructors, records, and students in place, training events must be sequenced and scheduled to advance students through the training progression. Classroom education or skill training for activities which occur routinely will probably not be limited by operating or maintenance schedules. But on-the-job training for non-routine activities requires careful planning so that it can be incorporated into the schedule of operating and maintenance events.

Planning and scheduling of training activities clearly will not transpire without someone designated to manage it. Nor will it occur smoothly without cooperation between the operating, maintenance, and training groups. Therefore, a training organization structure is usually necessary to achieve the objectives of the training plan.

Training Priority

One of the most important functions of the training plan is to establish the value and priority of training events. Unless training, operating, and maintenance groups are committed to the training process, "crises" will inevitably displace training. Accordingly, once a schedule of training events has been coordinated and published, the events must be punctually conducted.

Great organizational discipline is necessary to ensure that training is carried out. When personnel scheduled to conduct training are, on short notice, assigned other "more important" duties, students soon lose faith in the value and certainty of the training process. Obviously, circumstances sometimes arise that require events to be rescheduled; but, such occurrences should be minimized.

Training Evaluation

No training plan is complete without describing a means for evaluating individual training events and the program as a whole. Every training program, every lesson, and every instructor can be improved. Therefore, training program policy must provide for evaluation of training presentation, lesson content, and program effectiveness.

Effective training comes from thorough preparation. But, don't expect it to be flawless on the first attempt. Improvement is the result of constant evaluation and refinement. One of the most effective means of appraising the quality of the program is through instructor round table discussions. Well-chosen and well-trained instructors are usually very capable of identifying the strengths and weaknesses of the individual lessons and of the training program as a whole.

Another important source of performance evaluation is from the students themselves. Evaluations can be obtained through questionnaires and interviews. If the students have not been intimidated by the instructors, they will provide useful suggestions and criticism. The instructors, however, must acquire an attitude that values and listens to criticism, even when it is not complimentary.

A third source of evaluation involves critical plant, process, or facility leaders. Plant managers, as well as managers of operations, maintenance, and training, should observe training events, including classroom training and OJT, frequently. They are, after all, the recipients of what the training program produces. Their reviews should include informal interviews with students and instructors to determine whether the training process is effective, challenging, and interesting. When they find flaws, they must provide constructive criticism and recommendation for change.

Leaders should be particularly sensitive to inconsistencies between training and field operations. Whenever they hear the words, "You can forget what they taught in training, because that's not how we do it out here", they must get to the root of the problem. If field operations are inconsistent with training, either the operation is wrong or the training is wrong. One or both must be brought quickly into line.

Ultimately, the most important form of training evaluation is an assessment of how well former students perform the job assignments for which they were trained. Consequently, ongoing performance evaluations ranging from routine daily observation to formal, annual readiness reviews are necessary to judge training effectiveness.

Phases of Training

The training plan must also clearly describe the phases of training. Operator training typically progresses through three general phases for each position requiring qualification. The first is initial training. This phase comprises the body of education and skill training necessary to bring a candidate to the point at which he or she is prepared to take a certification examination for the rating being sought.

After initial training, operators progress to a certification testing phase in which they must demonstrate their competence and ability to operate on their own. *Qualification* skills usually must be demonstrated to leaders who have the authority to grant *certified* status.

Once certified, operators enter a never-ending training phase during which they must periodically demonstrate proficiency in their assigned tasks, upgrade their knowledge and skill with new equipment and technologies, review procedural changes, and undergo casualty drill training as well as operational readiness reviews. This third phase of training is, as a result, sometimes referred to as *continuing training*.

Rickover described similar training phases before Congress:

> *The programs to train personnel for engineering duty aboard naval nuclear-powered ships are centered around four major phases—formal academic instruction, operational training at one of the*

Department of Energy land-based naval reactor prototypes, training and qualification [certification] as a watchstander aboard an operating naval nuclear-powered ship, and continuing shipboard training. Each of these four phases is essential in the satisfactory training of an operator and providing assurance that only those who are mentally and emotionally capable, and who have demonstrated ability as a competent nuclear propulsion plant operator are assigned duty aboard nuclear-powered ships.

Initial Training Phase

The first training phase (*initial training*) usually includes classroom instruction, individual study, small group seminars, and, most importantly, rigorous on-the-job training. The **initial training** phase must ensure that graduating operators not only have the skills to meet the challenges of routine operations, but also that they are prepared to manage abnormal or emergency situations.

The initial training phase incorporates education and practical training in plant components and systems, the purpose and use of operating procedures, actions for likely situations not addressed by procedure, and the margins and boundaries of the plant's safe operating envelope. Just as the captain of an airliner must understand the design, construction, and functional operation of aircraft systems under his or her control, so must facility operators acquire the ability to safely operate facility equipment, especially under abnormal or emergency conditions. Without such training, facility operators are deprived of the knowledge and skills necessary to control their equipment responsibly under all conditions.

Classroom Training

Classroom instruction is a critical part of each phase of training. It is necessary to impart theoretical understanding of the components and systems that an operator must control. Without sound, theoretical understanding of the function and operation of the equipment, operators lack confidence in their equipment and in themselves. They tend to become blindly reliant upon procedures. Then, if situations arise that are outside the scope of operating procedures, or if there are no procedures to address the current conditions, improper actions are bound to be taken.

OJT (On-the-Job Training)

Practical, on-the-job training must follow theoretical training to ensure that operator candidates learn hands-on equipment operation, procedure usage, and response to abnormalities. Educational studies indicate that on-the-job training, when effectively performed, yields 80–90 percent retention. Classroom training, on the other hand, can yield, at most, 50 percent retention. Further, classroom environments usually cannot provide the realism of on-the-job training.

In addition to individual on-the-job training, it is important that operators learn to function as cooperative operating team members. Success or failure often depends on the coordinated actions of the team. Operators must understand their primary function with relationship to the mission and how to prioritize their actions within the context of the team. Consequently, evaluated performance under the control and instruction of a senior operator in the role of an operating team member is as important as the evaluation of individual knowledge and skills.

Well-written qualification standards that clearly address what a student must know or be able to do for each task will assist OJT instructors in maintaining training integrity. In the Big Bayou Canot bridge accident, Warrior & Gulf's on-the-job training was incomplete. It did not address in detail the tasks and skills necessary to perform well as a tow operator under normal and abnormal conditions. The NTSB investigation concluded that:

> *W&GN's evaluation form is a check-off sheet rather than an in-depth assessment form for assessing an operator's skills and abilities. Whether management, using this form, could accurately evaluate an individual's abilities is questionable. The criteria for the four rating levels are not listed on the form, and the six areas of performance evaluated are too general to allow meaningful assessment of an operator's skills. For example, "knowledge of position" is not defined. The Safety Board concludes that operators should be evaluated on their proficiency in use of wheelhouse equipment such as radar (under various visibility conditions and circumstances, including finding a suitable place to tie off), the swing meter, and rudders (including backing rudders) and engines in high water and high current conditions. The Safety Board also concludes that W&GN's written evaluation form did not fully identify and assess those skills critical to vessel operation, thereby limiting its value as a management tool for ensuring safe vessel operations.*

It is probable that better qualification standards, improved training guidance, and more thorough OJT evaluation forms would have improved training for their tow operators.

Certification Phase

When an operator candidate has successfully completed the classroom instruction and on-the-job training of the initial training phase, he or she often must undergo final written, oral, and practical examinations to demonstrate overall knowledge and skill. These examinations comprise the certification phase of training.

The terms *qualification* and *certification* are sometimes differentiated. **Qualification** refers to the training process which prepares an operator to perform independently. **Certification** refers to the examination process by which an authority validates the knowledge and skill of an operator. All operating personnel must be qualified, but some positions are so important that policy, regulation, or law requires certification. Driver's education, for example, is a qualification process, whereas successful completion of the written and practical driver's licensing examinations constitutes certification.

Continuing Training Phase

Individual and team training should be a continuing process. Just as in the domain of sports, no player or team ever gets good enough that practice is no longer necessary. Constant evaluation and improvement are required. They are the hallmarks of a vibrant team.

Such is the purpose of the continuing training phase of the training process. Constant performance improvement is sought through drill training, innovative refresher training, and analyzing accidents and undesirable events from which applicable lessons can be drawn.

Just as initial training must be well-planned and executed, continuing training must also receive high priority. It cannot be relegated to a second-rate function if a team is to reach its potential. It must be a well-planned and surgically executed process. Bill Walsh, former head coach of the San Francisco '49ers, says:

> *The way I coach, I know ahead of time how I am going to run the whole season's worth of practices. I have established the priorities for what we need to accomplish and allocated the time in which to teach the necessary skills. I establish the program long before we take the field so I can use most efficiently the time available for learning and so the players do not get bored or distracted. The players must know clearly and at all times exactly what it is that they have to get out of any given drill.*

Industrial teams should be just as well-schooled. Every training drill or class should have a specific purpose which fits into an overall plan that is identified long before the "season" begins. Each team member should know the purpose of each training activity.

Classes and drills should be planned and executed with enthusiasm and imagination to eliminate distraction and boredom—sitting in a classroom after working a graveyard shift is not training; it's an exercise in staying awake.

Fitness for Duty

If the training plan is well-conceived a viable training process is likely to result. But, highly knowledgeable and well-skilled operators and leaders are not enough. Operating personnel must also arrive for work on every shift in a proper state of mental and physical fitness.

Education and training can be seriously compromised by poor mental and physical fitness. Operators must be rested and mentally attuned to their tasks to safely operate. In fact, the requirements for safe equipment operation are predicated upon the machinery being controlled by *fit-for-duty* operators. **Fit-for-duty** implies that an operator not only has the skills and knowledge necessary to perform during routine and emergency situations, but also that the operator is in the proper physical and mental state to utilize those capabilities.

There are many ways for an equipment operator to be unfit for duty. The use of drugs or alcohol will certainly impair the skills of an otherwise excellent operator. On January 3, 1987, a Conrail train ran six warning signals at a speed of 62 mph and pulled in front of an Amtrak train that had been cleared for right-of-way outside the city of Baltimore. The Amtrak train slammed into the Conrail unit at a speed of nearly 100 mph resulting in a disastrous accident. Drug testing revealed that members of the Conrail crew had been using marijuana. After denying in testimony the use of marijuana, they later acknowledged that they had been using the drug at the time of the accident.

Drug use need not be illicit to cause a fitness problem. Just as in driving, prescription drugs may result in lapses in attention, drowsiness, or distraction. Equipment operators bear a duty to themselves as well as to the others to analyze their own physical and mental states and take the actions necessary to preclude unsafe acts.

Fatigue is another common cause of impairment. It may be the result of excessive overtime or team members who stay out too late. Regardless, it is unreasonable to believe that inadequately rested operators will exercise the judgment and skill necessary to safely manage routine operations, let alone emergency situations. Following the *Exxon-Valdez* accident, the investigation team from the State of Alaska concluded that fatigue of the crew was a significant contributor to that event. The National Transportation Board drew the same conclusion.

Impairment may also be caused by improper mental attitude. An equipment operator who is unusually distressed or distracted is unlikely to exercise the level of concentration necessary for consistent, safe operation. Yet, we know that the quality of *alertness* is essential to safe motor vehicle operation. How much more so to industrial "defensive driving"?

Whether from drug use, fatigue, or improper attitude, mental unfitness for operating tasks is dangerous. Just as poor maintenance or poor design will affect safe mission outcome, the safe operating envelope is also invalidated by impaired operating personnel. Though the definition of impairment may vary from facility to facility, a standard for fitness must be established and maintained based upon the risk associated with the hazards characteristic of the operating machinery.

Effective Management Oversight

One of the most important forms of continuing training is the feedback that comes from good coaches throughout individual and team performance periods. Even the finest players need oversight and feedback. As such, team leaders bear a grave responsibility to vigilantly observe for danger signs of inattentiveness, fatigue, or improper mental outlook. Leaders who don't know the normal characteristics of their team members are often unable to recognize when something is wrong.

The process of management oversight is an elemental part of coaching. It has as its purpose the identification of team problems in order that preventive and corrective measures can be instituted to prevent the growth of serious team deficiencies.

The best of teams on the best of days can still improve. Observant leaders acting as coaches are the principal instrument for team improvement. The interaction between team members and team leader is, perhaps, the most important of all human factors.

Topic Summary and Questions to Consider

After the TMI accident, Admiral Rickover summarized training deficiencies at that site:

> *There was an apparent lack of attention and devotion of resources to the training of operators. Site managers did not consider themselves responsible for operator training. The training department was undermanned and was staffed by instructors no more qualified educationally than their students. There was no training for engineers or managers at a level higher than that for control room operators, although during the accident operators turned to their supervisors for guidance.*
> *The training for newly-qualifying control room operators was done essentially on a self-study basis. The curriculum did not cover the principles of science and engineering necessary for understanding the operation of the power plant, nor was it reviewed and approved by people qualified to do so. The requalification program, which served as a continuing training program, was shallow and haphazard. It did not continually upgrade knowledge and understanding through reinforcement of principles and procedures. Course content was not reviewed and approved by management, nor did they monitor the conduct of formal instruction.*

Warrior & Gulf fell victim to the same errors. And, though tow navigation on the Mobile River is a far cry from reactor operations, no one died at Three-Mile Island.

Rigorous selection criteria and excellent training are very expensive. Almost without exception, however, the best in every industry attest to learning the hard lesson that poor selection and training in the beginning are *far* more expensive than doing the job correctly the first time. The Big Bayou Canot bridge accident resulted in material losses alone of nearly $20 million. The value of 47 lives is more difficult to determine.

Ask Yourself What is the state of *your* training program? Do you have a training plan? Are all of the phases of training in place and working? If you expect your operators to perform intelligently under both normal and adverse circumstances, you must educate and train them. Remember: **There is no substitute for the alert, well-trained operator controlling equipment parameters within specified operating bands in accordance with approved procedures.**

Part III

Vital Operating Skills

- Controlling Equipment and Processes
- Conducting Pre-Task Briefings
- Understanding and Using Procedures
- Monitoring Critical Operating Parameters
- Independent Verification
- Communicating Vital Information
- Keeping Logs and Recording Data
- Recognizing Abnormalities
- Combatting Emergencies & Casualties
- Overseeing Maintenance
- Isolating Energy Hazards
- Training On-the-Job
- Performing Shift Turnovers

Controlling equipment and processes within established boundaries of safe and efficient operation is the goal of the operating strategy. And people (alert, well-trained operators) are the single most important element in achieving that goal.

Part III, therefore, introduces and expands on the details of the twelve most important skills for every industrial operator, including: what they are, why they are important, and how to establish and use them.

Chapter 7

Controlling Equipment and Processes

As we just learned, no matter how well-designed or well-constructed the equipment, management of operating risk relies, ultimately, upon intelligent control by well-trained personnel.

In Chapter 3, we learned that any strategy for successful industrial operation has, as its objective, intelligent control of equipment and processes. Intelligent control means operating, maintaining, modifying, and testing equipment in accordance with the limits and practices dictated by design. If components and systems are not controlled within prescribed tolerances, they cannot be expected to function in accordance with design specifications.

It follows that *intelligent* operators, leaders, and support team members must thoroughly understand their equipment and its governing procedures. If they neglect, however, to control equipment through the wise use of procedures, observation of equipment status, and communication with other team members, disaster waits in the wings. Continuing our emphasis on the alert, well-trained operator as a fundamental element of the operating strategy, this chapter introduces the twelve vital skills essential for every industrial operator to master for effective, safe equipment and process control.

Fatal Gas Release at Bhopal, December 1984

In the early morning hours of December 3, 1984, a deadly cloud of gas was released from a Union Carbide-operated chemical plant four hundred and fifty miles south of New Delhi, India. The release resulted from a runaway reaction in a 15,000 gallon storage tank containing 13,000 gallons of methyl isocyanate (MIC), a chemical used in the production of pesticides at the site. It is conjectured that the violent exothermic reaction was triggered by a small amount of water inadvertently forced into one of the MIC holding tanks during a routine preventive maintenance piping flush. The end result was a cloud of toxic gases cascading out of a site vent stack into the atmosphere. The cloud drifted in a southwesterly direction into the adjacent city of Bhopal, a population center of 800,000, killing and injuring thousands of its citizens.

Plant History   In the early 1970s, Union Carbide India Ltd., a Union Carbide Corporation subsidiary established by agreement with the nation of India, built a pesticide and fertilizer production facility northeast of the city of Bhopal, the capital city in India's central state of Madhya Pradesh. Pesticides and fertilizers fulfilled a necessary role in India's ongoing effort to improve agricultural efficiency.

By 1981, the pesticide production process was improved by adding a plant for the manufacture and storage of a new compound, methyl isocyanate (MIC). MIC was to be produced and stored on-site, combined as a primary constituent with carbon tetrachloride and alpha-naphthol to formulate a proprietary pesticide.

The Bhopal MIC facility design was patterned after Union Carbide's Institute, West Virginia, plant, where the compound began use as a less toxic substitute for other chemicals several years earlier. Though Union Carbide's technical manual warned that it "may cause fatal pulmonary edema", it was thought to be primarily a skin, eye, and lung irritant in low concentrations.

Plant Description

Since MIC is unstable and highly reactive, a number of procedural controls and engineered safeguards were designed into the plant's safe operating envelope. The boiling point of MIC is 102.4°F. To reduce its volatility, MIC was to be kept at or below 41°F through the use of a chiller and heat exchanger adjacent to the MIC storage tanks. (See Figure 7-1.) This chiller system would cool and circulate the MIC in the tanks. A high temperature alarm was originally set at 52°F.

The compound was stored in any of three 15,000 gallon storage tanks buried in an earthen berm and covered with a six inch concrete cap. (See Figure 7-2.) The maximum volume of MIC stored in any of the three tanks was to be kept at 60% (9,000 gallons) or less. The vapor space above the liquid compound provided a buffer against a pressure excursion and, as a result, provided time to cool the tank in case of a reactive chemical excursion. One storage tank was always to be kept empty to allow transfer of compound to the empty tank in case of emergency.

A dry nitrogen head was to be kept continuously on the MIC compound contained in each tank. Dry nitrogen limited vaporization of the MIC compound and served as a spark and fire suppressant. Also, nitrogen pressure was used to transfer compound out of the tanks.

Frequent MIC transfers within the piping system dictated routine flushes of the piping inner walls (using fire main water). Without flushing, compound salts would build up and eventually obstruct the transfer piping. Since MIC reacts violently with water, piping flushes were to be preceded with an exclusionary valve lineup and installation of blank flanges (often called slip blinds or pancake flanges) to prevent water contamination.

Water isn't the only contaminant with which MIC reacts violently. Acids, bases, chlorides and other impurities (and contact with certain metals) were excluded to prevent uncontrollable excursion.

Each storage tank had a pressure relief valve set to relieve at 40 psig. Off-gassing from the tanks (both pressure relief and normal process venting) was routed through a caustic scrubber to strip

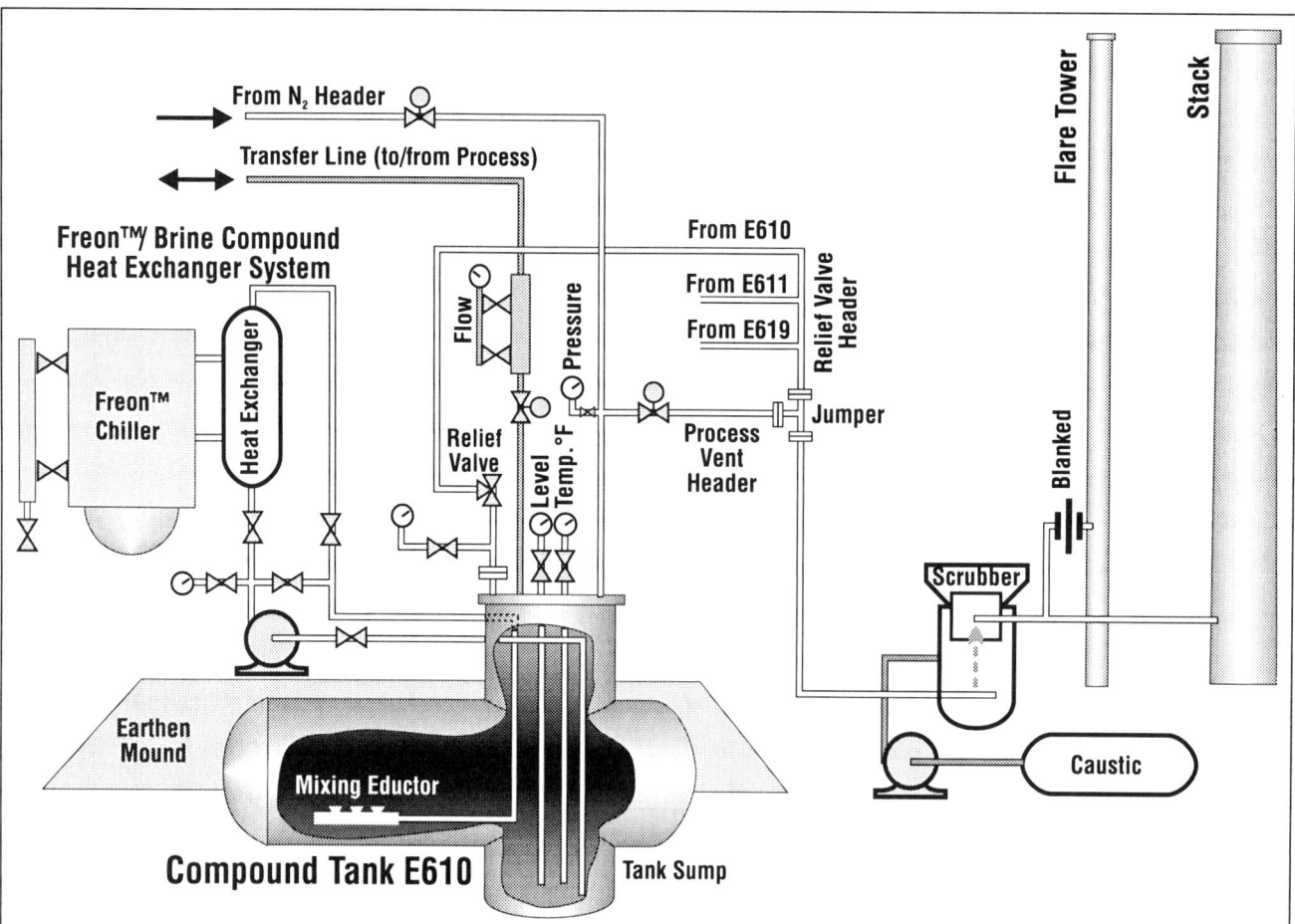

Figure 7-1: Details of connections and equipment associated with MIC storage tank E-610.

toxic components from the vapors. The remaining gaseous byproducts were routed to a ventilation stack which emptied to the atmosphere 120 feet above ground. If scrubbing was insufficient, a spool piece could be installed at the caustic scrubber exhaust and the gases routed to a flare tower to strip toxic components through combustion. The flare tower also exhausted 120 feet above ground.

Finally, fire main spray nozzles were located around the plant to raise a water curtain around the facility to help combat and contain a gaseous release in the event of a serious leak.

Operating Degradation

The MIC facility went into production in 1981, supervised and operated by a well-skilled and well-educated staff of East Indians. Selection criteria for supervisors and operators within the facility required that each have a university science degree and undergo a thorough initial training process. It was seen as a model industrial facility.

Little more than a year after the plant went into production, product sales declined and funding became limited. Cuts were made in operating and maintenance staff as well as in training and maintenance budgets.

As original staff members left the MIC facility for other jobs, the selection criteria for operators and supervisors were lowered. By 1984, only a high school degree was required for employment and training began to deteriorate. Detailed, integrated knowledge of the components and systems was originally required, and operators were cross-trained in other facility functions. As standards waned, however, the plant supervisors, technicians, and operators received only minimal levels of training in their own areas of expertise and little (if any) cross-training.

Maintenance Degradation

Corrective and preventive maintenance declined as well. Instrument and alarm maintenance became less rigorous. Most operators and supervisors had little confidence in the instrumentation reliability. Many plant components and systems were in a state of disrepair as well. As a cost-cutting measure, refrigerant had been drained from the storage tank chiller to be used in another part of the facility. Temperature in the MIC storage tanks ranged higher than 77°F during hot weather, nearly 40 degrees closer to the 102 degree boiling point. How much higher tank temperatures normally reached is unknown since 77°F was the highest readable by the instrumentation, and the temperature alarm had been either reset or disabled.

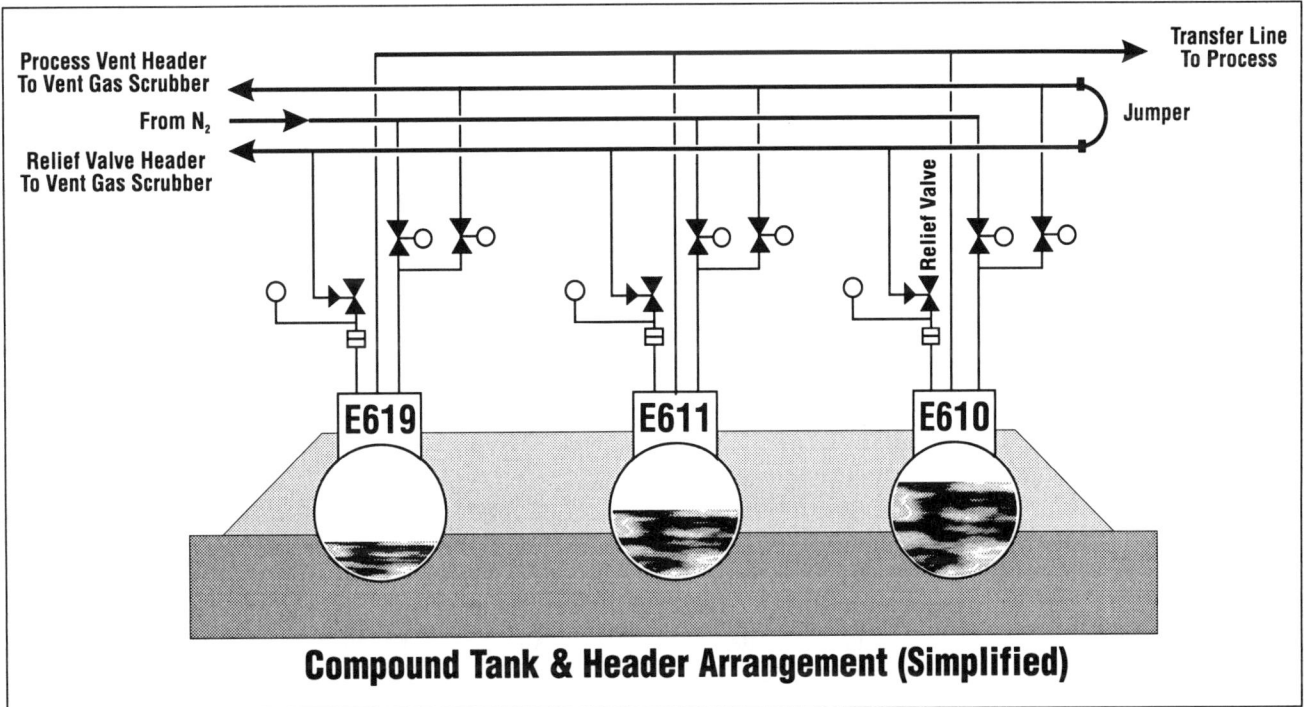

Figure 7-2: Simplified diagram of the MIC compound tank and header arrangement at Bhopal.

Controlling Equipment and Processes

The caustic scrubber had undergone maintenance nearly six weeks before, but was not functioning on the night of the event. The flare tower was also out of service because of a missing four foot pipe section. Minor MIC leaks had become commonplace as a dwindling maintenance staff struggled with a rapidly deteriorating facility.

In the weeks prior to December 2, 1984, one of the three MIC storage tanks (E-610) was allowed to fill to 87% of capacity or 13,000 gallons. A leak in the nitrogen system was preventing a sufficient pressure buildup to transfer MIC out of the storage tank. The location of the leak had not yet been identified. Contrary to procedure, *both* of the other tanks (E-611 and E-619) also contained MIC—tank E-619 was designed to be empty as an emergency dump. Its level gauge, however, indicated 3,300 gallons. [It was later found that E-619 actually contained only 437 gallons.]

A Routine Piping Flush

On the afternoon of Sunday, December 2, 1984, the MIC facility second shift crew came on duty at 2:45 P.M. At 9:30 P.M., an hour prior to completing their eight hour shift, the second shift supervisor directed an operator to begin a flush of one section of the transfer piping. The MIC operating manual directs that a flush be conducted with these restrictions:

> *"Isolate the equipment positively by inserting suitable blinds. Isolation by valve or valves is not to be relied upon."*

Piping flushes were normally overseen by the maintenance supervisor; but, the maintenance supervisor wasn't available that night. The second shift operator completed the valve lineup for the flush under the oversight of the shift supervisor; but no slip blinds were inserted. The flush commenced.

The third shift crew arrived at 10:30 P.M. The off-going control room operators noted pressure in tank E-610 as normal, approximately 2 psig. The shift change was completed at 10:45 P.M. and the piping flush continued to run.

Rising Tank Pressure

At 11:00 P.M. the third shift senior control room operator noticed tank E-610 pressure at 10 psig—five times its previous reading. Yet, he and his shift supervisor (because of past erratic instrument indications) believed the reading to be within the range of normalcy. Complicating the matter was the fact that the operating data record sheets had no column for recording tank temperatures. As a result, tank temperature trends were not tracked except in the memories of the plant staff.

At 11:30 P.M. operators began to sense the tell-tale smell and irritation caused by leaking MIC. Upon investigation, a small liquid and gas leak was noted in a pipe run 50 feet up in the superstructure. An operator reported the leak to the MIC unit shift supervisor at 11:45 P.M., but no investigation was initiated. The shift supervisor later testified that he thought the operator was referring to a water leak.

An Uncontrolled Leak

At 12:15 A.M. the shift supervisor and operators began their traditional tea break. Investigation of the previously reported leak did not commence until after the break at 12:40 A.M. But at 12:45 A.M., the senior control room operator noted tank pressure near the 40 psig relief setpoint. He went to the storage tank to investigate. Within seconds, the six-inch concrete cap on tank E-610 began to crack upward as the tank swelled. The operator heard an ominous rumbling sound beneath the cap. As he turned to run back to the control room, he saw a gaseous white cloud burst from the 120-foot ventilation stack in a ten foot high plume. Back at the control room, he noted that the pressure gauge for tank E-610 was off-scale high at 55 psig. The tank temperature reading had also exceeded scale.

With the chiller out of service and with the rate of the reaction, cooling the tank was an impossibility. Neither could MIC be transferred to the dump tank since it was partially filled. Regardless, the nitrogen pressurization system for material transfers had not been repaired for tank E-610, precluding a transfer. Though the caustic scrubber was on line, it had not been restored to its proper configuration after maintenance. The flare tower remained out of service and thus could not be used. Even if the scrubber and flare tower had both been fully functional, neither had the capacity—separately or together—to handle a release of this magnitude. One of the few remaining actions available was to secure all sources of water to the MIC tanks with the expectation that water contamination was causing the uncontrolled reaction.

The piping flush was secured around 12:45 A.M. and the fire brigade arrived on scene about 12:50 A.M. Water curtain sprays were initiated to combat the escaping cloud, but the curtain reached a height of only 110 feet. The gas cloud continued to jet out of the ventilation stack to a height of 130 feet or more and slowly drift in a southwesterly direction into the ghetto community outside the plant perimeter.

"Avoiding a Panic" The plant staff began a disorderly evacuation of the MIC installation at approximately 1:00 A.M. A few stayed to combat the emergency, but the control room was inundated with MIC fumes by 1:30 A.M. The staff activated a warning siren to alert the city to the accident, but the siren was the same one frequently used for other communications. In any case, the alarm was turned off after only 15 minutes in order that it not panic the community.

The gas plume spread slowly into the city on a mild breeze coming from the northeast—conditions allowing the cloud to remain concentrated. Some of the Bhopal citizens near the plant awakened and began to flee their homes; others died in their sleep. With the wind direction nearly indistinguishable, many ran downwind into the cloud. The city's hospitals and clinics filled to overflowing with patients suffering severe skin and eye irritation, a continuum of respiratory difficulties, and severe lung edema. Makeshift morgues accommodated the hundreds of dead.

Combined Accident Contributors

The mechanism of the runaway reaction in tank E-610 is still a matter of conjecture. Tank samples revealed impurities of acids, bases, and other contaminants. What seems likely is that the MIC in tank E-610 was already dangerously close to an excursion point through contamination and elevated temperature. The piping flush well may have been the triggering event that catalyzed the reaction to an uncontrolled state.

We must, however, look far deeper into the supervision, training, operations, and maintenance of the Bhopal plant to fully understand the causes of this accident, for this was a team that discarded nearly every boundary and margin of their safe operating envelope.

Comparison of the events at Bhopal with the nineteen common accident components presented in Chapter 2 delivers overwhelming evidence of an industrial team out of control.

Degradation of Operating Limits Investigation revealed the event to be the result of a series of small, interrelated problems that had compounded over several years of plant operation. Degradation of organizational structure, training, knowledge of plant components and systems, communications, and maintenance all contributed to the development of a dangerous operating environment. Worst of all, assuming the leaders, technicians, and operators had originally identified and implemented rigorous safe operating boundaries, those boundaries were no longer understood and certainly no longer observed.

Deficient Organizational Staffing and Structure The facility was originally staffed for construction and startup with specially selected, well-trained members of India's indigenous population, coached by an American cadre of Union Carbide employees familiar with the plant design. But, over time, personnel with lower qualifications were substituted as staff members were replaced. With degraded selection and training came limitations in operator and supervisor abilities. Stuart Diamond, investigative reporter for the **New York Times**, wrote in the January 28, 1985, issue:

> ...[W]orkers raised questions about lower employment qualifications. Methyl isocyanate operators' jobs, which once required college science degrees, were filled by high school graduates, they said, and managers experienced in dealing with methyl isocyanate were often replaced by less qualified personnel, sometimes transfers from Union Carbide battery factories, which are less complex and potentially dangerous than methyl isocyanate operations.

Furthermore, plant procedures and technical manuals were written in English which initially wasn't a problem since most of the Indian staff spoke English as their second language. By the time of the accident, many of the plant staff were not literate in English. Yet, the procedures were never translated.

Ignorance of Equipment Operating Characteristics Training proved to be insufficient to familiarize the supervisors and operators thoroughly with the theory and risks associated with the process. The third shift supervisor on duty the night of the event had only recently come to the Bhopal MIC unit from a job in a Union Carbide battery factory. The training provided this supervisor was far inferior to that originally provided to plant supervisors. Similarly, the training provided to plant operator and technician replacements was limited in scope and detail. Stuart Diamond:

> Some of the operators at the plant expressed dissatisfaction with their own understanding of the equipment for which they were responsible. M.K. Jain, an operator on duty on the night of the accident, said he did not understand large parts of the plant. His three months of instrument training and two weeks of theoretical work taught him to operate only one of several methyl isocyanate systems, he said. "If there was a problem in another MIC system, I don't know how to deal with it," said Mr. Jain, a high school graduate. Rahman Khan, the operator who washed the improperly sealed pipe a few hours before the accident, said: "I was trained for one particular job. I don't know about other jobs. During training they just said, 'These are the valves you are supposed to turn, this is the system in which you work, here are the instruments and what they indicate. That's it.'"

These operating and maintenance team members were not prepared to control their facility equipment during routine operations, let alone under emergency and casualty conditions.

Deficient Equipment Maintenance

Several maintenance actions and modifications to the plant occurred over the years that reduced margins of safety. The decision to incapacitate the MIC storage tank chiller system by draining refrigerant to use elsewhere dramatically reduced the margin of error by allowing MIC to exist almost 40 degrees nearer its boiling point. (The quantity of water introduced during flushing was probably sufficient by itself to cause the disaster, but this elevated temperature certainly caused the resulting reaction to be even more violent and uncontrollable.) Stuart Diamond:

> The methyl isocyanate operating manual in use at Bhopal, which was adapted by five Indian engineers from a similar document written for the West Virginia plant, according to a former senior official at Bhopal, says: "Keep circulation of storage tank contents continuously 'ON' through the refrigeration unit."

Also, a jumper had been installed in the MIC storage tank header system to connect the headers for ease of maintenance in flushing. (See Figure 7-2.) This jumper had not undergone a formal safety review before installation, and it compromised safeguards to prevent water from entering the tanks.

Deficient control of maintenance was evidenced by other failures as well. Failure to identify the source of a nitrogen pressurization system leak allowed tank E-610 to be filled 27% over its specified maximum limit, contributing greatly to the consequences of the runaway reaction. The Bhopal operating manual, however, cautions, "Do not fill MIC storage tanks beyond 60 percent level."

Failure to restore the caustic scrubber and flare tower (two vital pieces of safety-related equipment) to their design configurations prior to MIC unit operation—a serious breach of good maintenance doctrine—precluded their use to control an excursion (although, in this instance, they would have been of little use).

Finally, failure to rigorously maintain instruments led operators and supervisors to disbelieve the temperature and pressure warnings of impending disaster. Stuart Diamond:

> Instruments at the plant were unreliable, according to Shakil Qureshi, the methyl isocyanate supervisor on duty at the time of the accident. For that reason, he said, he ignored the initial warning of the accident, a gauge's indication that pressure in one of three methyl isocyanate tanks had risen fivefold.

With no confidence in their instrumentation, this operating team was deprived of critical information necessary to operate their equipment within design specifications.

Non-Compliance with Procedural Requirements

In the area of procedural compliance, the operating staff was grossly deficient as evidenced by the failure to insert the required blank flanges to isolate the MIC storage tanks prior to commencing a water flush. In the January 30, 1985, issue of the **New York Times**, Stuart Diamond recorded that:

> Mr. Khan said he noticed that the closed valve had not been sealed with a slip blind, a metal disc that is inserted into pipes to make sure that water does not leak through the valve…but "it was not my job" to do anything about it…. Valves were notorious for leaking at Bhopal, the workers said. They knew, they said, that water reacted violently with methyl isocyanate. Page 67 of the MIC operating manual for Bhopal says: "Isolate the equipment positively by inserting suitable blinds. Isolation by valve or valves is not to be relied upon."… "I knew that valves leaked," said Mr. Khan, who has a high school education and less training than the amount originally established for methyl isocyanate plant operators. "I didn't check to see if that one was leaking…. It was not my job."

Industrial experience proves routinely that isolation valves alone cannot be relied upon to prevent leakage in high risk situations.

Inadequate Emergency Preparedness

Finally, emergency preparedness was almost nonexistent. In the January 28, 1985, **New York Times**, Stuart Diamond wrote:

There were no effective public warnings of the disaster. The alarm that sounded on the night of the accident was similar or identical to those sounded for various purposes, including practice drills, about 20 times in a typical week, according to employees. No brochures or other materials had been distributed in the area around the plant warning of the hazards it presented, and there was no public education program about what to do in an emergency, local officials said. Most workers, according to many employees, panicked as the gas escaped, running away to save their own lives and ignoring buses that sat idle on the plant grounds, ready to evacuate nearby residents.

A ghetto community had been allowed to concentrate outside the fence line of the installation. Little thought seems to have been given to what would happen in the case of a leak of toxic chemicals, how the community would be warned, and how the installation and community should respond. The warning siren, turned off after 15 minutes to "avoid panic", was turned back on a few hours later—after the cloud had already dispersed.

The Cost

The cost of the tragedy at Bhopal is inestimable. The Indian government originally established the death toll at 1,754 people. However, early in 1992 the count was revised upward to nearly 3,000 dead. Independent studies have estimated the number of deaths as high as 10,000. In addition, tens of thousands remain permanently injured. Lung and eye damage are among the leading problems.

The costs to the Union Carbide company are also great. The company originally paid $470 million to settle wrongful death and injury suits. Yet, litigation continues. From a business perspective, the intangible damage to this company's reputation and resultant increased regulation and oversight costs are enormous.

Yet, as is so often the case, this was an accident that could have been easily prevented. Successful operation of the facilities at Bhopal dictated that operators and maintenance personnel preserve the integrity of plant operating and safety systems. Instead, this team violated important operating and maintenance procedures, neglected to monitor critical operating parameters, failed to restore important systems and components to their design configurations following maintenance, inadequately isolated high energy systems, and ignored independent verification of critical operating and maintenance tasks. Serious breakdowns in judgment in what might otherwise appear to be minor daily problems resulted, ultimately, in the loss of control of safe equipment configuration.

The large loss of life qualifies this accident as the most lethal single industrial disaster in history. It would be even more disastrous if we failed to learn the lessons of equipment control that Bhopal teaches.

Twelve Vital Operating Skills

The Bhopal accident presents a momentous lesson for all industrial operators: *Maintain configuration control!* **Configuration control** simply means operating, maintaining, and testing equipment and processes in accordance with the limits and practices imposed by design. Components and systems *must* be operated within prescribed tolerances if they are expected to function in accordance with design specifications. Further, whenever maintenance, modification, or testing is performed, equipment *must* be restored to a level equal to or exceeding design specifications unless a revised hazards analysis indicates that lower levels of performance are acceptable.

In Chapter 5, we learned that the objective of the strategy for operating success—equipment control—is supported by a simple cardinal rule: *Humans must remain in control of their machinery at all times. Any time the machinery operates without the knowledge, understanding, and assent of its human controllers, the machinery is out of control.* Yet, at Bhopal, the operators no longer commanded the situation. Through disregard for operating restrictions and proper maintenance practices, they allowed their equipment (and the chemicals contained therein) to far exceed the limits of design—clearly violating the first principle of operation. They failed to maintain control.

Yet, *wanting* to maintain control of equipment and *doing* it are not the same. The first is necessary for the second, but they are separated by a gulf of knowledge, experience, and good coaching.

If proper equipment configuration is to be maintained—whether for operations, maintenance, or testing—every operator must acquire fundamental skills of equipment control. In the U.S. Navy's Nuclear Power Program, each operator is taught *twelve vital skills*—a systematic approach to equipment operation—designed to arrest accident contributors in their infant stages. As summarized in Figure 7-3, every team member must learn to:

1. Conduct pre-task briefings. A *pre-task briefing* is a preparatory meeting conducted before performing an operating, maintenance, or testing task. It is designed to ensure safe and efficient work by causing team members to think through tasks before performing them.

Through the process of pre-task briefing, operators become acquainted with unusual or complicated work prior to its performance. Roles are established, lines of communication designated, resources identified, and task steps deliberated—a prescription for eliminating many of the problems that plague efficiency.

2. Understand and use operating policies and procedures. The skill of using operating policies and procedures is based, obviously, on an initial ability to read and comprehend. In days past, reading skills were taken for granted. No longer. Inadequate reading ability has become one of industry's greatest obstacles. Therefore, selection criteria for operator candidates must require (and test for) sound reading comprehension. Without it, an operator is at an almost unrecoverable disadvantage.

To basic reading skills must be added the ability to find, interpret, and apply technical policies and procedures. Just as understanding legal literature requires familiarity with special language, format,

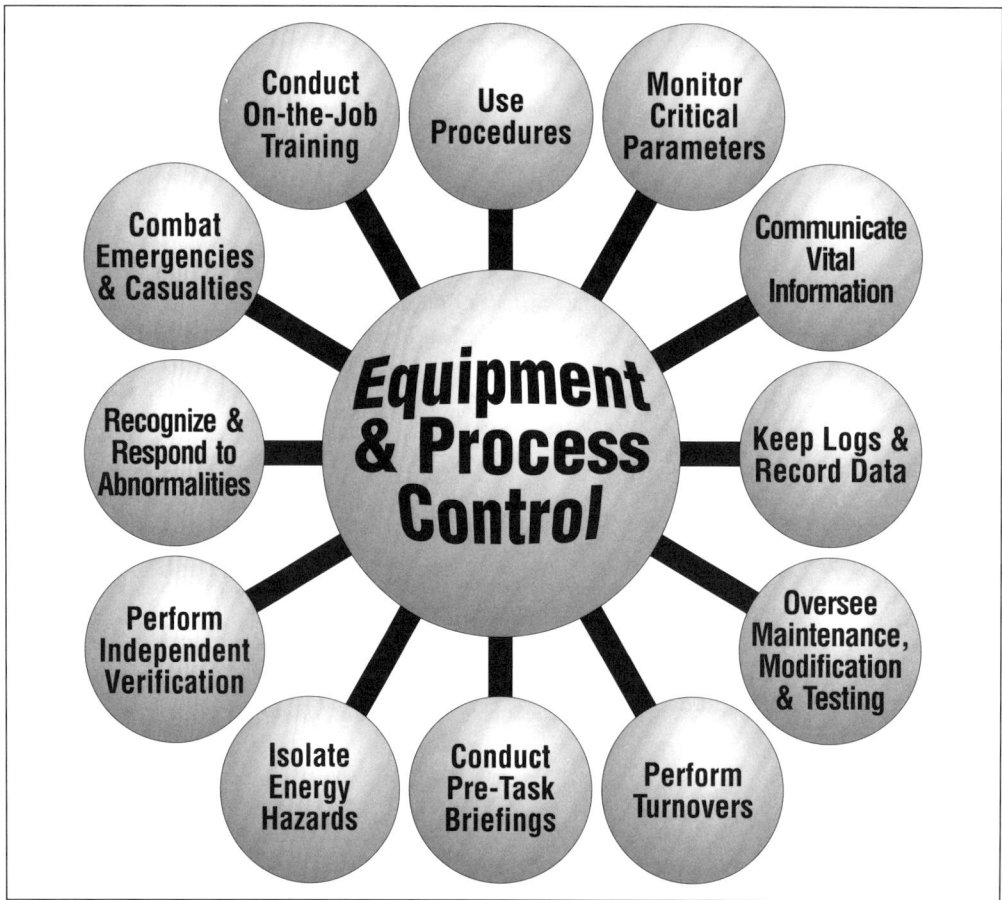

Figure 7-3: Twelve Vital Skills.

and conventions, operating policies and procedures (even if well-written) will contain language and conventions peculiar to the particular industry and technology. Knowledge of procedure format, for example, is not inherently part of basic reading skills. An operator must be taught where to find the section in an operating procedure which prescribes the initial conditions required for machinery start-up. He or she must also know where and how operating precautions and limitations are specified, and what the final conditions following task completion should be.

3. Monitor Critical Parameters. For equipment to be properly controlled, critical operating parameters—temperatures, pressures, levels, speeds, voltages—must be maintained within specified operating bands. *Configuration control*, as it is sometimes termed, is accomplished through detailed knowledge of components and systems, compliance with operating procedures, and frequent observation of the equipment. Accordingly, operators must learn to conduct careful inspection tours, monitoring critical operating parameters and symptoms as they proceed.

4. Perform Independent Verification. Some tasks are so important that failure cannot be tolerated. In such cases, separate proof of successful task completion, called *independent verification*, may be required. For example, if mispositioned valves or switches could lead to dangerous conditions, a second competent person may need to independently check valve or switch positions either during or after completion of valve or switch line-up. The second person would be fulfilling the task of independent verification, a skill critical to the success of industrial operations.

5. Communicate essential plant information. Industrial operations are almost always carried out in the context of a team. Communication, accordingly, fulfills a momentous role in orchestrating the efforts of team members during both routine and abnormal conditions.

Communications may take the form of face-to-face discourse (including directions, questions, and status reports), radio or telephone exchanges, and distinctive emergency signals. Regardless of form, the language must be understood, the communication method must be appropriate, and participants in the exchange must feel responsible to accomplish the objective of the communication.

6. Keep logs and record data. One way of communicating plant information is through records of vital plant data and history. Nearly all industrial machinery and processes are regulated by monitoring critical operating parameters. Therefore, many components and systems have data record sheets (or automatic recording devices) on which the critical parameters are registered.

Data record sheets (or automatic recording devices) provide a systematic means of checking critical parameter values and, by recording the values, a method of logging operating history. Operators must be capable of recording *and interpreting* these critical parameter values if the equipment is to be operated within design specifications.

Another important written communication common to industrial facilities is an operating log. Traditionally, logs are legal, narrative, chronological records of significant events at an operating station or facility. As with data record sheets, they provide a history of operation. They also help provide continuity of control by permitting operating personnel to review the events of previous shifts.

Other written communications which may require operator attention include entering information in machinery history, suggesting procedure changes, and submitting material deficiency reports.

7. Recognize and respond to equipment abnormalities. Part of competent equipment control requires an ability to recognize deficiencies or symptoms of abnormal equipment performance. Machines, as do humans, frequently display signs of their "illnesses". Component or system infirmities may be exhibited through abnormal sounds or smells, unusual changes in critical parameter values, or part failures. Operators who can recognize such symptoms are able to identify, report, and sometimes fix the deficiencies in their early stages.

Early detection and treatment, whether for human or equipment problems, is an important part of the operating strategy. Fortunately, undesirable events and accidents are frequently portended by prior events known as *accident precursors*. Precursors may take the form of close calls, comparable events which resulted in minor consequences, or habits and patterns which have been shown through accident studies to lead to mishaps. Operators who have been educated in precursors and abnormal condition response are far more apt to intelligently intervene in a questionable chain of operating events.

8. Combat equipment emergencies and casualties. Left unattended, abnormalities may lead to equipment emergencies or casualties. An *equipment emergency* is a threatening event that, if promptly and properly treated, will probably not result in equipment damage or personnel injury. An *equipment casualty* is a destructive event that *will* result in equipment damage or personnel injury—regardless of what is done to combat it. Actions prescribed to combat a casualty are designed to minimize the damage that occurs.

Unquestionably, the ability to detect and combat emergencies and casualties is a vital skill.

9. Oversee maintenance, modification, and testing. Control of equipment is important, not only while equipment operates, but also when it is in an abnormal state such as when shut down (or re-configured) for maintenance or testing. Accordingly, operators must coordinate with maintenance and testing personnel to secure, isolate, and remove equipment (or to simply place it in a reduced status), always with an understanding of how the configuration change affects on-going operation of other components and systems. Hence, the ability to monitor equipment maintenance, modification, and testing is essential for operating success.

10. Isolate energy hazards. Tools and machines use and develop energy—mechanical, electrical, hydraulic, pneumatic, thermal, and chemical—forms of energy which, if inappropriately regulated, can injure humans and damage equipment.

Because of unusual equipment configurations, risk of injury from energy hazards is often increased during maintenance and testing. Therefore, operators (vital members of maintenance teams) must work closely with maintenance and testing personnel to safely isolate hazardous energy sources before permitting work on their equipment.

Energy isolation is usually accomplished by interrupting energy paths through realignment of switches and valves, or insertion of physical blocks (such as blind flanges) in fluid flow paths. These *energy isolation devices* are then protected or flagged by installing locking devices or warning tags.

11. Train on-the-job. Team composition is seldom static. New team members (who need training, coaching and mentoring) demand experienced operators who are capable of teaching them how to control equipment and operate systems. Therefore, operating personnel must be equipped to train new team members, employing, in particular, the skill of on-the-job training.

On-the-job training (OJT) is a process in which an experienced operator/instructor demonstrates and explains operating tasks to an operator candidate. The candidate, after observing and discussing the task, must then perform and explain the task to the instructor's satisfaction.

On-the-job training requires special instructor skills. Not only must the operator/instructor be well-qualified on the equipment, he or she must also be able to teach and explain while performing operating tasks. As such, the skill of OJT instruction is one that usually requires special preparation.

12. Perform shift turnovers. *Turnover* is the mechanism for orderly transfer of responsibility for a task or process to a relieving work group. It is analogous to passing the baton in a relay race.

As with pre-task briefing, turnover is a unique communication skill important to the success of operating teams. Since most teams must *pass the baton* from time to time, the skill of shift turnover should be mastered by operators, maintenance technicians, testing personnel, engineers, and leaders at all levels in an organization.

Topic Summary and Questions to Consider

At Bhopahl, operators and managers alike failed to maintain control of their equipment and processes, and that failure resulted in yet another major disaster, the effects of which persist today, more than a decade later. Disaster could have been avoided if only they had been proficient and diligent in practicing the twelve vital skills presented in this chapter:
- Conduct pre-task briefings
- Understand and comply with procedures,
- Monitor important operating parameters,
- Independently verify critical task completion,
- Communicate vital information,
- Keep logs and record operating data,

- Recognize, prioritize, and respond to abnormalities,
- Combat emergencies and casualties,
- Oversee maintenance, modification, and testing activities,
- Isolate energy sources when required,
- Train others on-the-job, and
- Perform shift turnovers

These are the skills that constitute the defense and fundamentals for industrial teams. Practiced separately, they have little effect on the performance of the team. But, together, they create a formidable barrier to the development of accident conditions.

Ask Yourself Do your team members know more about their processes and equipment than the people at the Bhopal plant knew about theirs? Do your team leaders know the technology well enough to lead? Are your components and systems properly aligned before use?

You may not face the same risks as the Bhopahl plant, but the same principles of control apply.

Subsequent chapters explore each of these twelve vital skills in greater depth, demonstrating their power to improve operating efficiency and enhance safe industrial operations.

Chapter 8

Conducting Pre-Task Briefings

Many of the problems that occur during operations, maintenance, or testing can be anticipated and prevented. This chapter examines the purpose of and how to use one of the best techniques available to technicians and their leaders for problem pre-emption: the *pre-task briefing*.

Utility Crew Electrocution, April 1994

On April 18, 1994, two members of a city crew from a utility company's Electric Light Division sustained critical electrical injuries while moving an electric light pole to accommodate on-going road construction. While the crew foreman and one crew member endeavored to control the base of the pole during the lift, it swung out of control, contacting one phase of an energized 46kV phase-to-phase power line. The description of the following event is factual, but names and locations have been omitted to protect the privacy of the victims and their families.

Accident Description   On Monday morning, April 18, a six-man crew employed by a utility company's Electric Light Division was dispatched to replace a transformer. After completing the assignment and taking a lunch break, the crew responded to a radio interference complaint which subsequently required a guy rod installation.

While the crew members installed the guy rod, the division superintendent visited the work site and requested the crew foreman to accompany him to another job location. Road construction near a busy city intersection required removal of a 50 ft. steel light pole. After evaluating the new work location, the superintendent instructed the foreman to remove the pole before the end of the day.

The light pole to be removed was a 50 ft. tall steel pole with a 15 ft. davit, attached to which was a Luminaire street light fixture. (See Figure 8-1.) The Electric Light Division's accident report states that:

> *This type of pole (50 ft. height, 15 ft. davit) is significantly larger than our standard street light pole.*

Following the evaluation, the crew foreman returned to the guy rod installation site to inform the work crew of the street light removal task. The crew completed the work in progress and moved to the new location, positioning their line truck on the sidewalk beside the pole. (See Figure 8-2.)

The light pole to be removed was 40 feet north of a three-phase high voltage line (46kV phase-to-phase). The crew was to remove the pole from its base using the line truck's boom and tongs, electrically disconnect the pole from its supply cabling, and lay the pole horizontally on a grassy area to the south of the truck.

Before starting the job, communications regarding the work steps, attendant hazards of the task, and appropriate safety precautions appear to have been minimal. In particular, little consideration was given to the nearby power line hazard. The accident report states that:

> *As work in preparation of removing the light pole began, all crew members recall having little or no concern with the 46 kV power line, which was approximately 40 ft. from the street light location. However, not all crew members were aware of the direction the light pole would be moved.... None of the crew members were aware the line had been de-energized earlier that day, and subsequently re-energized.*

As the light pole removal progressed, they encountered some difficulty in disconnecting it and rigging it from its base. The lifting and handling operation had to be modified to allow a better purchase on the pole. The investigative report recounts the events as follows:

> *The boom was extended and positioned at the pole. A sling was attached to the pole. The sling slipped and the foreman directed that a longer sling be used to provide more wraps and better grip. The sling was attached approximately two thirds up the pole. the tongs were closed around the pole, above the sling. There was some difficulty removing the nuts at the base, requiring use of ground rod hammer with pipe wrench. After the nuts were removed, the pole was lifted slightly off the foundation to allow for access to the wiring. Crew members took turns steadying the pole as the wiring was disconnected. The pole did not spin or cause difficulty at this time.*

The line truck tongs provided a stabilizing force near the top of the pole while the lifting sling grasped the pole below the tongs. As stated in the report, members of the crew had to steady the base of the light pole as it was lifted.

After the pole was mechanically and electrically disconnected from its base, the crew began to move the pole to the south side of the line truck. The accident report describes the subsequent events:

> *After the wiring was disconnected, the line truck operator asked the foreman where to swing the pole. The foreman directed the operator to move the pole along the front of the truck, from right to left, to the sidewalk/grass area on the left side of the truck. It is believed the operator and foreman, and possibly [Worker 1], knew of this communication. The operator, from the passenger side controls, began to move*

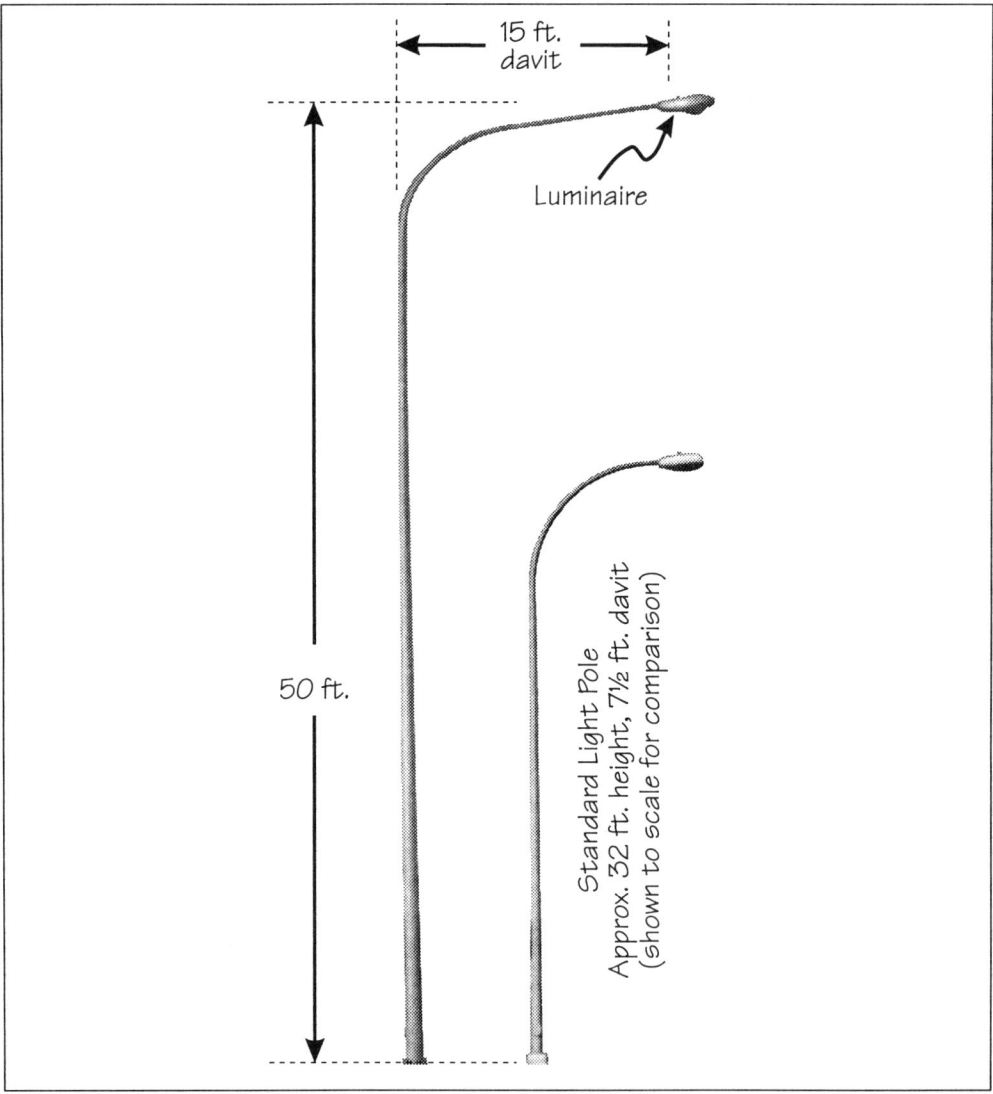

Figure 8-1: Illustrations of the involved light pole (left) and a standard light pole (right) shown to scale for comparison purposes.

the pole. The load appeared to be balanced and reasonably straight. The operator's vision from the passenger side controls was poor. He stopped boom movement, and stepped over to the driver side controls. He then continued to move the pole. At this time, the crew members were located in the approximate position shown on the attached map. [See Figure 8-3.] At this time, the crew recalls [the foreman] and [Worker 1] using great effort trying to guide the pole base. At some time prior to contact, two crew members began yelling. The pole then contacted the 46 kV (46 kV phase to phase) line. Contact was made on the north phase only, from 10 ft. to 14.5 ft. east of the power pole on the corner. Contact on the street light pole was on the luminaire and end of davit arm.

Witnesses and crew members reported an immense, bright flash accompanied by heat when the pole contacted the north phase of the power line. Electrical current traveled down the conductor and through the bodies of the foreman and crew member who were attempting to steady the butt of the pole. The report states that:

It is believed the pole butt was off the ground and not in contact with the truck at the time of contact.

The electrical fault opened the circuit breakers for a substation and a power plant on the grid. The breakers remained open as designed. With power now removed from the high voltage conductors, the line truck operator moved the pole clear of the power lines while emergency aid was summoned for the injured crew members. Both casualties were transported to the regional medical center and subsequently moved to a burn trauma center. The first succumbed to injuries. The second survived, but required amputation of both legs as a result of the burns.

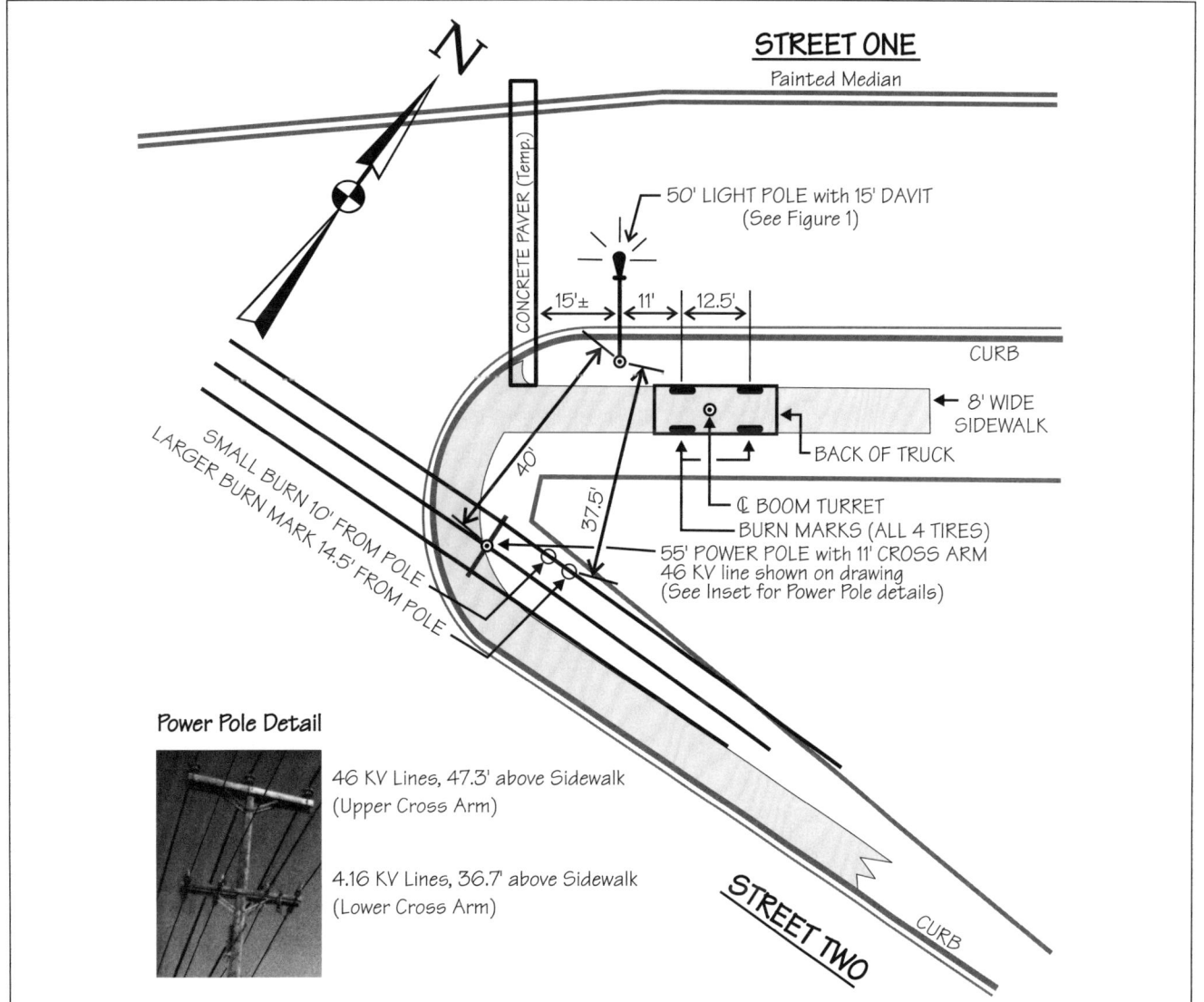

Figure 8-2: Diagram of accident location showing positions of boom truck, power lines, and light pole location before removal.

Conducting Pre-Task Briefings

Investigating Board Conclusions

An accident investigation board convened on April 21, 1994, to determine the cause of the accident and to establish corrective actions to preclude similar events. The investigation soon established that the light pole removal task had been conducted without the benefit of a pre-task briefing (known to the crew members as a *tailboard conference*). Consequently, potential hazards and attendant safety precautions were not formally discussed. The accident investigation board concluded:

> *A tag line on the pole was not considered. Use of rubber gloves was not considered. De-energizing the power lines, line clearances, and/or hot-line permits were not considered. Alternative positioning of the line truck was not considered.*

Attachment of a tag line near the top of the pole may have better assisted in controlling movement of the pole. De-energizing the nearby power lines, though inconvenient, would have eliminated the source of electrical energy which culminated in the accident. (The power line had already been de-energized once before during the day.) A properly performed rigging evaluation would have identified clearances to nearby obstacles, the potential for losing control of the pole, and the subsequent need for a different truck position. Finally, a well-conducted electrical safety evaluation and hot-line work permit would have led to consideration of electrical safety precautions including rubber insulating gloves.

In conclusion, the report stated that:

> *The task of removing the street light pole on 4-18-94 appeared to all involved to be very simple and routine. The possibility of contact with the power lines (46kV & 4.16kV) was not considered or discussed. Communication prior to commencement of work was inadequate; a tailboard conference was not conducted. A signal man was not designated. An observer was not designated.*

Investigating Board Recommendations

In light of the facts determined during the accident inquiry, the investigating board made six recommendations for future work:

- *An individual must be designated [whose] sole responsibility is to be in charge of the job.*
- *An individual crew member shall be assigned as signal man.*
- *An individual crew member shall be assigned as observer. The same individual may perform both signal and observer duties.*
- *Means of controlling unwanted pole movement, such as use of a tag line, shall be considered when handling poles.*
- *Tailboard conferences should be conducted prior to commencement of any work. The possibility of contact with power lines during the work should be specifically addressed during tailboard conference.*
- *First line supervisor training should be included in our training program.*

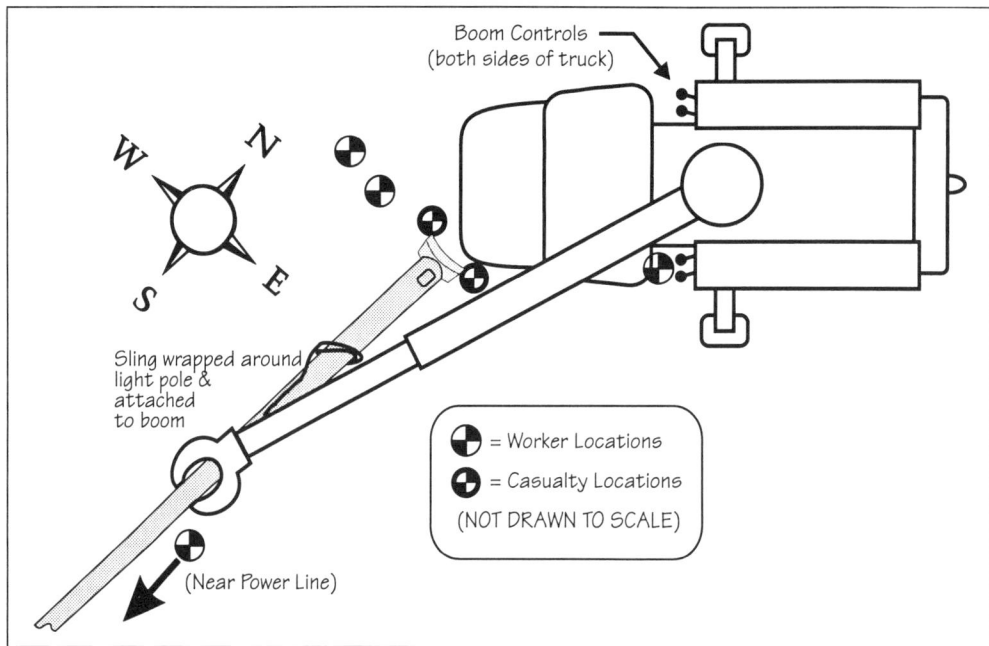

Figure 8-3: Illustration showing worker locations at time of accident. (See also Figure 8-2.)

Accident Prevention through Pre-Task Briefing

The errors committed by the Electric Light Division crew are not at all uncommon. They usually result from bad habits which have taken root over a long period. Unsafe habits are tolerated because they seem insignificant. The logic trap asserts that, "Since they *didn't* cause a problem with *yesterday's* job, they must not be necessary." Unfortunately, as this crew learned (and as you and I so often learn), poor habits aligned with non-conservative circumstances eventually create disaster. The solution? Don't tolerate bad habits—even when the risk is relatively low. Do it right every time.

As with so many accidents, this utility company mishap probably could have been prevented if the crew had performed a simple *pre-task briefing*. A properly conducted briefing would have addressed the objective of the task at hand, the major steps of the task, the roles of the team members in accomplishing the task, the characteristic hazards of the task, and the safety constraints employed to protect against the hazards. In this instance, it appears that none of these elements received appropriate attention.

By learning the basics of this simple anticipatory process, you can protect yourself and your colleagues from the majority of operating, maintenance, and testing hazards. So, the remainder of this chapter will explore the following topics:

- What is a pre-task briefing?
- When should a pre-task briefing be conducted?
- What are the elements of a properly conducted pre-task briefing?
- What guidelines will assist in conducting a pre-task briefing?
- What are some common pre-task briefing errors?

Definition of a Pre-Task Briefing

A **pre-task briefing** is a preparatory meeting conducted before performing an operating, maintenance, or testing task. The briefing is designed to ensure the safe and efficient execution of the task. Generally, it consists of a review of the task, the role of team members in task accomplishment, the hazards characteristic of the task, and the safety constraints employed to protect against the hazards.

Applicability of a Pre-Task Briefing

A pre-task briefing is valuable not only in high risk, infrequent, and complex tasks, but is also useful (and often necessary) for relatively ordinary tasks such as routine preventive maintenance. A thorough and time-consuming pre-task briefing may be essential for infrequently performed, high risk tasks. It may even require real-time walk-throughs to ensure that each participant understands the sequence of task steps and his or her role in accomplishing the steps. For simple and routine tasks, the pre-task briefing may last only a few minutes.

Normally, a pre-task briefing is conducted as a meeting with a task team. But, for a one-person task, it may simply be a written or mental checklist before starting the job.

Elements of a Pre-Task Briefing

Regardless of length, the form of the pre-task briefing should vary little. Each pre-task briefing should consider these ten elements:

1. Statement of Task Objective
2. Discussion of Team Member Roles
3. Identification of Necessary Tools and Materials
4. Determination of Initial Conditions
5. Review of Task Steps
6. Determination of Final Conditions
7. Analysis of Hazards and Risk
8. Review of Safety Constraints
9. Assessment of Emergency Preparedness
10. Review of Required Documentation

Conducting Pre-Task Briefings

Statement of Objective

Each pre-task briefing should begin with a statement of the task title and the objective of the task. All in attendance should obtain a clear understanding of the purpose of the task and the desired outcome of task completion. The briefing leader must avoid the error of bypassing a statement of objective as a trivial action. To assume that each team member already understands the objective can be disastrous.

Discussion of Roles

The second element of a pre-task briefing is the delineation of team member roles. This discussion ensures that the team members know how they fit together and how their responsibilities for task completion interrelate. This part of the briefing should identify the team leader and the roles of each member during both routine and abnormal conditions. This discussion should not normally be a monologue by the briefing leader. Each of the team members should be asked to state his or her role in the process.

Identification of Necessary Tools and Materials

An important function of the pre-task briefing is to ensure that all necessary tools, equipment, and materials are available to perform the task successfully. A little time spent during the briefing addressing this element will pay big dividends in time and, often, in safety as well.

Well-written work requests and procedures usually contain a *tools and materials* checklist as a standard part of the work package. Work teams, however, should not fall into the trap of bypassing a review of the checklist by assuming that it is complete. Seldom are two jobs exactly the same.

If no tool and material checklist is available, a few minutes spent in developing one during the briefing will be very beneficial in assisting the team to think through its task. Finally, the tool and material review should not ignore emergency equipment necessary for contingency response.

Determination of Initial Conditions

A fourth fundamental element of a pre-task briefing is determining the required initial conditions necessary for task performance. A review of the conditions that must pre-exist task performance helps to avoid the danger of starting a task before the supporting conditions are established.

One important reminder: Don't forget that, if the task is interrupted, the initial conditions may change. Significant work delays should be followed by a verification that the necessary initial conditions still exist before resuming work.

Review of Task Steps

The fifth essential element of the briefing is to review the steps associated with job performance and completion. Normally, this need not be a step-by-step reading of the task procedure. In fact, for most tasks this portion of the briefing is better conducted if the briefing leader asks team members to describe the task segments and their individual roles in accomplishing these steps. The active involvement of team members in this process raises participation, interest, and understanding. It also allows the team leader to evaluate the readiness of the task members.

Determination of Desired Final Conditions

Restoration of equipment to a desired final condition must also be addressed during the pre-task briefing. The question to be answered is, "How do we want to leave this equipment after finishing the job or the work for this phase of the job?." In particular, the work team should establish (or review) what post-work checks must be conducted to guarantee that the equipment has been restored to a safe and stable configuration. If the purpose of the task is to place the equipment back in service, the post-work checks must ensure that the equipment will operate in accordance with design specifications.

Analysis of Hazards and Risk

At some point in the briefing, the team must critically analyze the hazards associated with task performance and the probability (risk) of encountering a hazard. Particular attention should be given to hazardous chemicals that will be used, high energy systems that might be encountered, and precarious situations (for example, exposure to fall) that could occur.

As part of the analysis, the team should postulate scenarios for uncontrolled hazards that could be encountered. A discussion of past problems with the same or similar tasks involves the team in thinking through the event and how it could go wrong.

Review of Safety Constraints A review of the safety actions and constraints to prevent damage to personnel, equipment, or environment naturally follows the hazards and risk discussion. The review should consider electrical safety precautions, protective clothing, isolating (locking and tagging) hazardous energy paths, and independent verification of critical task steps as required.

Assessment of Emergency Preparedness The pre-task briefing should also consider the resources and actions necessary if an emergency occurs. Team members should understand how to get emergency assistance if required, where to obtain emergency equipment such as fire extinguishers or protective masks, and under what conditions to evacuate the work site. Pre-staging emergency equipment or response teams may be necessary for high risk events.

Review of Required Documentation Before the pre-task briefing is completed, the task team should review the documentation necessary to validate and record task completion. The review should ensure that team members have the appropriate data record sheets or recording devices.

Pre-Task Briefing Guidelines

Conducting a successful pre-task briefing is a skill that should be acquired by all team leaders. Here are some guidelines that will help to ensure a productive pre-task briefing:

- Ensure the personnel who will be involved in the actual task performance are chosen with regard to required training, qualification, or certification. Don't forget to include operating technicians! Sometimes the engineers talk the job through without consulting those who actually perform the work. The human engineering perspective gets lost in the shuffle and communication suffers.
- Ensure, if necessary, that the chosen personnel have been alerted to the event, that their work assignments have been appropriately coordinated, and that a study assignment has been issued, if necessary, in advance of the event to allow effective performance of the pre-task briefing.
- Assemble in a quiet location, if possible, free of unnecessary distractions. Sometimes, however, the briefing may be best accomplished at the location of the actual event in order to clearly visualize the equipment, systems, and components that will be manipulated or affected. In any case, make sure everyone can hear and understand.
- Have the requisite procedures, instructions, or technical manuals available and refer to them as necessary during the briefing.
- Have the briefing leader succinctly describe the task and the desired outcome.
- Discuss the initial conditions, precautions, and limitations associated with task performance. Include a discussion of what actions will be necessary to re-establish initial conditions should a delay in task performance occur.
- Determine the tools necessary to accomplish the task and check the status of the tools or ensure that the status will be checked prior to performance of the task.
- Have the task team participants describe (and demonstrate, if necessary) each of their individual roles in performance of the task, including timing of actions with other members of the team and methods of communication to accomplish the timing.
- Have the team leader lead an open discussion of the hazards associated with the task and the potential risk and outcomes of failed task performance.
- Have the team discuss the methods of combatting the emergencies or casualties associated with failed outcomes and the effect that each will have on the integrated operation of a facility or process.
- Discuss the process that will be used if the task cannot be finished and must be placed in suspension until return of the original task team members. Include a discussion of the process and problems associated with turning the task over to a relief crew.
- Review the requirements for safety equipment and safe work habits. Check the status of the equipment or ensure that it will be checked prior to initiation of the task.
- Review the documentation provided for validation and recording of the task and ensure responsibility has been assigned for recording required data.

All of these guidelines will not apply to all situations. *You* must use judgment in constructing each pre-task briefing. Risk should determine the comprehensiveness of the briefing.

Common Pre-Task Briefing Errors

There are several common errors that a briefing leader should understand and avoid when conducting a pre-task briefing. They include failure to conduct a briefing, failure to involve team members in the briefing process, and briefing the wrong people.

Conducting Pre-Task Briefings

Failure to Conduct a Briefing

When tasks are performed often or when the job seems to be relatively low risk, we all have a tendency to dispense with thinking through the process before starting it. Yet, remember what the Electric Light Division investigation committee said:

> *The task of removing the street light pole…appeared to all involved to be very simple and routine. The possibility of contact with the power lines…was not considered or discussed.*

In this instance, a high-hazard risk was located only 40 ft. from the light pole—and the light pole was 50 ft. tall with a 15 ft. davit extension.

The best professionals, whether in sports or industry, learn to take time before starting a task to think through the task and its attendant hazards. Pre-emption of mistakes is the purpose. Remember that even a routine preventive maintenance or operating task can be mentally briefed or briefed in a team environment in a few minutes. Time spent anticipating task steps and hazards usually offsets time lost through performance errors.

Failure to Involve the Team Members

Pre-task briefing is a skill. An unskilled briefing leader often talks without listening or providing an opportunity for team members to raise questions or objections. Briefing leaders who simply read the procedure steps while team members "listen" are unlikely to determine how well team members understand the job. Active involvement promotes interest and provides an opportunity to evaluate.

Briefing the Wrong People

Pre-task briefings are often prescribed as an initial step in a work document. When the briefing is conducted, the step is signed. If a delay prevents team members who have been briefed from performing the task, the briefing should be re-performed for those who will actually do the work. If the briefing, however, is viewed as an unnecessary inconvenience, leaders may be tempted to avoid briefing a new team since the package step is already signed.

Post-Task Debriefing

Though the theme of this chapter has been pre-task briefing, we would be remiss if we failed to mention her twin sister. **Post-task debriefing** is the simple process of evaluating task performance after completion. Questions should include, "Is the job complete?" "Was anything peculiar about this job?" "Were any lessons learned from this job that should be recorded and passed on to others?"

As with pre-task briefing, the post-task debriefing may be an involved and complex process, or, for routine tasks, little more than a mental check to ensure that everything has been properly restored. It may require a formal, written after-action report, or merely a word to a boss or colleague. Regardless, don't dismiss it as unimportant. It serves as a job-completion and lessons-learned checklist.

Topic Summary and Questions to Consider

The Electric Light Division accident was a tragedy of bad habits and poorly established work processes. Unfortunately, these are the same kinds of errors that *we* make every day in processes as "simple" as driving an automobile. Therefore, we have used this event, not to place blame or point fingers, but as all accidents should be used—to learn lessons.

Through the wise use of pre-task briefings, we can protect ourselves and our fellow team members from similar tragedies. The pre-task briefing is a simple tool that, properly conducted, causes a team to think through a job completely before jumping into performance.

Ask Yourself

Does your team regularly conduct pre-task briefings? Is a checklist used for planning and conducting the briefing? Have you observed a pre-task briefing lately? If so, did the team members actively participate, or was it just a boring lecture?

Be prepared for objections when you try to establish pre-task briefings as a work habit. Some will say that they take too long. Task briefings needn't take much time. In fact, the briefing usually saves far more time than it takes. Once the habit is established, you will find that work team members feel uncomfortable when a pre-task briefing is *not* performed.

So, take time before jumping into your next task. Remember the old adage: *Measure twice, cut once!*

Chapter 9

Understanding and Using Procedures

In Chapter 3 we learned that valid policies and procedures (and the discipline to follow them) constitute a vital part of any effective operating plan. We have already seen how conscious decisions to violate procedures (or ignorant neglect to follow them) played crucial roles in the *Exxon-Valdez* grounding, the Chernobyl reactor accident, and the Bhopal disaster. We will learn that procedure violations also seriously undermined safe operations at Three-Mile Island (Unit 2) and routine maintenance on Continental Express Flight 2574.

Through accident study, it doesn't take long to conclude that the ability to effectively use policies and procedures is a fundamental skill of the operating technician. Competent operators (and maintenance technicians) manage equipment within prescribed tolerances *in accordance with approved procedures*. This chapter reviews the process of developing and using good operating procedures.

Policy or Procedure?

In Chapter 3, we made a distinction between the terms *policy* and *procedure*. Policy was defined as written guidance describing the plans or general principles for performance. Policies usually prescribe philosophy and interactions between people and agencies. Operating and maintenance procedures, on the other hand, are instructions for running, preserving, and repairing machinery components and systems. They provide step-by-step guidance for starting, operating, stopping, and maintaining equipment.

When properly written and used, both form an important line of defense to ensure that equipment is operated and maintained within design specifications—within the boundaries of the safe operating envelope. Procedures, in particular, should be the constant companion of the industrial operator as he or she regulates equipment during the operating day. Therefore, *this* chapter is devoted specifically to the development, use, and revision of operating and maintenance procedures in an industrial facility.

Developing Procedures

When correctly developed, component and system procedures have a lineage that can be traced all the way back to the design basis of the equipment. Operators who are taught the lineage and usefulness of procedures are far more likely to respect and employ them competently.

Operating and maintenance procedures are usually derived from manufacturers' information such as component technical manuals that accompany the equipment. Technical manuals should reflect the design features, safeguards, and operating/maintenance standards necessary to insure safe, long equipment life.

Sometimes, technical manuals alone are sufficient for equipment operation. Usually, however, the contents of technical manuals must be carefully incorporated into plant-specific operating procedures. Especially where off-the-shelf components are incorporated into a customized arrangement, system operating procedures must be developed by the facility using component technical manuals, a task of great complexity.

In addition to manufacturers' information, resources for procedure development include assistance from vendors, facility engineers, and prospective equipment operators. Unfortunately, operating and maintenance technicians—the end-users of the procedures—are frequently not consulted until it is time to test the procedure. Failure to employ the knowledge and skills of the final users early in the procedure development process usually results in numerous review cycles.

Finally, an important step in procedure development is the process of testing the procedure. Validating a new procedure usually requires a step-by-step walkthrough of the procedure under controlled conditions.

Desirable Characteristics of Procedures

Effective procedures share a few common characteristics—they must be clear, concise, and easy to understand to a competent technician. It must be emphasized that the procedures must be written with the *competent technician* in mind—a technician who has been selected and trained in accordance with reasonable standards. It is a dangerous and fallacious notion to believe that detailed procedures can substitute for proper training. Good procedures are a vital adjunct to proper training, not a substitute.

Procedures are made user-friendly by selecting a format which emphasizes rather than obscures the initial conditions, precautions and limitations, and the action steps. Short sentences containing one action per procedure step help keep the sequence intact. Inserting pertinent precautions and warnings immediately before the step to which they apply is also a useful practice to remind the technician of hazards associated with the procedure.

Perhaps the most perplexing problem facing procedure writers is the question of what language level at which to write. Yet, if representative end-users are employed in the writing process, the language level problem is diminished.

Procedure Format

Before the procedure development process begins, it is necessary to establish a basic procedure format. Without such a procedure "style manual", many disparate and often incompatible formats will evolve—one for every organization involved in procedure writing.

The fundamental elements of an operating or maintenance procedure are:
- Purpose Statement
- Applicability Statement
- Precautions and Limitations
- Initial Conditions
- Procedure Steps
- Final Conditions

Purpose Statement One of the first paragraphs in an operating procedure should be a statement of its purpose. The purpose statement should clearly describe the task for which the procedure has been developed.

Applicability Statement Following purpose should be a statement of applicability. This section delineates the system and components to which the procedure applies and under what conditions the procedure should be used.

Precautions and Limitations Incorporating a precautions and limitations section early in a procedure alerts the operator to the general hazards and restrictions associated with performance. "Precautions and Limitations" may contain both general and specific warnings, although some organizations choose to include specific warnings or precautions immediately prior to the steps to which they apply.

Initial Conditions *Initial conditions* is a phrase used to describe the equipment status (e.g., valve or switch positions) that must exist before starting the procedure steps. Accordingly, an initial conditions section should appear early in the procedure format so that equipment status can be verified before starting the task *or after a long delay during task performance.*

Procedure Steps   The individual task steps are in this procedure section. They should prescribe in clear, sequential terms the actions necessary to perform the task for which the procedure has been written. A good guideline for composing procedure steps is to use a separate step for each action. Clearly, good judgment must be used in applying this guideline, for the question of "What constitutes a whole step?" can be taken to a ridiculous extreme. The overriding concern must be for clarity.

Final Conditions   A procedure is seldom complete unless the desired outcome is clearly described. Therefore, a section detailing *final conditions* is usually warranted. "Final Conditions" delineates the desired configuration of the system and its components upon successful completion of the procedure.

Review and Approval

Each facility must establish its own policy on the appropriate rigor of review for new procedures (or procedure changes). The ruling guideline for procedure review is that the rigor of review must correspond to the risk associated with incorrect performance of the procedure in question. If the subject procedure is associated with equipment or personnel safety, obviously, little room for error can be afforded. The procedure should be scrutinized closely.

Sometimes, charts, graphs, or other aids which aren't part of an operating or maintenance procedure are useful aids to operating technicians or maintenance technicians in performance of their tasks. These documents—sometimes termed *operator aids*—are usually posted near the equipment with which they will be used. If an operator aid is to be used, it should be reviewed, approved, and controlled in the same manner as the procedure it supplements.

Controlling Procedures

To avoid proliferation and use of invalid or outdated procedures, an industrial facility must establish a method to control procedures. A procedure control system should provide guidance regarding what procedures *require* tracking and control, criteria for review and approval of new procedures or procedure changes, repair or replacement of worn copies, distribution of new procedures, retrieval of old procedures, and reproduction of controlled procedures.

One useful control is to limit and individually number the copies which will be distributed to users. Numbered copies can be assigned to specific users or positions, making updates, changes, and recalls far simpler.

Copies of controlled procedures without an original copy number would be considered invalid for use. In cases where procedure extracts must be used in areas where they may be damaged or destroyed, portions can be copied from a controlled copy and verified as valid by an authorized team leader. A statement and signature of validity with the requirement to destroy the copy after use is normally sufficient to ensure proper control.

Revising Procedures

Procedures must not only be user-friendly, they must also be easy to revise when change is necessary. If the change process is excessively cumbersome, operating and maintenance technicians are apt simply to "work around" the procedure. This is a dangerous precedent to establish because it devalues *all* procedures. Also, unapproved changes made by technicians may have consequences that have not been considered.

The revision process must provide for both short-term and long-term changes. When a change is needed rapidly, the process must safely accommodate it. When many short-term changes have accumulated, a procedure revision is appropriate.

Using Procedures

Having well-written, valid procedures is an important element in the strategy for operating success—but it is not enough. You must also have people who follow them. When procedures aren't used, tasks soon become poorly and, in many instances, unsafely performed. Therefore, guidance regarding the use of procedures should be clearly established *and enforced*.

Procedural Compliance

The detail and complexity of procedures (as well as strict requirements for procedural compliance) are usually linked closely to the intricacy of equipment and the risk associated with operation. For some equipment, omitting steps or performing steps out of sequence won't affect the outcome and presents no operating hazard. For other equipment, step sequence and completion are vital to safe operation.

Because of this wide spectrum of operating complexity and risk, the meaning of the term *procedural compliance* varies broadly. Some interpret it to mean that general conformance is sufficient. Others believe it to mean exact, mechanical, step-by-step performance. Neither interpretation is correct.

Within this text, procedural compliance means *intelligent* compliance. Intelligent compliance requires a step-by-step application of the procedure, but it goes far beyond "cookbooking". For conformance to be intelligent, one who is performing procedure steps must understand the procedure's purpose and the effect of each step prior to performance. Further, unless the procedure allows for deviation from the sequence of the procedure steps, they must be performed in order.

Open and In Use

Must a procedure be open and in use for *every* task? No, a procedure is not necessarily required to be open and in use for some tasks. If not, under what circumstances *should* an operator have the procedure out? The answer to this question is not so easy. It depends upon the skill and knowledge of the operator, the complexity of the task, and the risk of failure. If the consequences of failure are low *and* the task is frequently performed, an operator may not need a procedure on hand and in use for task performance. Conversely, if the consequences of failure are severe, the task is complex or infrequently performed, a procedure should be used.

Some tasks are so critical that a procedure must be used in a "single-step, read-and-perform" manner. Such procedures often incorporate signature validation and verification for each step. Whenever the consequences of failure are this critical, it is clear that the procedure should be open and reviewed prior to *every* prescribed action.

Sometimes, performance of an operating or maintenance task needs to be controlled by a central control room. In some instances, it may be necessary for the control room to authorize the performance of the procedure a single step at a time (for example, in a critical valve lineup process). In other instances, the control room may release certain blocks of steps to be performed and reported. In still others, the control room may release the entire procedure for performance while requiring status reports at important checkpoints within the procedure.

Other tasks may not lead to disastrous consequences if incorrectly performed; but, because of their complexity, the prospect of failure is high. In such instances, the procedure should be open and reviewed frequently, if not before each action. When complexity is high, though the consequences of failure may not be great, the time required for re-performance is usually costly.

If a procedure is one with which the user is not familiar, a review prior to performance and at appropriate times during performance is warranted. (If you don't bake cookies very often, get the recipe out.) The likelihood of failure is great if familiarity is low.

When the technician is familiar with the task and the consequences of failure are not significant, it is normally not necessary for the procedure to be out and in use. An exception is when one technician is training another prospective technician in a task. In such an instance, the procedure should be open and in use as a training vehicle for the student.

When a plant emergency or casualty is indicated, the cognizant technician must have the requisite skills and knowledge to initiate immediate corrective actions from memory. Some events occur so quickly that there is no time to review procedures before responding. Just as an emergency medical technician does not review the procedure for CPR when confronted with a cardiac arrest, neither may an industrial technician delay immediate response actions while reacquainting him- or herself with the proper procedure. Nevertheless, at the earliest opportunity, the technician should review the procedure to ensure that all appropriate actions have been completed or initiated.

Determining when a procedure must be open and in use is obviously a subjective decision. Technicians should simply remember that if they choose not to use a procedure, they had better be right. One would also do well to remember that airline pilots—a group better trained and more experienced than nearly any other group of operators—always use checklist procedures before takeoffs, landings, or other critical events. As professionals, they have learned not to rely on memory alone.

An Example of Procedure Use

As we have seen, the procedure governing a task should be reviewed before, during, and after performance of any complex or infrequently performed task. Similarly, any task which will result in dire consequences if improperly performed should begin with a procedure review. Only those frequently performed, low risk tasks with few steps can be considered simple enough to dispense with a procedure review. Professionals seldom trust their memory when consequences can be serious.

As an illustration, let's examine a typical situation in which an operator must perform an important task that must be directed and coordinated by a central controlling station. Note that the sequence described is for a closely controlled process which potentially affects more than one operating station. For routine actions initiated by the station operator that do not require the direction or control of a control room, the equipment operator must use judgment and follow existing policy for coordinating with other stations and the central control room. In general, however, the same steps apply.

Receive Notification Communication from the central control room directing performance of the evolution has just been received by a station operator. (The term *evolution* is sometimes used to describe a discrete portion of a larger operation. For example, changing the oil in an automobile is an evolution in the process of engine maintenance.)

The operator acknowledges the control room's communication through a verbatim or paraphrased repeat-back. If the task is complex, warning well in advance of desired performance is appropriate to allow the equipment operator time for a thorough procedure review.

Review the Procedure Since the task has serious consequences if performed incorrectly, the operator opens the procedure and reviews the purpose, applicability, required initial conditions, and performance steps. In particular, a thorough review of the initial conditions is conducted to ensure that the equipment status will support the evolution.

Establish Initial Conditions Once the procedure steps have been reviewed, the equipment operator establishes (or verifies) the initial conditions prescribed by the procedure. For example, an initial condition requirement might state: *Ensure that the pump is being supplied by cooling water.* Creation of the appropriate initial conditions may require coordination with other equipment stations and operators.

Notify Others Having prepared the equipment for task performance, the operator calls the control room to advise them of impending commencement. Since the evolution we are considering will affect (and must be supported by) other stations within the facility, the central control room announces over the public address system that the evolution is commencing. The announcement alerts the affected stations to monitor important indications and to be prepared for emergency or casualty actions which may be necessitated by the performance of the evolution.

Commence Procedure Steps Following the plant-wide announcement, the operator initiates the task and informs the control room. For complex or high-risk tasks, the operator may need to consult the procedure before performing each step. For extensive tasks, the operator should inform the control room of task status at important junctures in the procedure. If, on the other hand, the evolution will be completed quickly, it is usually sufficient to simply report completion of the task. The length and complexity of the task and the consequences of failed performance should dictate when and how frequently status is reported to the control room.

Regulate the Process Once the process has begun, the equipment operator performs periodic tours of the machinery affected by the evolution to ensure that the equipment is responding as expected; if an abnormality is detected, the operator may find it necessary to stop the process. In such a case, the equipment operator should initiate the required immediate actions for the situation from memory, promptly notify the controlling station, and check the governing procedure at the earliest opportunity.

Upon completing the task, a check of the final conditions section of the procedure assures the operator that the task has been successfully performed. The operator reports completion of the evolution to the control room, and, if necessary, the control room announces to the entire plant that the task has been completed.

Update Logs At the conclusion of the task, the operator makes necessary narrative log and data record sheet entries. In addition, the operator updates any equipment status boards affected by the evolution.

Reporting Status Changes

For evolutions which affect more than one station, close communication and coordination are necessary. Consequently, to expedite procedural compliance, operators must report major equipment status changes to central controllers or to other affected operating stations. Here are a few guidelines governing when equipment status changes should be reported:

- At the start of major evolutions. Recall that *evolution* sometimes describes a discrete portion of a larger operation. (For example, starting a turbine-generator in a power plant would certainly be classified as an important evolution.) Before starting up a system or major component that affects other stations, the cognizant operator should normally report the impending event to the central control room. (In some cases, operators must first obtain permission to perform the task because of its impact on other systems.) By alerting the control room, controllers have an opportunity to announce the event to other equipment stations. The announcement forewarns affected stations to monitor important indications and to be prepared to act if an adverse situation arises.
- During important stages of an evolution. Plant controllers need to be kept informed of the status of evolutions in progress. Plant configuration can then be better monitored. Also, if the evolution in progress causes problems at other operating stations, controllers have a much better view of where troubleshooting should begin.
- At the completion of an evolution. When a major operating evolution is completed, the cognizant operator should report the task as complete.

Status reports are equally important during equipment emergencies and casualties. As with planned events, status reports for equipment crises should usually be rendered at the onset, periodically during, and upon conclusion of an anomalous event that affects other stations.

Training to Procedures

Well-written procedures are, next to excellent instructors, the best training tools. Whenever on-the-job training is being performed, the governing procedures should be open and in use. When training is performed in this manner, several important benefits accrue. First, the student learns where to find the procedure. Second, the student sees exactly how the steps of the procedure are related to the components and systems in use. Third, the student has an opportunity to ask clarifying questions of the instructor (and vice versa).

Not only is this process good for the student, it is also an important review for the instructor. In fact, procedure defects that would otherwise go unnoticed are often identified by questions from students.

Enforcing Procedural Compliance

Reliable procedures alone cannot guarantee safe, efficient operations and maintenance. In the hands of inadequately trained personnel, good procedures will still lead to failure. Assuming that a procedure is well-written, it must be intelligently applied to be effective.

As we have seen, the goal of procedural compliance is *intelligent* compliance rather than *blind* compliance. Operators who understand the reason for each step of a procedure (and how performance of every step affects the equipment) should be treasured as an invaluable commodity. Not only will they properly control equipment during routine operations, but their expertise also prepares them to respond appropriately to abnormal conditions.

Sometimes, however, procedures are viewed as an encumbrance to equipment operation. In such cases, procedural compliance may degrade. Facility leaders must detect such deterioration quickly and act to correct it. Uncorrected, a habit of non-compliance is a serious liability to safe operation.

The first step in correction is to determine *why* a procedure (or procedures in general) are not being used.

One reason is that procedures are sometimes poorly written. Non-compliance may simply be the result of an invalid (or "unfriendly") procedure. If a procedure is not understandable or, worse, not correct, it probably won't be used.

A second fault is that equipment operators often are trained *how* to use equipment with no explanation of *why* the equipment should be used in the prescribed manner. Without understanding the reasons for rules, operators will neither value nor respect them.

In some organizations, procedures are ignored because team members have been indoctrinated with the philosophy, "Any operator worth his salt shouldn't *need* the procedure." Such an argument clearly falls short when one considers that, regardless of how many takeoffs or landings an airline cockpit crew has made, the crew always uses the task checklists.

Another common problem is outdated procedures, usually resulting from an unresponsive procedure change system. When procedure changes fail to keep pace with equipment modification or legitimate changes in the operating conditions, operators lose faith in the procedures. Procedure change systems can become so cumbersome that they no longer respond to the operators' needs. As a result, procedures lose credibility and fall into disuse.

For procedures to work, the equipment must be properly maintained. When equipment maintenance is not sustained, workers devise ways to accomplish their tasks by operating equipment outside prescribed specifications. Some equipment may tolerate such misuse; other equipment may malfunction and injure or kill the operator.

Finally, if the value and practice of procedure use are not exemplified and enforced by the organization's leaders, procedure usage and compliance will deteriorate. If leaders (actively or passively) communicate that procedures are not very important, the procedures will not be used.

As with all coaching processes, team weaknesses (and strengths) are best monitored through frequent observation and coaching. Through close interaction, facility leaders are able to constantly reinforce this important operating value.

Topic Summary and Questions to Consider

If procedures for operating and maintaining an industrial facility are to be effective, guidelines for developing, controlling, using, and changing procedures must be established. Yet, no procedure or policy can substitute for well-trained personnel. As Admiral Rickover told Congress:

> *In final analyses, we must depend on human beings. No machine, including computers, can be more perfect than the human being who designed it, uses it, or relies on it.*

Proper procedure usage is a skill that must be taught and reinforced continuously. Without self-assessment and daily coaching in the process, the skill will atrophy. Once poor procedure usage habits are established, they are not only hard to correct, they may also become an accident contributor.

Ask Yourself Do your team members have and use reliable procedures when performing operating and maintenance tasks? Or, are they "flying by the seat of their pants"?

Do you have a system to formally change a procedure found to be incorrect or incomplete? Or, do you find lots of handwritten notes in the margins of the document?

Do you have the end-users help create and modify the procedures? Or, do you just let them complain about how bad the procedures are?

If your procedures don't support your activities, you're missing one of the eight essential elements of the strategy for operating success.

Chapter 10

Monitoring Critical Operating Parameters

In the Eastern 401 accident, three highly trained operators neglected to monitor aircraft altitude. Not until the aircraft was within 100 feet of the ground did they realize that altitude had changed. In the *Exxon-Valdez* grounding, neither the third mate nor the helmsman (by their own testimony) monitored rudder angle of the ship. Nor had the Coast Guard tracked vessel course as prescribed by regulation. At Bhopal, control room operators enjoyed tea while temperature and pressure in tank E-610 increased to off-scale levels. In the Big Bayou Canot Bridge accident, the towboat pilot failed to monitor radar closely enough to determine that his vessel was entering a tributary of the Mobile River rather than following the river channel. Finally, in Chapter 15, we will learn that control room operators at TMI-2 overlooked important valve position indications that would have informed them of the incorrect auxiliary feed pump stop valve status.

In each case, operators failed to monitor critical operating parameters. As discussed in Chapters 3 and 4, equipment is controlled by monitoring important indicators—physical characteristics such as temperature, pressure, speed, distance, fluid level, concentration, and purity—and making the necessary adjustments to keep these factors within specified operating tolerance bands. Critical to equipment performance, they are often termed *critical parameters*.

Yet, accurate information is worthless if not wisely acted upon by alert, well-trained team members. If critical parameter information isn't regularly monitored, analyzed when anomalies are observed, interpreted with intelligence, and used as a basis for regulating equipment, failure will soon follow. Accordingly, one of the most important of the vital operating skills is the ability to monitor critical operating parameters and act judiciously upon the information that they provide.

Why Monitor?

An imperative of equipment control is to frequently monitor (and record as necessary) critical operating parameters. **Critical parameters** are important operating properties that must be controlled to avoid equipment failure. For example, in a light aircraft, altitude, attitude, airspeed, fuel loading, oil pressure, exhaust gas temperature, cylinder head temperature, and a few other important parameters must be controlled to avoid unacceptable failure of one sort or another. Similarly, in an automobile, speed, fuel level, coolant temperature, oil pressure, and battery voltage are the prominent indications.

Critical parameter values provide important information about equipment performance with respect to design specifications. They help the operator to know how the machine is performing in relation to the boundaries of its safe operating envelope. Therefore, most critical parameters are sensed and indicated so that they can be readily observed.

It is incumbent upon equipment operators to frequently observe critical parameter indications, compare them with established boundary values, and respond to abnormalities indicated by the instruments. When critical parameter values approach boundary limits, equipment must be adjusted to maintain parameters within prescribed tolerances.

Timely adjustments to fluid flows, sump levels, voltage values, or other key operating elements allow smooth, continuous operation. Whereas, if parameters are allowed to drift outside operating bands through negligent monitoring, equipment damage or personnel injury may occur.

When to Monitor

Frequency of monitoring a critical parameter depends on the parameter's importance, its related equipment status, and its sensitivity to other equipment changes. Here are a few guidelines:

At Prescribed Periodicity — Often, critical operating parameters must be observed at times prescribed by policy or procedure (e.g., on the hour or half-hour) and recorded on forms known as *data record sheets*. Monitoring frequency is usually suggested or required by the related component or system technical manual.

During Inspection Tours — But, monitoring is usually desirable more often than just "when required". As operators tour their stations, they should check operating parameters as a part of the tour. A quick glance at lubricating oil temperature, cooling water flow, main steam pressure, or regulated voltage gives assurance that equipment is operating as designed.

During Status Changes — Many critical parameters require that, as the load or state of the equipment changes, monitoring frequency must increase. For example, when you drive your automobile in mountainous regions, you must monitor engine temperature more closely as you proceed up steep and long hills. You increase monitoring frequency because the load has changed on the engine, and you know that temperature will increase. If you're not careful, you may overheat the engine, violating a safe operating limit (or worse, a failure limit).

The same concept applies for industrial operations. Whether for diesel generators, turbine-driven pumps, electrical switch-gear, steam condensers, or boilers; as load changes (either increases or decreases), monitoring frequency must increase until the equipment stabilizes with the new load. Once stable, monitoring frequency can usually return to normal.

How to Monitor

When monitoring critical parameters, operators should watch for at least two problems. The first is a parameter which has gone outside its designated operating band (either higher or lower). The second is a parameter value that has not exceeded its limit but is on a trend to do so.

In the Band? — Most critical parameters have established upper and lower operating limits. The area of tolerance between these limits is often referred to as an *operating band*. In fact, the upper and lower boundaries are usually the normal operating limits. The acceptable band is the zone of normal operations for the parameter under consideration.

In the Eastern 401 accident, the crew established a target altitude of 2,000 feet. Their instrumentation, once set, allowed an operating band of 500 feet—from 1,750 feet to 2,250 feet. When either the upper or lower limit was reached, an alarm would sound. As Eastern 401 descended, that alarm sounded in the cockpit, but the crew took no action. Either they didn't hear it or they ignored it. They were too pre-occupied with the nose wheel indicating light to notice.

Temperatures, pressures, levels, speeds, altitudes—all must be maintained within assigned limits for design specifications to be met.

Undesirable Trend? — "In the band" is not the only concern for an operator who is monitoring indications. Though a parameter is not out of its prescribed operating band, if it is drifting in a direction that will soon take it out of the band, the responsible operator should take action to stabilize the equipment.

Vigilant operators can often detect undesirable parameter trends long before the parameter leaves its operating band. Eastern 401's altitude alarm should have been the *second* warning—not the first—to the cockpit crew.

Recording Parameter Values

At times, critical parameter values must be recorded. Data record sheets provide operators with a structured means to record those operating parameters. Through accurate recording, an historical record of operating values is established. The record is valuable for determining whether components and systems are functioning reliably. Critical parameter data may also be useful for troubleshooting problems and making preventive repairs.

Interpreting Parameter Values

Reading and recording instrument indications implies more than just seeing and writing. Interpreting and analyzing values and (especially) trends is a more complex and difficult task.

All too often in industrial operations, an operator will read and record an indication, only later to realize that the value was out of specification (perhaps as a result of an independent data review). Operators cannot be "brain-dead" during data recording. They must know what they are looking at, and whether it is right.

Scanning Control Panels

During pilot training, aircraft operators, if well-trained, are taught to develop an instrument **scan pattern**—the habit of glancing at and critically evaluating important operating parameters in a logical sequence. It is called a pattern because the eyes are directed, usually in a predetermined progression, from one instrument to the next.

In close quarters—be it control room, cockpit, or submarine maneuvering room—the instrument scan serves as an on-going operator inspection tour. Operators continuously monitor important indications to "check the pulse" of their equipment.

The instrument scan becomes, after a time, a matter of routine, requiring little conscious will. Yet, for professionals, it is not a thoughtless act. Reliable operators know what to expect—what is right for the current conditions. If they see something different than what they expected, they can redirect their attention and focus on the problem.

Efficient scan patterns are enhanced by well-engineered control panels. If related instruments are located near one another, cross checks can easily be performed. If, on the other hand, control panel layout is poorly designed, scanning is difficult (and probably neglected).

Actions for Abnormal Values

As noted a few moments ago, when a parameter value is drifting on a path that will exceed an operating limit, operators shouldn't wait until a limit has been violated before taking action. Equipment operation should be adjusted to maintain the parameter within design specifications.

If, however, an operating parameter is found outside of its limits, three actions are normally required. First, take prompt corrective action to return the value to its operating band. (But remember that if a parameter is beyond a failure limit, some situations dictate that no rapid changes be made that could lead to cataclysmic failure.)

Second, report the out-of-specification value to the team leader, stating the actions being taken to correct the problem. By reporting the problem, the team leader is kept informed and engineering review can be initiated if appropriate.

Third, record the out-of-specification reading on its related data record sheet, circle the value in red, and enter a note on cause and corrective actions in the narrative logs. Red-circling the out-of-specification reading highlights the problem for others who must review the logs and data record sheets.

Special Concerns

Two special concerns are worth noting before we close this chapter. The first is instrument validity and the second is computer-controlled equipment.

Validity of Indications

Monitoring critical parameter indications is of little value unless the indications are valid. At Bhopal, instrument indications had little credibility because sensing and indication equipment had not been properly maintained. In the January 28, 1985, issue of the **New York Times**, Stuart Diamond recorded that:

Employees at the plant recalled after the accident that during the evening of Dec. 2 they did not realize how high the pressures were in the system. Suman Dey, the senior operator on duty, said he was in the control room at about 11 p.m. and noticed that the pressure gauge in one tank read 10 pounds a square inch, about five times normal. He said he had thought nothing of it. Mr. Qureshi, an organic chemist who had been a methyl isocyanate supervisor at the plant for two years, had the same reaction half an hour later. The readings were probably inaccurate, he thought. "There was a continual problem with instruments," he said later. "Instruments often didn't work."

Instrument calibration was clearly ineffective at Bhopal. Operators didn't believe their indications.

Instrument and alarm maintenance is critical if operators are to observe and respond to equipment abnormalities. If instrument indications are not believable, they will not be heeded.

Instrument credibility requires that instruments (both sensing and indicating) be regulated by a calibration recall system—a preventive maintenance program for instruments that ensures reliability through periodic accuracy checks.

Computer-Controlled Equipment

Using computers to assist in equipment control is valuable and desirable for many applications. But, operators must understand that they are not relieved of any responsibility to monitor critical parameters, even though the equipment is on "autopilot".

In the Eastern 401 accident (Chapter 1), the captain, first officer, and second officer thought that the autopilot was flying the aircraft. As a result, they stopped monitoring critical parameter indications. The NTSB concluded that:

The flightcrew did not monitor the flight instruments during the final descent until seconds before impact. The captain failed to assure that a pilot was monitoring the progress of the aircraft at all times.

The inescapable inference is that humans are *always* responsible for equipment control.

While driving on a highway with the cruise control engaged, you *still* check the speedometer once in a while. Don't you?

Topic Summary and Questions to Consider

Critical parameters are the vital signs of the health or illness of equipment and processes. They require constant attention of alert, well-trained operators if they are to be used to control equipment within design specifications. Accordingly, the skill of monitoring critical parameters must be taught and coached by skilled industrial leaders for the team to consistently succeed.

Ask Yourself

Do your team members know which operating parameters are critical? Are those parameters regularly reviewed and recorded?

Do team members know what to do if a parameter exceeds established limits?

Chapter 11

Independent Verification

In Chapter 2, we learned that failure to independently verify completion of critical tasks is a common component of accidents. Whether it's neglect to separately verify rudder indication on a turning tankship, altitude in a passenger airliner, valve positioning in a power plant, or pressure and temperature indications in chemical storage tanks—failed *independent verification* increases the likelihood of mishap.

Some tasks are characterized by so much risk that separate and independent confirmation of correct task completion (or problem diagnosis) is warranted. If afflicted with serious disease, we solicit second opinions from doctors before making important (perhaps, life-threatening) choices about treatment. Before submitting tax returns, many of us engage another knowledgeable person to check data and calculations. Within the military training environment, artillery sight settings and propellant charges are independently verified—*second-checked*—by range safety officers before permission to fire is granted, thereby limiting the chance of rounds impacting outside authorized areas. It is because we cannot afford failure that independent verification is employed for these tasks.

Similarly, the ability to perform independent verification for critical tasks in industrial operations is a fundamental discipline. The training of an industrial operator is incomplete unless it has imparted this vital operating skill.

What Is It?

Independent verification is the process of *separately proving* that a task has been correctly performed. The skill is applied anytime one operator, maintenance technician, or team leader checks the work of another. For example, critical tasks performed by an aircraft mechanic are required to be verified by another master mechanic. Airline captains validate the observations and actions of first officers and vice versa. Electricians employ the "buddy system" when working in hazardous conditions. Even when driving a car, passengers perform independent verification by checking traffic along with the driver before changing lanes, entering a busy thoroughfare, or backing the vehicle. Whether by visual check, calculation, or testing, completed work (or important stages of work) is confirmed by another person to be correct.

Building code inspections are a good example of independent verification. If you have ever built a house, you know that, at certain points in the process, building inspectors must check the quality of the work—structural concrete, framing, plumbing, electrical wiring—to validate that the craftsmanship meets state building code requirements.

In industrial operations, independent verification is used in the same way to make sure important work has been correctly performed. Critical calculations are repeated by another qualified engineer before proceeding in the engineering process. Important valve lineups, established, for example, in preparation for hydrostatic testing, are second-checked by a separate qualified operator. Lockouts and tagouts, once attached at prescribed locations to isolate energy sources, must be verified by a separate party. There are dozens of ways in which industry uses the independent verification process.

For All Tasks?

Obviously, not all operating or maintenance tasks require independent verification. Separate validation should, however, be performed for those tasks (such as isolation of high energy systems) which, if performed incorrectly, are apt to result in serious injury to team members or in unacceptable damage to equipment. Independent verification is also indicated if incorrect performance would result in other unacceptable consequences such as an inordinate loss of time.

At Bhopal, the task of establishing the valve lineup and inserting blank flanges to perform a piping flush merited independent verification. Allowing water into a MIC tank had a serious potential to cause a runaway reaction. Yet, not only was there no independent verification of the task, blank flanges to isolate the flush from the tanks weren't even installed.

In the cockpit of Eastern 401, monitoring altitude was a critical task, requiring a second-check by the captain (or first officer, depending upon who was flying the aircraft). Independent verification was neglected, however, since the captain failed to designate who was flying. An independent check by the air traffic controller failed as well.

Aboard the *Exxon-Valdez*, the third officer should have independently verified the ship's heading rather than assuming that the helmsman was responsibly monitoring rudder indication. And the Coast Guard failed to fulfill their regulatory duty to provide course monitoring.

These were all critical tasks, worthy of independent verification. If the necessary second-checks had been properly performed, the resulting accidents could have been averted. Furthermore, in each case, second-checking was taught and required by policy. It simply wasn't executed.

Who Decides?

Clearly, the decision to use or not to use independent verification is a risk-based decision. As such, the decision will ultimately rely on human judgment. Therefore, people responsible for deciding whether independent verification is necessary should be qualified or certified on the components or systems involved.

To assist in determining when independent verification should be performed, some organizations develop a list of tasks which *always* require independent verification, and another list of those which *never* require it. Such lists serve as beneficial guidelines for those who make the decision whether to independently verify. But, lists should never substitute for a critical thought process; complicating circumstances may indicate a need for separate validation—regardless of what the list says.

Process or Point Verification?

Is independent verification required for the entire duration of a task, only at important junctures during the task, or only at task completion? The type of task dictates the answer to the question. Some tasks require full-time, on-scene verification because every step in the process is so important. Others can be verified simply by observing the final product.

Continuous Verification If an error in task performance (or an uncorrected material deficiency) would have unacceptable consequences, the entire process must be observed.

Continuous observation is indicated whenever only a final inspection would be insufficient to verify that all portions of the task were performed in accordance with prescribed standards. For these situations, qualified team members, often quality control inspectors, are required to observe every step of task performance. For example, work on a turbine, engine, or pump which will be inaccessible after the task is complete is likely to require full-time observation.

Multiple-Point Verification Somewhere between process (continuous) verification and verifying after task completion is what we will call *multiple-point verification*. By that term, we mean that for a given task, more than one verification is required (e.g., after each step or group of steps during performance, but not a continuous "over-the-shoulder" inspection or verification).

Verifying After Completion Other times, an inspector may need only to observe the finished product. For example, code inspections for plumbing or wiring compliance during the building of a structure are usually performed only after the entire wiring or plumbing task is completed. The inspector is able to judge the quality of the job by observing the finished product.

Guidelines for Independent Verification

Methods and techniques of independent verification are so numerous that they cannot (and need not) be delineated one by one. Here are a few fundamentals that capture the *spirit of the law*:

Independent Person Independent verification usually means that a separate, qualified person observes or checks the work of another. Different people (most fortunately) bring different perspectives to problems. Deficiencies to which I might be blind are readily observed by another who is separate in thought and vision.

This does not mean that self-checking is not important as well. If the task is crucial enough to require an independent check, it *certainly* is worthy of a self-check first.

Independent Time or Location For a check to be independent, it must be performed separately (physically and often chronologically) from original performance. For example, the second-check of a valve lineup is best performed by another qualified operator moving behind the first, separated by time. Otherwise, independent perspective is tainted if one operator looks over the other operator's shoulder as a valve is repositioned. (In numbers, there is security—also dilution of responsibility.) Therefore, the second-checker should trail after a delay such that he cannot observe the first operator's performance.

A second example would be a process operator reporting to the central control room that he had shut a particular valve. The control room operator independently verifies this action by observing the remote valve position indicator on a control room panel. In this case, the verification is at an independent location, but is done at the same time that the process operator shuts the valve.

With Intelligence As with all rules, there are exceptions. For some tasks, verification by directly observing the first operator is the wisest choice. For example, when repositioning a throttle valve, it would make no sense for the first operator to open the valve 2¾ turns with no one watching. To verify, the second operator would have to close the valve and open it again to the same position. Therefore, to verify positioning of a throttle valve, the second operator should observe the first. Another common example is when an airliner pilot and copilot simultaneously have their hand on the throttle and together increase power for take-off; each is checking (and can override) the performance of the other at the same time.

Good sense must also be applied when checking the position of a valve which is danger-tagged in the shut position. The purpose of the danger tag is to isolate a high energy system. Opening, even slightly, a danger-tagged valve and then shutting it again is a serious breach of high energy boundary isolation. The result could be a pressurized system that will later injure someone. Therefore, valves which are danger-tagged in the shut position may only be checked in the shut position.

Informal Verification

Most of the situations discussed to this point have involved the application of *formal* independent verification. Yet, *informal* independent verification is just as important and more often needed.

By Vigilant Team Members Informal verification should occur continuously on any work team. On the flight deck of an aircraft carrier, for example, though many formal checks are necessary, many more informal checks are continuously exercised. Members of flight deck teams are taught to *continuously* look around the deck and the sky. There are so many hazards in the environment that team members cannot afford a moment's complacency. Through this abnormal and continuous vigilance, the awareness of the team is kept at peak level.

The "buddy system" developed from this thought process. Divers use it, mountaineers use it, electricians use it—wherever the hazard potential is high, team members should be watching one another.

By Vigilant Leaders Leaders set the standards (and the example) for teams. High standards and attention to detail become the norm when leaders vigilantly coach, praise, and correct their team members.

Accordingly, team leaders should frequently perform "over-the-shoulder" checks of their personnel. In control rooms, for example, the lead control room operator should monitor important parameters and actions just as is supposed to occur in aircraft operations. The issue is not one of trust. Rather, the issue is verification. Everyone makes mistakes. Over-the-shoulder checks should not be shunned, but rather, welcomed.

How the checks are performed has everything to do with how they are received. If you act like a coach, your advice will probably be accepted. If you act like a tyrant, the response will be quite different.

Topic Summary and Questions to Consider

Clearly, independent verification is an important skill for every operator to acquire. But, be very careful in applying it, for it is not a panacea. It can never substitute for thorough training. Without proper training, you can never safely turn your team members loose in the workplace to perform on their own.

On the other hand, independent verification, in addition to providing assurance of correct task performance for critical tasks, is also a valuable training tool. Federal Aviation Administration rules (as well as airline policies) require check-ride captains to observe the performance of cockpit crews at prescribed intervals. Though cockpit crews are highly trained and experienced, they, like everyone else, are constantly learning. Individual proficiency (as well as team proficiency) can deteriorate when not practiced and evaluated. If it's important for the airlines, isn't it important for your industrial team?

Ask Yourself Do you have a clearly-worded policy that specifies when independent verification is required?

Are second-checks performed for placement of danger tags and energy isolation devices? Are second-checks *really*—and *independently*—performed?

Chapter 12

Communicating Vital Information

Among the requisite skills for industrial team success, the ability to communicate effectively is, perhaps, the most critical. Whether face-to-face, via radio or telephone, through written communiqué, or with emergency signals—mastery of communication skills often spells the difference between fortune and failure.

Sinking of R.M.S. Titanic, April 1912

At approximately 11:40 P.M. on the night of April 14, 1912, the *Royal Mail Steamer Titanic* struck an iceberg on her maiden voyage from Southampton to New York. Ignoring six iceberg danger warnings from other ships, *Titanic* had steamed ahead at almost full speed, in darkness. The vessel went down in calm seas approximately 400 miles off the coast of Newfoundland at 2:18 A.M. on the morning of April 15, a little over two-and-one-half hours after the accident. Despite a series of radio and rocket distress signals and a ship not over eighteen miles to the north, more than 1,500 of the 2,228 passengers on board were lost.

Background The 46,328-ton *Titanic* (Figure 12-1.) was one of two mammoth passenger liners built at the Harland and Wolff shipyard in Belfast, Ireland, for the White Star Line. She and her sister ship *Olympic* had been designed and constructed through the funding of J.P. Morgan's International Mercantile Marine to compete against the British Cunard Line and the German liners for the rich passenger trade across the North Atlantic. This nearly 900 foot passenger liner was the largest vessel ever constructed when she was slipped from her Belfast dry dock on May 31, 1911, and towed to the fitting-out basin for completion. She boasted a steam power plant with both turbine and reciprocating engines supplied by twenty-four double-ended steam boilers housed amidships in six huge boiler rooms. The ship's hull was protected with double-bottoms and was separated by fifteen "water-tight" bulkheads that extended over half way up the hull. These bulkheads could be sealed by activating water-tight hatches that would compartmentalize the hull into sixteen different sections. As a result, she was thought to be nearly unsinkable.

Within a little over ten months, the *Titanic* had been fitted out and left Belfast on April 2, 1912, for sea trials. (Sea trials are designed to prove capabilities of equipment, crew, and procedures during

Figure 12-1: RMS Titanic. [From 1912 news photo]

both normal and abnormal conditions.) After little more than eight hours of sea trials, she was declared seaworthy and began preparations for her maiden voyage. Only one life boat drill was conducted and several members of her crew of nearly 900 would not report on board until the day of departure.

Accident Description

Titanic departed Southampton on April 10, 1912, destined for New York on a scheduled seven-day voyage across the North Atlantic. The ship was commanded by Captain Edward J. Smith, Commodore of the White Star Line. He expected this to be his last voyage and crowning achievement with White Star. From the onset, however, the *Titanic*'s maiden crossing was unusual. As the *Titanic* eased from her Southampton dock toward the River Test, she drew alongside the liner *New York*. Suction created by the *Titanic* passing at a distance of 85 feet pulled the *New York* from her berth, breaking her hawsers, and swinging her dangerously near the *Titanic*. Rapid action by the *Titanic* and a tug maneuvering the *New York* narrowly averted disaster. A similar event occurred just seven months earlier when *Titanic*'s sister ship *Olympic* passed the British cruiser *HMS Hawke*, drawing it into its side, damaging both vessels.

After picking up passengers at Cherbourg, France, on the afternoon of April 10 and Queenstown, Ireland, on April 11, the *Titanic* headed to sea in beautiful weather, expecting to arrive in New York on Wednesday, April 17, 1912. The first days of the voyage were uneventful, except that *Titanic* altered course to a more southerly route. The arctic winter was unusually warm and ice had been reported at lower latitudes than normal.

By Sunday night, April 12, *Titanic* was steaming at over twenty-two knots into an area of reported ice. The air temperature had been dropping and was below freezing. Marconi operators Jack Phillips and Harold Bride had received ice warnings from four separate vessels during the day. The warnings had been passed to Captain Smith, but only one was posted in the chart room for the ship's officers' information. The sea was a flat calm and the sky was clear and starlit but moonless. Radar and sonar were unknown in 1912. Icebergs could be detected only through vigilant observation by the crew. Normally, light reflected from the surface of a berg or waves lapping against the base of the berg were the only signs of ice in the water. With no moon and a dead calm sea, these indicators were absent.

At 9:20 P.M., Captain Smith left the bridge under the command of Second Officer Charles H. Lightoller. His instruction to Lightoller was, "If in the slightest degree doubtful, let me know." At 9:30 P.M., Marconi operator Jack Phillips received a wireless message from the steamer *Mesaba* warning of pack ice and large icebergs in the area toward which *Titanic* was steaming. Phillips, whose passenger message traffic had backed up due to a failed relay in the Marconi set earlier, merely thanked *Mesaba* and did not forward the message to Captain Smith or to the bridge.

Watch relief occurred between the lookouts in the crow's nest at approximately 10:00 P.M. and on the bridge between Second Officer Lightoller and First Officer William Murdoch at the same time. Lightoller advised Murdoch that they would come into the reported ice about 11:00 P.M. A short time later, the British steamer *Californian* sent a wireless message to *Titanic* stating, "Say, old man, we are stopped and surrounded by ice." Phillips merely replied, "Shut up! Shut up! I am busy. I am working Cape Race." The *Californian*'s report was never forwarded.

Despite the absence of moonlight, a clear demarcation between sky and sea was visible from the crow's nest. At 11:35 P.M., crow's nest lookout Frederick Fleet observed something unusual ahead—the stars were momentarily disappearing and then reappearing. In a moment he realized that the phenomenon was caused by a huge iceberg. Immediately, he pulled the bell rope to the bridge and reported, "Ice right ahead". At nearly the same time, First Officer Murdoch, now in charge of vessel navigation, saw the berg and ordered "Hard a'starboard!" and "Full speed astern!"—orders intended to slow the vessel and steer to the left of the ice. [Note: In 1912, a steering order to starboard—turning the wheel clockwise—turned the ship to the *left*.] At 11:40 P.M., as the bow swung slowly to port, *Titanic* grazed the berg, showering ice onto the forecastle and well deck. First Officer Murdoch then ordered "Hard a'port!" in an attempt to swing the stern clear. As the vessel grazed the ice, hull plates ripped open along a 250 to 300 foot section, opening the forward six compartments to the sea. First Officer Murdoch's actions were contrary to **Knight's Seamanship**, the ruling seamanship manual of the day, which advised that striking head-on with the bow was far more desirable than an oblique strike, should a collision be unavoidable. Furthermore, First Officer Murdoch's action to stop and then reverse the engine had the effect of reducing steerage, the ability to rapidly turn.

Perhaps no more than twelve square feet of hull surface was opened, but the damage was sufficient to begin dragging the *Titanic* down by the bow. The watertight doors between compartments were closed, but as the forward sections filled, each compartment spilled over the top into the next.

Captain Smith soon ordered the ship's chief architect, Thomas Andrews (aboard for the maiden voyage), below to inspect damage. Andrews, after learning of flooding in six compartments, reported that the ship had only a short time to live.

At 12:10 A.M. on April 15, Captain Smith, ordered Marconi operator Jack Phillips to send out a "CQD", *Calling Distress*. Yet, many on board the *Titanic*, including some of the officers, were still not aware that the ship had struck a berg. The general alarm had not been raised. Most of those who *were* aware remained unconcerned.

Titanic's first distress signal was answered almost immediately by the British steamer *Carpathia* from approximately 60 miles away. Communications from the German steamer *Frankfurt*, nearly 150 miles from *Titanic*, were rudely rebutted by Marconi operator Phillips. But the *Californian* failed to answer—she had shut down for the night in heavy ice and the Marconi operator was no longer at his station. Her position is conjectured today to have been only 5 to 18 miles from *Titanic*.

Lifeboat operation on the *Titanic* progressed lethargically and with significant confusion. Only half enough lifeboats—sixteen regular boats and four collapsibles—were on the *Titanic* for the passengers and crew. Though the *Titanic* had a capacity for sixty-four boats, British Board of Trade regulations did not require vessels to carry enough lifeboats for all on board. Passengers were slow to realize the danger and go to lifeboat stations. Ship's officers were unfamiliar with the lifeboat davits and their capacities. Most lifeboats were less than half full when lowered into the water. No lifeboat drills had been run with passengers and the single drill run during sea trials was insufficient to familiarize crew members with emergency procedures. Many crew members assigned to the lifeboats were not familiar with the boats—many of the sailors, unfamiliar with small boat handling, had to be taught how to row by passengers.

Shortly after midnight, Fourth Officer Boxhall sighted the mast lights of another vessel, apparently a steamer. His morse signals went unanswered. At 12:45 P.M., Quartermaster Rowe began firing distress rockets at five minute intervals. Eight rockets went unanswered though sighted by the crew of the *Californian* and reported to her captain, Stanley Lord. Captain Lord, upon hearing of the signals, ordered the morse lamp manned. When no response was received, he neither investigated the rockets nor stationed the Marconi operator who had secured at 11:35 P.M. Even today it is unknown whether the mast lights sighted were those of the *Californian* or another ship in the area—possibly the Norwegian *Samson*, operating illegally in the area for seals.

As the *Titanic* settled, the bow sank and the stern rose into the air. The ship broke in half (perhaps a result of brittle fracture) with the stern settling back into a horizontal position for a few moments. The bow drifted to the bottom, two-and-one-half miles below, followed shortly by the stern. Most of the passengers still on board either jumped or were thrown into the freezing waters of the North Atlantic (approximately 33°F) and died within minutes from hypothermia.

By 3:30 A.M., after a heroic rescue response, the *Carpathia* reached the area of *Titanic*'s foundering and began picking up lifeboat survivors. She remained on scene until 8:50 A.M. at which time she made for New York harbor. The *Californian*, having learned of the accident, had come on scene an hour earlier and remained in the area searching for survivors. No others were found. The *Carpathia* arrived in New York with the survivors on Thursday evening, April 18, 1912.

Questions of Cause

Soon after the *Carpathia* docked in New York harbor, a Senate Investigating Committee, headed by Michigan Senator William Alden Smith, began an inquiry into the causes of the accident. Surviving *Titanic* passengers and crew (including White Star Line president J. Bruce Ismay) were interviewed within hours. The government of Great Britain (through the British Board of Trade led by Lord Mersey) also began an official inquiry on May 3, 1912. The investigations raised many questions about routine shipping operations and contingent emergency planning, some of which remain unanswered today.

Reasonable Caution? In maintaining nearly full speed (22½ knots), had Captain Smith applied *reasonable caution* given reports of unusually heavy ice? Second Officer Lightoller, the only surviving *Titanic* officer, testified that every captain he had sailed with during his twenty-five years at sea would have made the same decision.

Yesterday's successful strategy, however, does not guarantee today's victory—especially if conditions have changed. Four ice warnings had been forwarded to the captain portending the danger of icebergs. Further, without the benefit of moonlight and wave action, bergs were difficult to detect. The situation was complicated by the fact that the lookouts hadn't been issued binoculars. Frederick Fleet testified that, rather than detecting the berg at a distance of one mile, he probably would have seen it from three to five miles away with binoculars.

It would seem that the question of "due caution" must be answered, not by asserting that "Everyone else does it!", but rather in accordance with prevailing conditions.

Sufficient Sea Trials? Were only eight hours of sea trials sufficient to test the skills of the officers and crew as well as the reliability and characteristics of the ship and her equipment? The *Titanic*'s near collision with the *New York* and the *Olympic*'s earlier collision with the *HMS Hawke* under remarkably similar circumstances would seem to warrant an investigation of the ship's handling characteristics and its effect on other objects near it. (Some conjecture that the drafting effect forewarned by these two incidents may have contributed to the *Titanic* striking the iceberg.)

Regardless, each ship's officer assigned the duties of conning the vessel should have had opportunity during sea trials to maneuver the vessel during both normal and abnormal conditions. The trials, however, were so brief (and conducted in broad daylight) that little more than simple familiarization was possible.

Further, only one lifeboat drill was conducted during the trials, probably with little rigor. Many of the officers manning lifeboat stations on the night of April 14 were apparently unsure of the capacity of the lifeboats and the ability of the boat davits to sustain full loads.

Wise Leadership? Why wasn't Captain Smith on the bridge when the ship entered the most critical phase of her voyage? Successful leadership relies in part upon an ability to bring good judgment, wisdom, experience, and confidence into play during difficult times. Warnings had indicated that *Titanic* would enter ice packs about 11:00 P.M. Would not prudence dictate that the most senior and experienced officer be on the bridge at that critical juncture?

The Lesson of Communication

Each of these questions has been argued (and will continue to be argued) for decades. But there is one issue over which there is very little argument—failed routine and emergency communications contributed immeasurably to the accident and subsequent response. Deficiencies included:

- Captain Smith posted only one of four ice warnings received during the day of April 14 in the chart room where the information was needed by the ship's officers. His inaction may have fostered a sense of complacency about the seriousness of ice in the region.
- Wireless operator Jack Phillips failed to pass two critical ice warnings—from the *Mesaba* at 9:30 P.M. and from the *Californian* at 11:30 P.M.—to Captain Smith because of a higher perceived priority of sending passenger messages to the relay station at Cape Race. (The Marconi wireless was a relatively new invention and was a convenience rather than a requirement. Marconi operators were responsible only indirectly to the ship's captain. As a result, there seemed not to be a proper attitude of accountability and responsibility on the part of the wireless operators. Wireless operator Phillips' communications to other ships warning of ice and responding to the *Titanic*'s distress call were, at times, highly unprofessional and unresponsive.)
- The general alarm was never sounded on the *Titanic*, lending to the lethargy of passengers reporting to lifeboat stations.
- Captain Lord, master of the *Californian*, failed to station the wireless operator after receiving reports of rockets launched from another ship. Had the *Californian*'s Marconi operator been stationed, the *Californian* may have been able to assist in rescue prior to the sinking.

Not only for marine operations, but for leaders of all industrial endeavor, the *Titanic* should teach us one clear lesson: **Routine and emergency operations are doomed to fail in the absence of clear, accurate communication of vital information!**

Elements of Effective Communication

There are many forms of communication necessary for successful operation of an industrial facility. They include verbal dialogue between team members, written interchanges for giving instruction or soliciting information, and audible or visual signals designed specifically to warn people of hazards or emergencies.

Regardless, all effective communication is characterized by a few fundamental elements: (1) a sender, (2) a receiver, (3) a message, (4) a medium of communication, (5) an environment which supports a successful communication, (6) a feedback response mechanism, and (7) as many confirmation/correction responses as necessary to ensure that the message has been received as intended. (See Figure 12-2.)

Consider, for example, the process of ordering a pizza over the telephone. Let's assume that you want to order a twelve inch, thin crust, combination pizza with a side order of garlic bread from your local concessionaire, Antonio's. The conversation might proceed along these lines:

Antonio: (Phone rings) *Antonio's. Can I help you?*
Alicia: *Hi, Antonio. This is Alicia Richardson. I'd like to order a pizza.*

In this short segment, you have chosen the communication medium (the telephone with voice transmission) and have contacted the receiver to whom you wish to deliver a message. The receiver identifies himself upon learning of the need to communicate (when the phone rings) and you have identified yourself as well.

Antonio: *Hello, Alicia. What would you like?*
Alicia: *I'd like a twelve inch, thin crust deluxe, but hold the anchovies.*

You have now passed the message. Has it been received intact and as you intended?

Antonio: *I've got you down for a ten inch, thin crust deluxe, no anchovies. Do you want any garlic bread, salad, or drinks with your order?*

Antonio has just responded with a feedback communication and a clarification question.

Alicia: *No, that's a twelve inch deluxe. And I would like an order of garlic bread with it, but nothing else.*

You corrected Antonio's response and have responded to his clarification question.

Antonio: *OK, I've got you down for a twelve inch deluxe, thin crust, no anchovies, and an order of garlic bread. Where would you like it delivered?*

Figure 12-2: The Communications Cycle showing (1) the "Message" and "Feedback" elements common to all verbal communications and (2) the additional "Confirmation/Correction" element required for formal communications.

Communicating Vital Information

Antonio provided a confirmation of your correction and asked another clarification question.
Alicia: *Deliver it to 468 Pine Street here in town.*
Antonio: *468 Pine. Got it. That'll be $11.89 and it should be there in about forty-five minutes. What's your phone number in case I need to call back?*
Alicia: *The phone number is 379-5784. $11.89 you said?*
Antonio: *Yeah, $11.89. I've got your phone as 379-5784. Thanks for your order.*
Alicia: *That's correct. Thanks. We'll be looking for the delivery. Goodbye.*

In just this short exchange, you have employed a communication medium, identified sender and receiver, sent the message, confirmed some parts of the message, and corrected other parts. The same elements are necessary for face-to-face communication, radio/telephone communication, and even some specialized forms of communication such as paging devices.

Effective Verbal Communication

Verbal interchange usually is the most frequently used form of communication. It's quick, convenient, and familiar. It differs, however, from written communication in a very important way. It normally must be verified and understood at the time of transmission since it is ordinarily not recorded. Therefore, some special rules of communication apply for *formal* verbal interchange in an industrial facility.

Formal vs. Informal Communication

Because verbal interchange is such an integral part of every aspect of daily life, there must be a distinction made between formal and informal verbal communication. In this discussion, *formal* verbal communication is narrowly defined as a spoken interchange between equipment operating personnel, the purpose of which is to direct a change in the state of plant operating equipment or to obtain information regarding its status.

Formal verbal communication has a long history in shipping, aviation, military operations, and military or commercial radio communication. In these disciplines, the outcome of poorly conducted communication can be disastrous. Therefore, strict communication patterns are maintained in which sender and receiver are identified, concise messages are transmitted, and responsive feedback contacts for confirmation and correction are employed. A short period spent in a disciplined airport control tower quickly educates an observer regarding the need for formal verbal communication.

Informal verbal communication, on the other hand, is interchange that is not intended to result in the operation of equipment or to obtain status for the purpose of plant operation. In other words, informal communication is verbal interchange employed for issues that do not carry great risk.

Though not as stringent as formal communication, informal exchanges should, nevertheless, be intelligently managed. Even though not related to plant equipment operation, one should consider using the same guidelines as for formal communication if an interchange has the potential for hazard if misunderstood.

Face-to-Face Communication

The most common form of verbal communication used in an industrial facility is face-to-face interchange. The rules for formal face-to-face communication should vary little, if at all, from the guidance just presented. The sender and receiver must be clearly identifiable to one another, the message must be clear and concise, and the environment for communication must support effective verbal interchange (e.g., reasonable background noise level). Once the message has been passed, there must follow as many feedback responses and confirmations or corrections as necessary to ensure that the message has been received intact.

Face-to-face communication between a control room and equipment operator might proceed as:
Control Room Operator: *Jim, start up Number 3 Turbine Feed Pump and idle it in standby mode.*
Equipment Operator: *Rick, I understand that you want me to start up Number 3 Turbine Feed Pump and idle it in standby mode.*
Control Room Operator: *Jim, that's correct.*

One who is unfamiliar with formal industrial or military communication might view this exchange as stilted or unwieldy; but, by ensuring that the message has been heard through repeating it back, the chance for mistaken communication is dramatically reduced, actually with very little effort.

Repeating back directions can be accomplished in a *paraphrased* format or a *verbatim* format. Each has special benefits. Acknowledgment of a direction through paraphrasing often provides better assurance to the message sender that the receiver truly understands the direction. Paraphrasing is desirable for most routine communications. Verbatim response, however, should be used when a message must travel through more than one communication link or when the message contains elements that could be easily confused such as component names or numbers.

Formal verbal communication seems very simple and straightforward; but, because it is so often used, face-to-face communication is routinely mismanaged. Identifiers are omitted when they should not be, feedback response is often omitted, and signals or body language are substituted for appropriate words. For example, the communication between a control room operator and an equipment operator might deteriorate to something like this:

Control Room Operator: *Start up Number 3 Turbine Feed Pump.*
Equipment Operator: *OK.*

Though the direction might be carried out properly, there is an unnecessary opportunity for error. The control room operator is deprived of an opportunity to verify that the equipment operator received the message *as intended*. There may be no second chance to correct a communication error.

The use of names or titles in face-to-face communication should always be observed if other people in the vicinity might mistakenly believe the communication is directed to them. This is especially important in an environment such as a control room where most operators have their attention fixed on instruments or controls. Lack of eye contact can lead to dangerous misunderstandings.

Radio/Telephone Communication

Formal radio and telephone communication should follow the same general guidelines as formal face-to-face communication. In few cases should sender and receiver identifiers be dropped since no eye contact is possible.

When a radio link or telephone line is a shared medium, identifiers should be used before every communication. Titles rather than names should be used to further limit the possibility of confusion.

Radios and telephones also form an important part of the emergency communications system. When an emergency is in progress, the lines or frequencies must remain clear of unnecessary message traffic. To notify others who might be using shared frequencies or lines, a phrase such as "Silence on the line!" may be adopted to indicate to others that the message that is to follow is of an emergency nature. Such a phrase also serves as a warning to limit unnecessary message traffic until the emergency situation has cleared.

Sometimes, personnel who are issued radios may engage in the frivolous practice of jamming their radio frequency or using the frequency for purposes other than for which it was designed. Such practices must be exterminated. The illicit use of the radio frequency, in addition to being illegal in many instances, deprives the operating team of a crucial emergency communication link—perhaps necessary to save the life of an injured fellow team member.

Written Instructions

Face-to-face verbal interchange carries the great advantage of seeing the sender's eyes and reading his or her body language. Studies indicate that over 50 percent of a message is contained in the gestures, eye movements, and body positions of a speaker. Sometimes, however, verbal communication is not enough. General George S. Patton, Jr., told his commanders:

> The best way to issue orders is by word of mouth from one general to the next. Failing this, telephone conversation which should be recorded at each end. However, in order to have a confirmatory memorandum of all oral orders given, a short written order should always be made out, not necessarily at the time of issuing the order, but it should reach the junior prior to his carrying out the order; so that, if he has forgotten anything, he will be reminded of it, and, further, in order that he may be aware that his senior has taken definite responsibility for the operation ordered orally.

It is probable that General Patton, as we ourselves, learned from hard experience that communication cannot be left to chance. It must be "nailed down" if it is expected to succeed.

Communicating Vital Information

Note the importance Patton attaches to *writing* an instruction as a backup to giving it verbally. Written instruction carries the advantage of accessibility. You can look at it again. When used in conjunction, verbal and written instruction together create a parallel information conduit (at little cost) which dramatically reduces the chances of communication failure.

Patton undoubtedly learned this lesson while serving as *aide de camp* for General John (Blackjack) Pershing prior to World War I during the U.S. Army's quest to capture Pancho Villa in Mexico. Patton had been given verbal instruction by General Pershing and a sealed letter of instruction to carry to one of Pershing's field commanders. After delivering the letter to the field commander, Patton learned that the written instruction conflicted with Pershing's verbal instruction. Knowing General Pershing's desire and convinced that the written instruction had misidentified a mountain chain that was to be reconnoitered, Patton countermanded the order. The field commander complied but threatened Patton with court martial if incorrect. Later, Pershing confirmed that Patton had made the correct decision. In this instance, written instruction alone would have failed.

As with verbal instruction, written instruction must be clear and concise. Use of language or phrasing not understood by the intended receiver leads to confusion and reduces the chance of successful compliance. The rule is "Keep it short and to the point!" Again, Patton:

> It is my opinion that Army orders should not exceed a page and a half of typewritten text and it was my practice not to issue orders longer than this. Usually they can be done on one page, and the back of the page used for a sketch map.

Additionally, no written instruction can be effective if it isn't delivered. Marconi operator Jack Phillips' failure to deliver two ice warnings deprived Captain Smith of information that could have changed his assessment of the situation and the speed of the *Titanic*.

Some organizations create special communications vehicles to ensure that written messages arrive. One common example is the *written instruction to backshifts* (known aboard British fighting ships in the mid-1800s as *night orders*). Organizations which work multiple shifts almost always need a method to communicate directions and instructions from the dayshift staff to backshift crews. Most adopt the simple expedient of writing instructions in a journal or book which is passed from crew to crew.

Written instructions to backshifts can be a very effective tool. The instructions must obviously carry the clear authority of the writer. The register of *who* may issue such instructions should be kept short to avoid confusion. Normally, only officials such as the plant manager, operations manager, maintenance manager, and training manager should have this authority.

One danger of a daily written instruction to backshifts is the temptation to use the book as a vehicle for changing equipment operating procedures. Specific policies should exist to control procedure changes. If they are bypassed, the sanctity of procedures is degraded. Unauthorized (and sometimes hazardous) equipment operation is sure to occur once this poor habit takes root.

Finally, written instructions should not be drafted if they won't be enforced. Patton put it this way:

> Avoid as you would perdition issuing cover-up orders, orders for the record. This simply shows lack of intestinal fortitude on the part of the officer signing the orders, and everyone who reads them realizes it at once.

Emergency Communications

Emergency communication signals represent a third vital form of communication within an industrial facility. Effective emergency response is contingent upon clear and timely emergency communications. When a plant casualty or emergency has been identified, the appropriate members of the plant team must be notified promptly so that team action can be rapidly initiated to combat the situation. Therefore, a system of signals and notification procedures which utilize appropriate communications equipment must be established *and rehearsed* to ensure effective response.

Emergency Communication Systems

Emergency communication systems are usually composed of multi-use communication equipment in combination with specially designed emergency transmission devices; i.e., telephones, radios, and public address systems are used in conjunction with fire alarms, warning sirens or horns, and visual indications to warn facility and off-site personnel of emergencies or casualties.

Regardless of the signals chosen for emergency communication, they must be clearly audible or visible to all who must respond, clear in their meaning, and functionally reliable. Emergency signals which have other meanings and are frequently used for other purposes will be ignored. During Union Carbide's 1984 methyl-isocyanate (MIC) gaseous release outside Bhopal, India, the emergency siren *was* sounded; but it meant little to the townspeople—it was the same siren used to announce shift change and lunch break.

Reliability of emergency communications must be proven through frequent testing. After responders are taught the meanings of the signals, preventive maintenance checks and personnel response tests should be incorporated as a part of maintenance and emergency response programs.

Finally, if established emergency signals are not used, they do no one any good. On board the *Titanic*, the general alarm was never sounded.

Emergency Notification

Most plant emergencies or plant casualties necessitate notification of senior plant leaders who are ultimately responsible for the health, safety, and welfare of all team members within an operating complex. For some events with great potential hazard, the need for assistance often extends far beyond the facility's own team. Aid from local, state, or federal agencies may be necessary. Rapid, organized response can make the difference between a controlled event and exposure of the public and the environment to peril.

To ensure effective response, an industrial facility should have an emergency response plan that delineates events requiring notification. The plan should define the types of events that require notification and incorporate a reliable process to ensure that the notifications are completed in a timely manner.

An effective emergency notification system must include a list of *who* to notify, *who makes* the notifications, the *means* of notification, and guidelines for *when* to make and complete the notifications.

Leadership and Communication

This discussion of communication in the industrial facility would be incomplete without noting that *no* form of communication is sufficient in the hands of an incompetent leader. Patton wrote:

> *Commanders must remember that the issuance of an order, or the devising of a plan, is only about five per cent of the responsibility of command. The other ninety-five per cent is to insure, by personal observation, or through the interposing of staff officers, that the order is carried out. Orders must be issued early enough to permit time to disseminate them.*

There is no substitute for leaders who know both their technologies and their people.

Topic Summary and Questions to Consider

The *Titanic* disaster was a watershed event in the world of 1912. The resulting harsh feelings between Great Britain and the United States were not resolved until the sinking of the *Lusitania* in 1915. The women's suffrage movement was divisively rent as the policy of "Women and children first" was reconsidered.

Yet, the tragedy of *Titanic* resulted in some positive outcomes as well.
- The requirement for routine, mandatory lifeboat drills aboard passenger vessels was established.
- A requirement to carry sufficient lifeboats for both passengers and crew was established.
- A twenty-four hour radio watch on board vessels of this class was established.
- Searchlights (previously used only on warships) became standard equipment for passenger liners.
- The International Ice Patrol was established in 1914.
- The United States Coast Guard was also an outgrowth of the accident.

Though over eight decades have passed since this terrible tragedy occurred, it is gratifying to consider that others may avoid wanton injury or loss of life by learning the lessons of *Titanic*—especially the lesson of communication. Have you?

Ask Yourself

At your facility, are frequent errors caused by miscommunication? Is the repeat-back communication technique routinely used for critical operations?

Does your facility have distinct emergency signals with specific meanings that everyone understands? (Remember, Bhopahl didn't.)

Are written instructions to back-shift personnel clear and concise?

Check your communications closely. The quality of communication can "make-or-break" your operations—regardless of how good your other skills are.

Chapter 13

Keeping Logs and Recording Data

Not only must facility team members be able to communicate with others in verbal and written form, they must also be able to clearly record vital aspects of plant data and history. These very specialized forms of written communication play an indispensable role in controlling plant equipment both in the near- and long-term.

Data Record Sheets

Data record sheets (sometimes called equipment logs or round sheets) normally are preprinted forms designed for recording the current values of important equipment operating parameters (e.g., temperatures, pressures, tank levels, etc.). A data record sheet is different than a narrative station log which is a chronological narrative record of significant station events and conditions.

During each shift, an operating technician must periodically check the critical parameters of equipment operation if the machinery is to be safely and reliably controlled. Critical parameters are instrumented or indexed to ensure that the facility equipment can be operated and maintained within design specifications. (As discussed earlier, critical equipment parameters are chosen and instrumented to provide the operator with indication of the margin to failure of that particular parameter. Neglecting to monitor critical equipment parameters at prescribed intervals seriously erodes an important barrier of the safe operating envelope.)

For systematic industrial operations, the values of those parameters are normally recorded on data record sheets. The function of data record sheets is to provide a structured process of observing, recording, and analyzing the values of critical equipment parameters. Observing and recording machinery parameter values not only provides a real-time check of the values of important parameters, but also provides an important record of the operation of the equipment from which historical performance and trends can be extracted.

Here are some common guidelines for developing and using data record sheets:

Purpose and Format

Data record sheets are designed with three purposes in mind: (1) to create an operating history of the value of critical equipment parameters, (2) to provide a tool for ensuring that critical equipment parameter values remain within the control bands specified by the designer, and (3) to provide a tool for analyzing short- and long-term critical parameter trends so that preventive or corrective actions can be taken to prevent damage.

Data record sheets usually have two parts. One part is the data recording section in which parameter values are recorded on a prescribed periodicity in recording columns. The other is a narrative section for explanation of abnormal values or for recording explanatory notes. Often, data record sheets also have a section for indicating completion of periodic supervisory reviews.

The required parameter recordings should have the maximum and minimum acceptable limits of their specified bands preprinted on the data record sheet where applicable. Maximum and minimum limits provide a ready reference to determine whether a value is approaching or has exceeded an out-of-specification condition.

Frequency of Recording

The required periodicity for recording critical equipment parameter values should be specified on the data record sheet. Data recording should be initiated at or near the specified recording periodicity. The purpose of this guideline is to ensure that recorded parameter data is representative of the actual machinery conditions at prescribed intervals.

A time bracket for recording the values should be specified by facility policy. One common guideline is that parameter values should be recorded within a range of time not more than 15 minutes before the specified periodicity and not more than 15 minutes after. For example, if recordings are required to be done each hour on the hour, values should be recorded within a one-half hour bracket around the hour mark. An 8:00 A.M. reading should commence no sooner than 7:45 A.M. nor any later than 8:15 A.M.

Sometimes, however, a situation arises in which parameter value recording cannot commence at the prescribed time. In such an instance, the equipment operator responsible for recording critical parameter values should either request assistance for data recording or request and receive from the central control room permission to delay the recording.

Late recordings should not be allowed to become a common practice. If frequently permitted, the data loses much of its value. Equipment operators, therefore, must necessarily manage the tasks at their station to avoid omitted or late entries.

Recording Data

A data record sheet is usually considered to have the force of a legal record since it tracks critical parameter values associated with equipment operation. Parameter recordings, therefore, should be entered legibly in a timely fashion. Falsification of the record by entering values without observing them is a dangerous practice, and also a criminal offense under some circumstances.

The equipment operator is to observe each specified equipment parameter value, record each in its appropriate block on the data record sheet, and compare each parameter value with its specified maximum and minimum limit.

If the values are within specification, the operator should compare the value with those previously recorded to determine if the parameter is tracking as expected considering the function currently being performed by the equipment.

If a value is out of specification, the operator must determine whether an equipment casualty or emergency is in progress. If the value is not indicative of a casualty or emergency, the operator should take action to return the parameter to its specified band, inform the control room of the out-of-specification value, its cause (if known), and the actions being taken to return the parameter to its specified control band. The operator should then circle the out-of-specification value in red ink, and describe the problem in the narrative section of the data record sheet noting its cause (as soon as discovered) and the actions taken to return the value to its control band.

If a value indicates a plant casualty or emergency in progress, the operator must initiate immediate action to combat the casualty or emergency, announce the casualty or emergency to the central control room while informing the control room of the out-of-specification value, log the out-of-specification parameter at the earliest convenience, circle the value in red ink, and make an entry in the narrative section of the data record sheet regarding the symptom, its cause (as soon as discovered), the actions taken to correct the out-of-specification value, and the current status of the equipment.

Correcting Errors

When an error is made while recording data, the error should be corrected by drawing a single horizontal line through the data to be corrected, initialing and dating the lineout, and recording the correct data next to it. This method of error correction is characteristic of handwritten corrections for any legal document. If there is likely to be a question about the correction, a note may be entered in the narrative section or in the margin explaining the correction.

Recordings for Out-of-Service Equipment

During shutdown periods (when the equipment is not operating), parameter value recording may not be necessary. In such instances, recording may be suspended by order of an authority in the operations hierarchy. If complete suspension is undesirable, a decrease in the frequency of recording may be appropriate. In any case, the data record sheets should be clearly annotated to show the reason for the deviations and the status of the equipment.

Data Record Sheet Review

Data record sheets should be reviewed periodically by a senior member of the plant operations staff (such as the operations manager) as an independent check for trends or anomalies. Such reviews should not imply distrust of the station operator. Rather, the oversight process is a team function necessary when dealing with complex equipment or processes. Such a review also alerts the operations manager to any deterioration of the logkeeping process within the facility.

Storage and Retrieval

Data record sheets should normally not be removed from their stations for at least 24 hours after completion. They must remain on station so that oncoming crew members may review them prior to assuming the duties of the station. A system of retrieval, filing, and storage should be established so that the data record sheets can be used as an historical resource.

Narrative Logs

Narrative logs are chronological written accounts of the events which occur at operating stations or control centers in an industrial operation, military venture, or commercial enterprise requiring a legal record of occurrences. Logs have been used for centuries and are firmly rooted in commercial and military shipboard operations.

Traditionally, logs are handwritten documents. Just as data record sheets, they should be considered to have the force of a legal record. The language, accuracy, objectivity, and detail should reflect such consideration.

Purpose and Format

The purpose of a narrative log in an industrial facility is to provide a chronological record of prominent events and actions that occur at an equipment station or control room. They are useful to others who must later determine important event sequences and evidence of past performance.

Logs are often contained in a bound journal, often with sequentially numbered pages. Sometimes, however, they are integrated as narrative sections to data record sheets for stations at which data recording is required. Regardless of format, log pages must not be removed or destroyed if the log is to be considered complete, accurate, and reliable.

The log format should ensure that the dates and times of the entries are unmistakable. Some logs include a place to record the date at the top of each page, and many have instructions requiring log entries for each new day or shift to start at the top of a new page. Unused page space may be voided by an indication after the last entry of the shift or day that no further entries follow.

Recording Log Entries

Log entries should be clear, legible, and concise written statements. Each entry should begin with the time of the event. Whenever possible, entries should be made at the time that the event occurs. Delayed entries are often incomplete entries.

Items that should be logged include major routine events and most abnormal events which affected the equipment status. All plant casualties or emergencies should be documented with log entries explaining the initial symptoms of the casualty or emergency, the major actions taken to combat it, its cause, and the status of the equipment following the casualty or emergency.

A series of log entries would proceed as in this example:
- 0400 A loss of feedwater occurred to the Number 2 Boiler. Immediate actions were initiated in accordance with Operating Instruction Number 12.
- 0408 Main steam isolation valve was shut as a result of rapidly lowering water level in the Number 2 Boiler.
- 0410 Feedwater was restored to Number 2 Boiler using the auxiliary feed system.
- 0412 Number 2 Boiler was returned to normal feedwater supply.
- 0415 Cause of the loss of feedwater to Number 2 Boiler was discovered to be a main feed pump trip due to loss of control air to the feedwater regulating valve.

Spaces should not be left between lines. Filling in log entries in open spaces at later times diminishes the credibility of the log.

Out-of-Sequence Entries

When an out-of-sequence entry is necessary (e.g., because of failure to log an important step in an event), a *late entry* should be made by entering an asterisk (or other identifying mark) at the point in the log where the missed entry would fit chronologically. A corresponding mark should then be placed at the point where you enter the out-of-sequence item—including the time of the missed event with a parenthetical "Late Entry" note. An example of such an entry is illustrated as follows:

   0400   A loss of feedwater occurred to the Number 2 Boiler. Immediate actions were
     *   initiated in accordance with Operating Instruction Number 12.
   0410   Feedwater was restored Number 2 Boiler using the auxiliary feed system.
   0412   Number 2 Boiler was returned to normal feedwater supply.
   0415   Cause of the loss of feedwater to Number 2 Boiler was discovered to be a main feed pump trip due to loss of control air to the feedwater regulating valve.
 *0408   (Late Entry) Main steam isolation valve was shut as a result of rapidly lowering water level in the Number 2 Boiler.

Baseline Equipment Status Entry

A baseline equipment status entry is used to record initial station equipment status. It is usually recorded at the beginning of each new day and is the first entry on a shift that takes over at midnight.

The purpose of the baseline equipment status entry is to provide a clear picture of the starting status of plant equipment. The entry should state the condition (e.g., running, standby, out-of-service) of key equipment, the alarm conditions at the start of the shift, and any unusual conditions or anomalies present at the start of the day. A standardized baseline status entry checksheet can be very useful when the station has a sizable amount of equipment and controls.

If the station data sheet doesn't have a narrative log section, it may be desirable to use a separate equipment status check sheet (to be completed at the start of each shift day) as the baseline entry.

Acronyms and Abbreviations

Acronyms and abbreviations can be a source of confusion in logkeeping. Some organizations find it helpful to include an index of common acronyms and abbreviations with logkeeping instructions. Acronyms or abbreviations not included in the index should not be used.

Correcting Errors

Logs should be corrected in the same manner as data record sheets. A horizontal line should be drawn through the entry or portion of the entry determined to be in error. The correct entry should be entered legibly beside the error or in the margin. The lined out portion should be authenticated by the initials of the person making the correction, including the date and time of correction. If necessary, an explanatory note should be added.

Erasures and obliterations must be avoided. It may, at some later time, be important to see what was originally written. And, again, they have the effect of discrediting the validity of the log.

Log Review

Narrative logs are an important tool in the turnover process. Log and data review will assist oncoming crew members to gain a clear understanding of the events of the previous shifts.

Requirements for log reviews before station relief should be the same as for reviewing data record sheets. Logs should normally be reviewed for the previous 24-hour period or since the reviewer last controlled the station, whichever is less.

Supervisory logs should be reviewed by facility managers on a prescribed periodicity just as supervisors review data record sheets.

Storage and Retrieval

Retrieval, storage, and filing of logs should normally adhere to the same rules as prescribed for data record sheets. When properly archived, logs are an important historical resource and can be invaluable for reconstructing abnormal events.

Specialized Formats

We would be remiss if we neglected to include two specialized formats for data records and narrative logs in this discussion: automated data recording (for data) and voice or video recording for narrative logs.

Automated Data Recording

In situations where critical parameter data must be recorded continuously or at such frequent intervals that manual entry on data record sheets is impractical or impossible, automated data recording formats are often used. Typical automated formats include: continuous analog chart recorders, digital printers set to record values at fixed intervals, and digital or analog recording direct to magnetic media (tape or computer disk). One example is the "black box" data recorder on commercial aircraft (which records both digital and analog data to magnetic media). Another example is a seismograph which records analog earth tremor data (as a function of time) on a strip chart recorder. (Note that analog chart recorders have one distinct advantage: data trends as a function of time are immediately apparent.)

Whatever the format of such automatically recorded data, it is useless unless it is periodically reviewed. Hence, similar review, storage, and retrieval requirements (as discussed for data record sheets) should be established for automated data records. Remember in the *Exxon-Valdez* grounding that VTC watchstanders abandoned the practice of plotting tanker traffic on navigation charts after automated tracking equipment was installed. But the automatically recorded location and course data was essentially useless—it was rarely, if ever, reviewed.

Voice/Video Recording Less common in industrial situations is the use of voice and/or video recording to supplement or substitute for narrative logs. But these formats are especially useful for situations where events occur so quickly that handwritten logs cannot be reliably or comprehensively prepared or (for video) where a visual record is necessary to supplement narratives.

If you stop to think about it for a moment, the use of audio and video recordings for this purpose is not as uncommon as you might expect. Commercial airliners are equipped with a cockpit voice recorder (CVR); police and emergency dispatchers tape all "911" telephone calls (as well as all radio communication traffic) for later review and (if necessary) transcription into written log format. Banks, department stores, and convenience stores routinely use video cameras and recorders to preserve a visual record in the event of a robbery.

While the use of such continuous voice/video recording is usually not necessary or desirable (because of legal issues) in an industrial setting, strong consideration should be given to having those *capabilities* available for use during an emergency or casualty situation.

Topic Summary and Questions to Consider

Narrative logs and data record sheets are no more (and no less) than specialized communication formats. Just as for all other communications in the industrial environment, they must be clear, understandable, and unambiguous. Well-kept logs and data records are characteristic of organized and disciplined teams.

Ask Yourself Does your operation use carefully designed data record sheets for recording critical operating parameter values? Do your data record sheets include pre-printed upper and lower limits for each critical parameter?

Do your team members have—and use—narrative logs to chronologically record important events of the shift? Do they make their log entries throughout the shift (as the events occur), or do they wait until the end of the shift to update logs?

Do oncoming team members review the previous shifts' logs and data record sheets *before* performing station reliefs? Do team *leaders* review logs and data records for accuracy, completeness, and trends? Frequently?

Do you have a reliable method for archiving (and retrieving) logs and data record sheets?

Keeping Logs and Recording Data

Chapter 14

Recognizing Abnormalities

During industrial operations, abnormalities in equipment or operator performance inevitably occur. The manner in which team members deal with abnormal events and conditions is based, primarily, on how well they have been trained and how wisely they are led.

The best industrial teams learn to recognize, respond to, and report abnormalities to gain two important advantages. The first is prompt repair of machinery or swift correction of error in team performance. The second is an opportunity to learn from the abnormalities in order to prevent their recurrence. Hence, ability of operating personnel to recognize, prioritize, and respond to abnormalities is a vital operating skill that must be taught to each operator.

Crash of Air Florida Flight 90, January 1982

On January 13, 1982, at 4:01 P.M. Eastern Standard Time (EST), a Boeing 737 (N62AF) operated by Air Florida crashed shortly after takeoff from Washington National Airport, killing 74 of the 79 people on board including 4 of the 5 crew members. The aircraft had attempted a departure with only three-fourths the necessary thrust in both engines, a situation complicated by snow and ice accumulation on the wings and fuselage. Because of a communication failure between the captain and first officer, the engine anti-ice system had not been activated, resulting in ice-blocked pressure ducts for the engine power sensing system. Though other instruments indicated a problem, the captain chose not to abort the takeoff. The aircraft stalled after takeoff, striking the barrier wall of the northbound span of the Fourteenth Street Bridge three quarters of a nautical mile past the end of runway 36. The aircraft then plunged nose down into the frigid waters of the Potomac. (See Figure 14-1, Aircraft Impact Attitude.) Besides those aboard the aircraft, four people in vehicles on the bridge were killed and four more injured.

Delayed Departure   N62AF, a Boeing 737-222 operated by Air Florida, Inc., was scheduled to depart Washington National Airport as Air Florida Flight 90 at 2:15 P.M. en route to Ft. Lauderdale, Florida, with a layover in Tampa. Departure, however, had been delayed because of a wet (and, at times, heavy) snowfall at the airport, a condition which had necessitated a temporary closure to clear snow from the runways. Temperature at the time was around 24°F and visibility on runway 36 varied from 1,800 to 3,800 feet. Wind was out of the north at approximately 11 knots and the ceiling hovered around 200 feet.

Between 2:00 and 2:30 P.M., Air Florida's crew (captain, first officer, and three flight attendants) boarded passengers on Flight 90, expecting the airport to reopen at 2:30 P.M. During the boarding process, an American Airlines ground crew, at the captain's request, began de-icing the aircraft to remove accumulated snow on the wings and fuselage. Shortly thereafter, the captain terminated the de-icing operation, after learning that the 2:30 P.M. airport opening time was no longer valid (and that eleven other aircraft had departure priority). Only a small portion of the left side of the aircraft had been de-iced.

Deficient De-Icing and Anti-Icing   Removal of ice and snow accumulation from commercial aircraft is usually accomplished using high pressure spray equipment which mixes hot water and ethylene-glycol. The equipment used to de-ice N62AF was a Trump Model D40D vehicle with a proportioning nozzle for selecting the desired mixture of hot water and anti-freeze. During *de-icing*, accumulated snow and ice are removed from the aircraft surface using primarily hot water with some anti-freeze (e.g., 70% hot water and 30% anti-freeze). Once snow and ice have been removed, the aircraft is *anti-iced*, by spraying with

100% anti-freeze. The anti-ice over-spray hinders water freezing on the surfaces after the accumulation has been removed.

At approximately 2:45 P.M., the captain requested the de-icing process to be resumed. The de-icing truck operator stated in testimony that he sprayed the left side of the aircraft with a heated solution of 30–40% ethylene-glycol and 60–70% hot water in accordance with American Airlines guidance, but didn't apply an over-spray. Then, sometime between 2:45 and 3:00 P.M., the de-icing truck operator turned the task over to his relief, having informed his relief that he had de-iced the left side of the aircraft. The relief operator de-iced the right side of the aircraft and then over-sprayed the right side with a mixture of 70–80% hot water and 20–30% glycol.

Guidelines for de-icing and anti-icing mixtures were ambiguous. Manufacturer instructions for the de-icing fluid used (Union Carbide Deicing Fluid II PM 5178) states:

> *Do NOT use diluted deicing fluid for anti-icing treatment of ice-free aircraft.*

But the Trump Vehicle Operator's Manual recommended anti-icing with a mixture of 35% hot water and 65% glycol. And the American Airlines anti-icing procedures were even more lenient, recommending (for temperatures in the range of 20–25°F) anti-icing could be performed with 75% hot water and 25% glycol.

During the de-icing and anti-icing process, no covers or plugs were installed over the engines and other openings in the airframe. Yet, the Air Florida Maintenance Manual stated that:

> *Prior to application of solution, covers and plugs will be installed. In applying the deicing solution around openings in the airplane care must be taken to limit the application to the amount required for anti-icing. Avoid directing the fluid stream into openings or the use of excess solution in ducting or appliances served by the openings.*

Further, the Boeing Manufacturer's Maintenance Manual required that airframe and powerplant covers and plugs should be installed whenever the aircraft was expected to be exposed to heavy snow or ice, even for short periods.

The de-icing and anti-icing was completed at approximately 3:10 P.M. By then, heavy, wet snow had accumulated on the ground surrounding the aircraft to a depth of two to three inches. Prior to closing the cabin door, the captain of Air Florida Flight 90 asked the Air Florida station manager (standing near the cabin door) how much snow was on the aircraft. The station manager replied that a light dusting of snow had accumulated on the left wing outboard of the engine. A few minutes later, the cabin door was closed and the aircraft was prepared for departure.

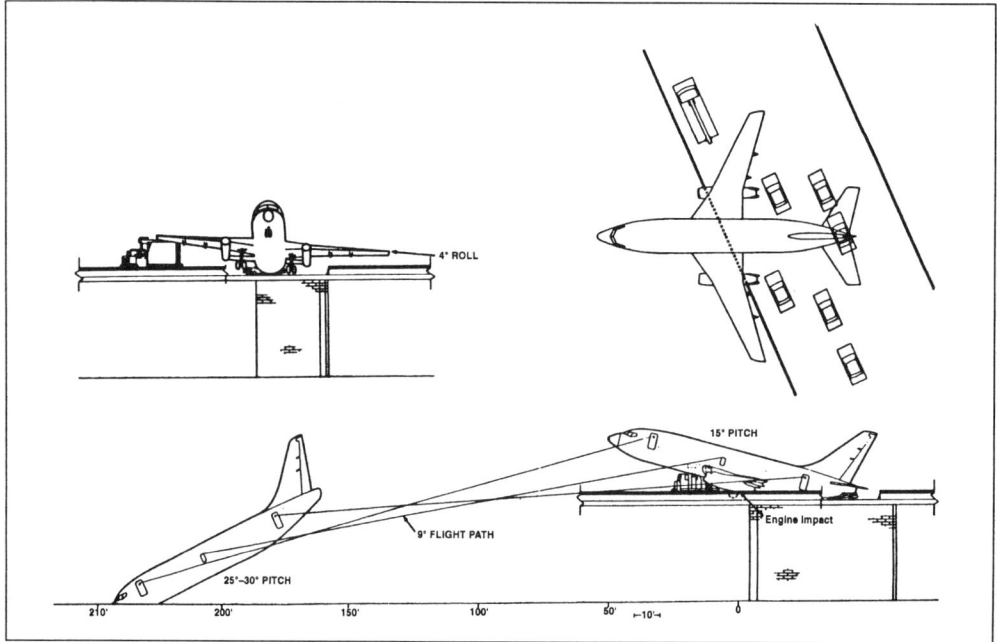

Figure 14-1: Air Florida Flight 90 aircraft impact attitude. [From NTSB Report]

The NTSB accident report states that:
> *No witnesses saw the flightcrew leave the aircraft to inspect for snow/ice accumulations while at the gate. Departing and arriving flightcrews and others who saw Flight 90 before and during takeoff stated that the aircraft had an unusually heavy accumulation of snow or ice on it.*

Yet, Air Florida operating procedures stated that:
> *No aircraft will be dispatched and no take-off will be made when the wings [or] tail surfaces have a coating of ice, snow or frost.*

Further, FAA Regulations "prohibit takeoff when frost, snow or ice is adhering to the wings, control surfaces or propellers of the aircraft." Air Florida Flight 90 was in an aerodynamically clumsy condition, preparing to take off in violation of both operating procedures and federal regulations.

Improper Use of Thrust Reversers

By 3:25 P.M., Flight 90 had received permission to push away from its gate and enter the sequence for takeoff. But, because of ice and snow on the ramp (as well as de-icing fluid), Flight 90's tug didn't have enough traction (without chains) to push the aircraft away from the gate. The flightcrew subsequently recommended deploying the thrust reversers (devices that redirect engine exhaust forward) to help push the aircraft away from the gate. The tug operator responded that deployment of reversers at the gate was prohibited by American Airlines policy. Nevertheless, Flight 90's crew started the engines and deployed the reversers. The Boeing 737 Operations Manual states:
> *A buildup of ice on the leading edge devices may occur during ground operations involving use of reversers in light snow conditions. Snow is melted by the deflected engine gases and may refreeze as clear ice upon contact with cold leading edge devices. This buildup, which is difficult to see, occurs in temperature conditions at or moderately below freezing. Crosswind conditions can cause the ice buildup to be asymmetrical, resulting in a tendency to roll at higher angles of attack during subsequent takeoffs.*

The NTSB accident report states:
> *Witnesses estimated that both engines were operated in reverse thrust for a period of 30 to 90 seconds. During this time, several Air Florida and American Airlines personnel observed snow and/or slush being blown toward the front of the aircraft.*

After an unsuccessful attempt to back away from the gate (assisted by engine power), N62AF's engines were shut down (with thrust reversers still deployed) and another tug with chained tires was positioned for the push back. At 3:35, N62AF was successfully pushed away from the gate, the engines were re-started, and the reversers were stowed.

Engine Anti-Ice "Off"

Preparing for taxi to their assigned runway, Flight 90's flightcrew performed an after-start check. One important checklist item was the engine anti-ice system status. The captain, responding to the first officer's checklist query, *Anti-ice?*, replied, *Off*. And it remained off.

Both engines were equipped with a thermal anti-ice system which bled exhaust gases from the turbine and routed them to, among other components, the sensing probes for the critical operating parameter called *engine pressure ratio* (EPR). (See Figure 14-2.)

EPR is a comparison of turbine discharge pressure to engine inlet pressure. If the pressure ducts for EPR are allowed to become (and remain) blocked with ice, they will provide a false indication of engine thrust. By activating the anti-ice system, valves are opened which route hot exhaust gases to keep the probes clear of ice.

The Boeing 737 Operations Manual *requires* use of the engine anti-ice system during conditions such as Flight 90 was experiencing:
> *Engine anti-icing should be turned on during all ground operations, takeoff and climb when icing conditions exist or are anticipated…. The Pt_2 probe will ice up in icing conditions if anti-icing is not in use; therefore, the first noticeable indication could be an erratic EPR reading.*

But, for unknown reasons—probably complacency or distraction—this flightcrew failed to turn on engine anti-ice, a fact confirmed by analysis of the wreckage. NTSB's investigation report states:
> *One function of the engine anti-ice system is to maintain a flow of heated air at the Pt_2 probe to prevent ice formation and blockage. Strong evidence from the accident investigation—namely the post-impact closed position of the anti-ice valves, valves which are electrically motor driven and thus not susceptible to position changes by impact loading—indicates that the engine anti-ice system was*

off at the time of impact. The CVR [cockpit voice recorder] recording substantiated that the engine anti-ice system had not been used during the pretakeoff ground operation.

For Flight 90 on this snowy day—especially after de-icing and anti-icing without plugs and covers installed over openings—activating the engine anti-ice system was a neglected necessity.

Increasing Uneasiness

As Flight 90 prepared to taxi at about 3:40 P.M., ground crews made a final attempt to de-ice the aircraft. All the while, the captain and first officer recorded concerns about icing conditions. Their comments, continuing for nearly fourteen minutes, voiced increasing uneasiness:

Captain: *…go over to the hangar and get deiced.*
1st Officer: *Yeah, definitely.*
Captain: *Tell you what, my windshield will be deiced. Don't know about my wings.*
1st Officer: *Well, all we need is the inside of the wings anyway. The wingtips are gonna speed up on eighty anyway. They'll shuck all that other stuff.*
Captain: *Gonna get your wing now.*
1st Officer: *D'they get yours? Did they get your wingtip over 'er?*
Captain: *I got a little on mine.*
1st Officer: *A little. This one's got about a quarter to half an inch on it all the way.*
1st Officer: *Boy…this is a losing battle here on trying to deice those things. It [gives] you a false feeling of security. That's all that does.*

They recognized that their margins of safety were being eroded. They knew that they were operating outside the zone of normal operations—as were probably many other aircraft at this airport that day. But, because of schedular pressures (compounded by an already lengthy delay), they were willing to operate with reduced margins. Unfortunately, as so often is discovered in accident studies, other factors, unknown to the operators, had already greatly reduced their margin.

Anomalous Engine Indications

At 3:48 P.M., the first officer asked the captain, *"See the difference in that left engine and right one?"* The captain responded, *"Yeah"*. The first officer replied, *"I don't know why that's different, 'less it's hot air going into that right one. That must be it—from his exhaust. It was doing that at the chocks awhile ago…."*

The first officer was observing a stack of ten engine instrument indications (five for each engine) in the center console. (See Figure 14-3, Engine-Related Instruments.) From top to bottom, these are:
- N_1 RPM, the turbine low pressure compressor speed in percent of RPM,
- EPR (Engine Pressure Ratio), which is directly proportional to developed thrust,
- EGT (Exhaust Gas Temperature) in °C, measured by thermocouple,
- N_2 RPM, the turbine high pressure compressor speed in percent of RPM, and
- Fuel Flow to the engine, measured in pounds per hour

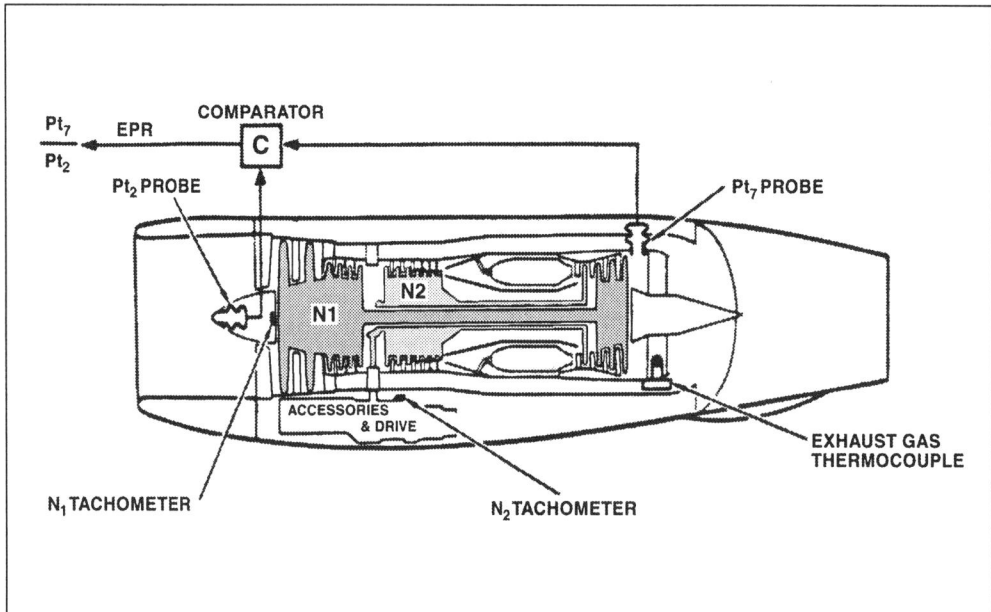

Figure 14-2: Engine Sensor Locations. [From NTSB Report]

Though the EPR setting was the primary indication of thrust being developed by its engine, the other four indications also characterized critical engine parameters. They changed in relationship to one another. For example, as EPR increased, so should fuel consumption, exhaust gas temperature, and both compressor speed indications.

Just before 3:38 P.M., Flight 90's aircrew performed the pre-takeoff checklist which included setting the EPR for both engines. (Prescribed EPR was 2.04 for takeoff.) Then, at approximately 3:59 P.M., Flight 90 was cleared to taxi into a hold position in preparation for entering the active runway. Just before 4:00 P.M., Flight 90 received a takeoff clearance and the captain said to the first officer, *Your throttles*, a direction to the first officer to perform the takeoff.

Then, over the next minute, during the takeoff roll, the first officer called the captain's attention to anomalous engine indications four more times:

3:59:45 P.M. Captain: *Holler if you need the wipers.*
3:59:56 P.M. Captain: *Real cold, real cold.*
  [Probably indicating that exhaust gas temperature was lower than he expected.]
3:59:58 P.M. 1st Officer: *God, look at that thing. That don't seem right, does it?*
4:00:05 P.M. 1st Officer: *Ah, that's not right.*
4:00:09 P.M. Captain: *Yes it is. There's eighty [knots].*
4:00:10 P.M. 1st Officer: *Naw, I don't think that's right.*
4:00:19 P.M. 1st Officer: *Ah, maybe it is.*
4:00:21 P.M. Captain: *Hundred and twenty [knots].*
4:00:23 P.M. 1st Officer: *I don't know.*

Despite the first officer's insistence that something was wrong with the engines, the captain chose not to abort the takeoff. (The Air Florida Training Manual states, "The captain <u>ALONE</u> makes the decision to 'REJECT'.")

Thrust Too Low, Stall, Impact

Almost as soon as Flight 90 left runway 36, it was in trouble. At 4:00:39 P.M. the sound of the stick-shaker, "a device which activates to warn the flightcrew of an impending stall", is recorded on the cockpit voice recorder tape. Flight 90 had experienced an unusual pitch-up of the nose as it rotated off the runway, placing it in a nose-up attitude that, without more engine thrust, would severely limit the lift of the aircraft.

Center Instrument Panel Functions (Boeing 737-222)

In this illustration the right engine instrument indications are shown for normal engine function and the left engine instruments show indications observed with the Pt_2 probe blocked—as was the case for both engines of Air Florida flight 90 (N62AF) on January 13, 1982.

Although (in this illustration) both engines show an EPR setting of 2.04 (set by the throttles), the left engine indicators show significantly lower turbine compressor speeds (N_1 and N_2), exhaust gas temperature, and fuel flow.

With the Pt_2 probe blocked and throttles set to a 2.04 EPR indication, the actual EPR is about 1.70 which is much lower than required for takeoff.

N_1 RPM
Indicates low pressure compressor speed as percentage of RPM.

Engine Pressure Ratio (EPR)
Indicates ratio of turbine discharge pressure (Pt_7) to compression inlet pressure (Pt_2). Primary thrust setting since EPR varies in proportion to the developed thrust.

Exhaust Gas Temperature (EGT)
Indicates turbine exhaust gas temperature as sensed by thermocouple.

N_2 RPM
Indicates high pressure compressor speed as percentage of RPM.

Fuel Flow
Indicates fuel consumption rate in pounds per hour (PPH).

Figure 14-3: Boeing 737-222 Center Panel Engine Instruments.

The captain, warning the first officer to control the pitch, urged him to push forward on the yoke to lower the nose:

4:00:45 P.M. Captain: *Forward, forward. Easy. We only want five hundred [feet above sea level altitude].*
4:00:50 P.M. Captain: *Come on, forward. Forward. Just barely climb.*

Undoubtedly surpised by the unusual upward pitch of the nose, Flight 90's crew response was ineffective. The NTSB accident report states:

> *The Safety Board believes that the crew probably reduced nose attitude at first, but later increased it to prevent descent into the ground. It should have been apparent from the continuation of the stickshaker and the continuing decrease in airspeed that the aircraft was not recovering from a serious situation.*

But Flight 90's situation was neither new nor unique. Abrupt nose pitch-up caused by ice on leading edge surfaces had been experienced and documented by other Boeing 737 crews. Twenty-two such incidents since 1970 prompted Boeing to issue Operations Manual Bulletin 74-8 on October 24, 1974. The NTSB accident report states:

> *The bulletin advised operators of the incidents in which asymmetrical clear ice had built up on the leading edge devices during ground operations involving the use of thrust reversers in light snow conditions with cross winds. It appeared that the snow melted due to hot engine gases and refroze on contact with the cold leading edge devices. The presence of the ice resulted in a tendency to roll at higher angles of attack during ensuing takeoffs. The bulletin cautioned flightcrews to assure compliance with all ice and snow removal procedures prior to takeoff under suspected icing conditions and to avoid maneuvers requiring unnecessary "g" loads immediately following takeoffs in weather conditions under which icing might be suspected. This bulletin had been incorporated into Air Florida Flight Manuals.*

A similar warning was issued on February 23, 1979, in Operations Manual Bulletin 79-2. Whether Flight 90's crew was aware of the pitch-up tendency (and its cause) is a matter of conjecture. What *is* certain, however, is that they were not prepared to respond to the event.

As Flight 90 approached the Fourteenth Street Bridge, it was still at too steep an angle without sufficient thrust to provide the necessary lift. The NTSB's accident report states:

> *Ground witnesses generally agreed that the aircraft was flying at an unusually low altitude with the wings level and attained a nose-high attitude of 30 [degrees] to 40 [degrees] before it hit the bridge.*

They needed more thrust. But with the EPR settings at their indicated limits, they were apparently reluctant to increase engine power, since over-powering (exceeding critical parameter limits) would require the engines to be dismantled and inspected—a costly process. NTSB:

> *The Safety Board is concerned that pilots are so indoctrinated not to exceed engine parameter limitations that they will withhold the use of available thrust until it is too late to correct a developing loss of control. Pilot training programs should be reviewed to ensure that they place proper emphasis on adherence to engine limitations, but that they also stress the use of available thrust beyond those limits if loss of an aircraft is the other alternative.*

But thrust was far less than EPR readings indicated. Following tests by Boeing, the NTSB concluded that actual engine power was only three-quarters of that indicated—corresponding to an actual EPR setting of 1.70 instead of the 2.04 indicated setting.

Flight 90 was in crisis. The choice had become one of exceeding engine limits or losing an aircraft, its crew, and passengers. NTSB:

> *…with the aircraft near to stall and close to the ground, the crew should have responded immediately with a thrust increase regardless of their belief that EPR limits would be exceeded. Furthermore, in this case the crew should have known that all other engine parameters—N_1, N_2, and exhaust gas temperature—were well below limit values.*

Tragically, the crew failed to increase engine power. In the final few seconds, the cockpit voice recorder registered a chilling account of their confused world:

4:00:59 P.M. Captain: *Stalling, we're falling!*
4:01:00 P.M. 1st Officer: *Larry, we're going down, Larry!*
4:01:01 P.M. Captain: *I know it!*
4:01:04 P.M. [Sound of impact]

An Eye Witness Account

At 4:01 P.M., N62AF crashed into the crowded northbound span of the Fourteenth Street Bridge, striking six vehicles and a boom truck. (See Figure 14-4, Flightpath.) Eight people on the bridge were injured; four died.

One eye witness described the crash sequence:

> *I heard screaming jet engines.... The nose was up and the tail was down. It was like the pilot was still trying to climb but the plane was sinking fast. I was in the center left lane...about 5 or 6 cars lengths from where [the red car] was. I saw the tail of the plane tear across the top of the cars, smashing some tops and ripping off others.... I saw it spin...[the red car]...around and then hit the guardrail. All the time it was going across the bridge it was sinking but the nose was pretty well up.... I got the impression that the plane...seemed to go across the bridge at a slight angle and the dragging tail seemed to straighten out. It leveled out a little. Once the tail was across the bridge the plane seemed to continue sinking very fast but I don't recall the nose pointing down. If it was, it wasn't pointing down much. The plane seemed to hit the water intact in a combination sinking/plowing action. I saw the cockpit go under the ice. I got the impression it was skimming under the ice and water.... I did not see the airplane break apart. It seemed to plow under the ice. I did not see any ice on the aircraft or any ice fall off the aircraft. I do not remember any wing dip as the plane came across the bridge. I saw nothing fall from the airplane as it crossed the bridge.*

At 4:00:33 P.M., air traffic control advised Flight 90 to *"Contact departure control"*, but got no reply. Then, at 4:01:22 P.M. (twenty seconds after impact), the controller realized that the aircraft had disappeared from radar. By 4:04 P.M., the airport fire department had been notified of a probable crash and, by 4:06 P.M., the Washington Metropolitan Area Communication Circuit of the Defense Preparedness Agency had learned of an aircraft emergency.

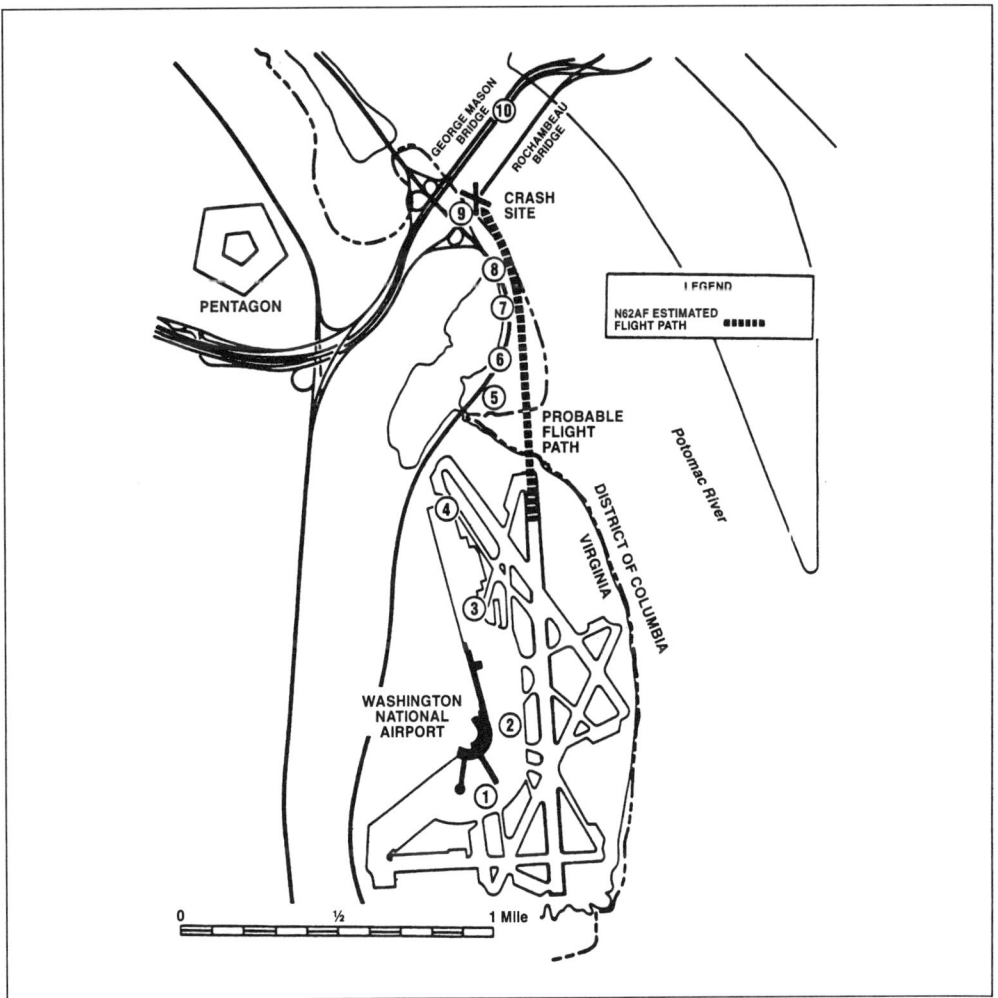

Figure 14-4: Air Florida Flight 90 flight path. Numbers indicate witness locations. [From NTSB Report]

Recognizing Abnormalities

Among those alerted by the area communication circuit were the U.S. Park Police. The Park Police's Bell Jet Ranger helicopter responded promptly and, by 4:22 P.M., was on-scene, aiding rescue efforts by heroic bystanders. But only five people were pulled from the water alive.

An Accumulation of Abnormalities

The tragic crash of Air Florida Flight 90 illustrates the classic accident pattern we have seen repeatedly. No single error or factor was sufficient to cause this mishap. Rather, a series of "minor" deviations from approved standards and practices (in the context of non-conservative circumstances) accumulated to create a problem, with effects far greater than the sum of individual contributors.

In their statement of probable cause, the National Transportation Safety Board determined that:

> *...the probable cause of this accident was the flightcrew's failure to use engine anti-ice during ground operation and takeoff, their decision to take off with snow/ice on the airfoil surfaces of the aircraft, and the captain's failure to reject the takeoff during the early stage when his attention was called to anomalous engine instrument readings. Contributing to the accident were the prolonged ground delay between deicing and receipt of ATC takeoff clearance during which the airplane was exposed to continual precipitation, the known inherent pitchup characteristics of the B-737 aircraft when the leading edge is contaminated with even small amounts of snow or ice, and the limited experience of the flightcrew in jet transport winter operations.*

The contributors:
- A long schedule delay,
- Incomplete de-icing, and failure to apply an appropriate anti-ice overcoat,
- A slick ramp, leading to an inappropriate decision to deploy thrust reversers,
- Ice-contaminated leading edges,
- A missed checklist item (engine anti-ice system), leading to ice-blocked pressure ducts,
- Departure after exceeding time limits requiring additional de-icing/anti-icing,
- Ignored anomalous engine indications on takeoff roll.

Each contributing factor was a human, equipment, or environmental *abnormality*. Further, each was easily recognizable (and controllable) before the accident.

Lamentably, as is so commonly discovered in accident investigation, proper recognition and response to almost any one of the abnormalities would probably have prevented this waste of lives. Accordingly, one of the most important of the vital operating skills for equipment operators to acquire is the ability to recognize and respond to abnormal human, equipment, and environmental conditions.

Sources of Abnormalities

Abnormalities, for our purposes, are conditions or events which deviate notably from those which are desired or expected. They usually result from one (or a combination) of three factors: equipment defect, human error, or environmental anomaly. Either something is wrong with the equipment, something is wrong with those who are running it, or something has changed in the operating environment. In the Air Florida crash, it was all three.

Equipment Defects Equipment deficiencies are sometimes attributable to flaws in design or construction. In Flight 90's case, the Boeing 737 had an unusual design characteristic:

> *The reports since 1970 by...operators who have experienced abrupt pitchup or rolloff immediately after liftoff of B-737 aircraft indicate that the B-737 may have a greater known inherent pitchup characteristic than other aircraft in this regard as a result of small amounts of frost, snow, or ice on the wing leading edge. The Safety Board could not determine whether the aerodynamic design makes the B-737 more sensitive to pitching or rolling moments when the wing is contaminated, or whether more frequent operation of these aircraft in environmental conditions conducive to snow or ice accretion during ground operations, coupled with the near to the ground wing placement, accounts for the higher number of reported B-737 pitchup/rolloff incidents. Regardless, the Safety Board concludes that the pitchup tendency of the aircraft because of leading edge contamination contributed to the accident.*

The B-737 was not unsafe for flight, but operators needed to understand this peculiarity.

Other times, equipment deficiencies result from normal wear during proper usage. And, on occasion, they occur because of improper use or maintenance of equipment—failure to use a part, component, or system in the manner prescribed by design.

Upon examination of the Trump D40D truck used to de-ice and anti-ice Air Florida Flight 90, the NTSB learned that the proportioning nozzle had been replaced with a commercial "equivalent":

The replacement of the nozzle on the Trump deicing vehicle with a nonstandard part resulted in the application of a less concentrated ethylene glycol solution than intended.

As a result, the ground maintenance crew may have been *icing* rather than *de*-icing Flight 90.

Regardless of cause, if equipment defects are not quickly recognized and addressed, they may lead to degraded equipment reliability, serious equipment damage, or personnel injury.

Human Error

Though as important as the equipment deficiencies they often cause, human deficiencies are often far more difficult to identify. Machines don't have personalities and feelings to worry about. Nor do they rationalize actions or make excuses when something goes wrong.

A human defect may take the form of unpreparedness, unfitness, or neglect. In the *Exxon-Valdez* accident, for example, at least five people knew that the master had been drinking prior to reporting back on duty. Moreover, his alcohol problem was well-known to the shipping company. Yet, the deficiency was apparently not considered serious enough to warrant action. The master was guilty of unfitness (leading to unpreparedness) and several others were guilty of neglect for failing to report or otherwise act upon his problem.

In the Air Florida disaster, neglect was demonstrated by failure to follow known, prescribed standards. The NTSB accident report states:

Regardless of the many factors which may have influenced the flightcrew, the Safety Board concludes they should not have initiated a takeoff with snow visible on the aircraft wings. The Federal Aviation Regulations are very specific, requiring that "no person may take off an aircraft when frost, snow or ice is adhering to the wings, control surfaces, or propellers of the aircraft." Since the snow or ice on the wings of Flight 90 degraded the aerodynamic performance significantly, the Safety Board concludes that the flightcrew's decision to take off with snow or ice on the wings was a direct cause of the accident.

Environmental Anomalies

Sometimes, even though both equipment and controller are performing properly, an unexpected environmental anomaly—too hot, too cold, too wet—can cause equipment or process upset. In the Air Florida accident, the true effect of prevailing environmental conditions was clearly not recognized by the flightcrew. NTSB:

The Safety Board believes that the flightcrew accepted the fact that snow or ice had accumulated on the aircraft and believed that while such accumulation may have some deteriorating effect, it would not affect significantly the aircraft's takeoff and climb performance.

In industry, environmental anomalies result more often from material defect or human error than from an Act of God—for example, flooding (environmental anomoly) resulting from a burst pipe (material defect) because an operator forgot to activate heat tapes to prevent the pipe from freezing (human error). Regardless of cause, operators must respond to such environmental changes.

Recognizing Abnormalities

Proper response to an abnormal condition is predicated first upon the ability to *recognize* it. Without recognition, there can't be effective action.

Recognizing a problem is the first step in all complex problem-solving. For example, the skill of cardio-pulmonary resuscitation is useless unless accompanied by knowledge of the signs of cardiac and pulmonary arrest. Similarly, in industrial operations, if abnormal conditions or events remain undetected (or ignored), they usually grow into more difficult problems.

In the Air Florida tragedy, perhaps the most conspicuous example of failure to recognize abnormalities occurred when the captain ignored the first officer's concerns about anomalous engine indications. Following the accident, the NTSB commented that:

> *Although the first officer advised the captain of his concerns several times, the captain apparently chose to ignore his comments and continue the takeoff.*

On five separate occasions the first officer attempted to direct the captain's attention to them. We do not know why, but the captain seemed to be blinded to the first officer's concerns, perhaps because of an overwhelming desire to get back on schedule. He used his position of authority to overrule the first officer's implied recommendation to abort the takeoff.

Knowing What's Right

In morals or machines, you have to know what's right before you can determine what's wrong. Therefore, industrial team members must have the knowledge and skill to discern when a condition is incorrect or when a process or event is not proceeding normally. If Flight 90's captain was able to discern a problem with engine indications, he certainly failed to exercise that ability.

Failure to "know what's right" was also demonstrated in the *Exxon-Valdez* grounding. The Coast Guard VTC radar operators believed that the radar was malfunctioning since targets faded so quickly on the screen. Actually, the operators just didn't understand how the range selector switch functions differed between the master and slave radar units.

"Knowing what is right" comes from a learning process that never ends. Rudimentary knowledge of equipment operating characteristics is, of course, acquired during initial training; but, mastery of abnormality recognition comes through extensive experience and continuous, superior coaching.

When Abnormal Becomes Normal

It's very easy to create an environment in which deficiencies are allowed to exist as "normal". All you have to do is ignore them when they present themselves. Pretty soon they will be accepted. During the Air Florida investigation, NTSB recorded that:

> *…the flightcrew was aware that the top of the wings were covered with snow or slush before they attempted to takeoff. The captain and first officer continued to discuss the weather conditions until they were first in line for takeoff clearance. There is no evidence that the flightcrew made any last minute visual assessments of the amount or character of the snow or slush on the wings before taxiing into the takeoff position. The lack of increased background noises indicating that windows were opened or pertinent conversation on the CVR are consistent with the conclusion that neither crewmember left the cockpit to observe the wings from the cabin nor opened the cockpit windows to enhance observation from the cockpit…. The flightcrew was probably influenced by the prolonged departure delay and was thus hesitant to forego the takeoff opportunity and return to the ramp for another cycle of deicing and takeoff delay. The flightcrew was probably also influenced by their observations of other aircraft departing ahead of them and successfully completing the takeoff and climb.*

The fact that "everyone else is doing it" leads many to believe that observed behavior is "normal" and acceptable. But, "normal" may not be right. This is an insidious human tendency.

When the abnormal is allowed to become normal, it creates a *very* dangerous operating environment. Within the Prince William Sound Vessel Traffic System, it became common practice for masters to deviate from the traffic separation scheme lanes (though the regulations expressly forbad deviation without permission of the Officer of the Deck). At Bhopal, MIC leaks were common because the maintenance function degraded. In the Challenger disaster, the solid rocket motor joint sealing problems were accepted (at least by some) because there had been twenty-four successful previous flights.

To avoid similar errors, mature operators must recognize that performance varies from machine to machine, operator to operator, and environment to environment.

If operators have been well-equipped through training to recognize abnormalities, they must then exercise continuous, vigilant evaluation of conditions and environment—in other words, they must be in constant search of events and situations that are not suitable. Unless leaders set the example, aggressive identification of deficient conditions will not follow.

Performing Inspection Tours

Identifying abnormal equipment conditions is very difficult to do if you never look at the equipment. You've got to be willing to get out of the car, look under the hood, and crawl under the vehicle. You've got to perform *inspection tours*.

Checking for Abnormalities

Inspection tours are the most valuable technique available to operators and leaders for detecting deficiencies. An **inspection tour** is usually a walking tour of an operating space during which a cognizant operator checks equipment for abnormalities. Besides observing instrumentation, skilled operators check for fluid leaks, unexpected vibrations, noises, or odors which might indicate a malfunction.

In the Air Florida accident, failure of both operations and maintenance personnel to perform an inspection tour contributed to the accident. In their accident report conclusions, NTSB noted that:

> *Neither the Air Florida maintenance representative who should have been responsible for proper accomplishment of the deicing/anti-icing operation, nor the captain of Flight 90, who was responsible for assuring that the aircraft was free from snow or ice at dispatch, verified that the aircraft was free of snow or ice contamination before pushback and taxi.*

Inclement weather probably influenced both maintenance and operations to forego the tour.

As in this event, inspection tours may sometimes be inconvenient or downright unpleasant. Personal comfort, however, is no reason to neglect the duty.

Usually, all the equipment for which an operator is responsible is at a single operating platform or station. To observe the equipment and instrumentation, the controlling operator need only walk around the station.

Control Panel Tours

In other cases, an inspection tour may be performed simply with the eyes. In a control room, for example, the equipment (and related instrumentation) for which an operator is responsible is usually located in a compact space. The panels and indications can be observed with little effort, often from a single location. Control room operators learn to scan their critical parameter instrumentation on a fixed intervals, increasing the frequency for equipment with changing loads or rates.

Finally, in some circumstances, a single operator is responsible for equipment at several locations or spread over a large area. In such cases, vehicle transport may be necessary to check equipment operation.

Combined with Data Recording

Since most operating stations require the cognizant operator to record critical operating parameter data on a prescribed schedule, inspection tours are a natural part of the data recording process. But, inspection tours should not be restricted only to those times at which data record sheets must be completed. Usually, tours should be performed more often—especially to check on equipment or processes that need to be monitored more closely.

Supervisory Tours

Inspection tours are not just for operators. They should also be a significant part of each supervisor's shift routine as well. Though a supervisor's tour of each station will probably not be of the same depth as that conducted by the controlling operator, each station should, nevertheless, be inspected at least once per shift.

Occasionally, supervisors should also accompany operators on routine inspection tours. Such accompanied tours provide an opportunity for supervisors to evaluate operators in the performance of their duties and to stay abreast of current station problems. Accompanied tours (conducted in the right spirit) can be an excellent learning experience for both supervisor and operator. The team relationship formed establishes and reinforces a cooperative, self-evaluative attitude on the crew.

Prioritizing Abnormalities

Inspection tours *always* turn up abnormalities. Knowing which ones are important is more difficult.

Abnormalities, by nature, are an indication that something is wrong. Though you may not be able to clearly define the underlying problem, ignoring it is an invitation to failure.

In discussing the captain's neglect to respond to the first officer's concern for engine abnormalities, the NTSB wrote:

> *It is not necessary that a crew completely analyze a problem before rejecting a takeoff on the takeoff roll. An observation that something is not right is sufficient reason to reject a takeoff without further analysis. The problem can then be analyzed before a second takeoff attempt. On a slippery runway, a decision to reject must be made as early as possible.*

Flight 90's situation was filled with risk. It required action. But, not all abnormalities have the same significance. As we noted earlier, some material deficiencies need only be identified for evaluation and future correction. So how do you decide what's important and what's not? In final analysis, it's a risk-based decision. Risk (the product of probability and consequences) is the determinant for judging significance—for deciding whether to stop or to continue. Therefore, in the face of abnormalities, mature operators always ask, "What is the potential outcome of this event if things go wrong?"

Dangerous abnormalities may not, at first glance, appear serious or worthy of prompt attention. Sometimes, serious abnormalities reveal themselves through modest symptoms which a distracted or inexperienced operator might not recognize. For example, in diagnosing disease, a fatal malady may have only a sore throat as its first symptom. In industrial operations, an impending disastrous failure may similarly be foretold by a new or unusual sound or vibration.

Therefore, like doctors, well-trained and experienced operators don't dismiss operating abnormalities, especially when they begin to appear in groups. They continuously ask, "How are the symptoms that I see related?" and "What do they mean?" Accordingly, one distinguishing characteristic of skilled operators is the ability to assess significance of a potential accident contributor *in the context of other contributors*. Alone, a single abnormality may be of little significance at all. Always remember, though, that accident contributors build on one another. In combination with other factors and other circumstances, they may be deadly. For Flight 90, the flightcrew was so preoccupied with departure, they did a poor risk analysis in the face of many abnormal indications.

Close Calls

Serious abnormalities sometimes reveal themselves through *close calls*. A **close call** is an event or condition which *almost* became an accident. Some people call them *near misses*. Perhaps the best description of all is the term *near hit*. Whatever they are named, they must be recognized as significant abnormalities requiring prompt attention.

The Boeing 737 Operations Manual warned against the use of thrust reversers during ground operation. The Operations Manual warning was a result of events in which the thrusters had caused moisture to melt and refreeze on the leading edge surfaces, inducing erratic aircraft performance—close calls—on takeoff.

Most of us know when we've had a close call. For example, has it been long since you changed lanes in traffic without checking your blind spot, only to be rudely awakened by the blaring horn of the car you nearly hit? That's a close call—a precursor of a future accident if you don't correct your driving habits.

Reporting Close Calls

What should you do when you have a close call? In driving, it's usually enough to think about what you've done and determine how to ensure that it doesn't happen again. But, in a team industrial operation, that may not be enough. Reporting the near miss gives opportunity to investigate it more fully and determine if others on the team are having similar problems.

Boeing's Operations Manual Bulletins, 74-8 and 79-2, were outgrowths of reported close calls. They warned of unusual pitch and roll characteristics when ice or snow contaminated the leading edges of B-737 wing surfaces. Twenty-two such events had been well-documented. The bulletins were issued to warn others of event potential.

Close call reporting and tracking requires great effort. Further, the reporting and tracking process does little for those who have experienced the close calls. If they are wise operators, they have already learned from their mistakes. Why, then, should close calls be reported and tracked?

The reason is simple. Studies show that accidents are usually portended by *many* near misses. Unless reported and tracked, they have no value as a tool in interdicting future events. The Federal Aviation Administration's *Callback* program is based on just this premise. If an aviator violates FAA rules but self-reports within ten days of the event, no action is taken against the violator. Rather, the event is screened by NASA for valuable lessons and those events with important lessons are published.

Close call reporting can fail for a couple of reasons. One is failure to report and the other is failure to heed the reports.

One of the greatest barriers to improvement in industrial operations is the fear of operating personnel to report their own errors. Yet, without open admission of operating problems, the underlying causes of the problems cannot be determined and corrected. Accordingly, it is important for every industrial facility to encourage a communication environment where all participants feel confident to acknowledge error.

Finally, even when reporting is well-performed, if the lessons are not considered thoughtfully by others, they have little value. Flight 90's crew certainly had not heeded the lessons of other B-737 operators regarding thrust reverser deployment and the B-737 pitch-up characteristics.

Responding to Abnormalities

Recognize, prioritize, respond. Once recognized, abnormalities which threaten immediate consequences require action. Those that do not may be able to wait. In any case, none can be ignored.

Responding to Material Defects

When defects result from degradation, prompt maintenance must be performed to restore the equipment to design specifications. Accordingly, an important tenet of sound maintenance is that material deficiencies must be recognized and reported in a timely manner so that they can be promptly repaired (or scheduled for repair when conditions permit). Primary responsibility for that action devolves upon equipment users.

Though some discrepancies can (and should) be corrected by operators at the time of discovery, others, for want of time, parts, or capability must wait. In those cases, deficiencies must be noted and reported for future scheduling and disposition. Therefore, in addition to knowledgeable technicians and leaders, an industrial complex must have in place a reliable and user-friendly system for reporting deficient conditions.

For deficiencies that can't be fixed immediately, most facilities employ equipment deficiency tags to record and report material deficiencies. The tags usually have a carbon copy and a cardboard tag. When a deficient condition is identified and verified as a valid deficiency, a deficiency tag is completed. The carbon copy is routed to the maintenance department and the cardboard copy hung on the equipment as near to the deficiency as safe and practical. The maintenance department can then locate and evaluate the deficiency, immediately correct it, or schedule it for repair.

For deficiencies that threaten equipment damage or personnel injury, prompt operator action is required. Such a condition usually constitutes an *equipment casualty* or *emergency*. We will learn more about them in the next chapter.

Responding to Human Performance Problems

The consequences of allowing deficient performance by team members and leaders are apt to be greater than a missed material deficiency. Such was the case in the grounding of the *Exxon-Valdez*. Accordingly, operating excellence requires mature team members and leaders who recognize that correction of performance errors is in everyone's best interest.

Human deficiencies have many causes and can display themselves in a variety of ways. They may result from inadequate training, lapse of attention, incomplete communication, or lack of the discipline necessary to dutifully perform an assigned task.

Unfortunately, deficiency tags don't work for human performance problems. They need to be addressed by wise leaders and colleagues soon after discovery. Usually, just a word from a respected co-worker or leader is sufficient. But ignoring performance problems is a sure invitation to future difficulties. The rule is, "Take care of it now".

Topic Summary and Questions to Consider

After the Air Florida Flight 90 accident, the NTSB investigating team concluded that:

The aircraft could not sustain flight because of the combined effects of airframe snow or ice contamination which degraded lift and increased drag and the lower than normal thrust set by reference to the erroneous EPR indications. Either condition alone should not have prevented continued flight.

A combination of circumstances, events, and operator actions (or omissions) led to the crash of N62AF and the great loss of life.

Further, this was a preventable accident. The Safety Board stated that:

Continuation of flight should have been possible immediately after stickshaker activation if appropriate pitch control had been used and maximum available thrust had been added. While the flightcrew did add appropriate pitch control, they did not add thrust in time to prevent impact.

The circumstances experienced by this flightcrew were not new or unique. A more vigilant crew could have regained control of the aircraft in time to avoid the crash.

From the Air Florida accident, it is again clear that the alert, well-trained operator is an essential element of systematic industrial operations. In order to effectively find and report abnormal equipment or operating conditions, team members must thoroughly know their equipment and its normal operating characteristics. Then, when abnormalities present themselves, prompt, intelligent action can be initiated.

Performing frequent inspection tours is a vital technique for finding abnormalities in their infant stages. At times, however, even the best and most experienced have questions about what is abnormal and what is not. When you find yourself in that dilemma, ASK SOMEBODY!

Ask Yourself Do your team members know what conditions are appropriate/normal? With various operations?

Do your operators know how to recognize abnormal indications? Can they corroborate abnormal indications using other instruments or methods? Will they pursue a "gut feeling" that something "isn't right" before proceeding with critical operations? (Frequently that "gut feeling" is an unconscious recognition of an abnormal indication or condition—the kind of recognition that only comes from excellent training, long experience, and thorough understanding of the equipment or process.)

Are deficiencies identified and scheduled for correction (or corrected on-the-spot)? Are deficient conditions allowed to persist? Has the abnormal become "normal"?

Do team members feel confident to point out deficiencies to their leaders? Are they *encouraged* to identify deficiencies?

Chapter 15

Combatting Emergencies and Casualties

Industrial crises are an unavoidable part of manufacturing and process operations. Though proper design and construction (in conjunction with disciplined operations and maintenance) can prevent most crises from occurring, material failure and human error are inevitable. Accordingly, industrial team members who plan and train for abnormal events are better prepared to intelligently respond to them, thereby, limiting their consequences.

The TMI-2 Accident, March 1979

In March of 1979, the commercial nuclear industry was shaken by a series of events at Reactor Plant Unit 2 of the Three-Mile Island nuclear generating station near Harrisburg, Pennsylvania. Due to an incorrect diagnosis of a steam generator loss-of-feed problem, a manageable equipment emergency soon progressed to a nearly unmanageable industrial casualty. A chain of equipment failures, maintenance errors, and incorrect operator responses combined to create the worst commercial nuclear power generation accident in U.S. history.

The loss of this one billion dollar nuclear generating plant raised serious questions about operators' ability to maintain control of complex equipment on a daily basis. There were no deaths or serious injuries; yet, the future of commercial nuclear energy production in the United States was placed in jeopardy.

Loss of Secondary Feed

At about 4:00 A.M. on Wednesday, March 28, the isolation valves on the main feedwater supply lines to TMI-2's two huge steam generators inadvertently closed, resulting in a loss of the normal path for makeup water supply. (Figure 15-1, Point 1.) Within seconds, the turbines driving Unit 2's 880 megawatt electric generators automatically tripped off line to lower power demand on the reactor core and to conserve steam in the steam generators. At the same time, the auxiliary feedwater pumps activated as designed to supply water to the steam generators from an alternate feed source. But, unknown to the control room operators, the auxiliary feedwater supply line isolation valves (Figure 15-1, Point 2) had been inadvertently left closed following a routine preventive maintenance conducted a few days earlier. Though the auxiliary feed pump indicating lights on the control panel assured the control room operators that the pumps were running, they were pumping against closed isolation valves. As a result, the steam generators, though supplying steam at reduced rates, were not receiving any makeup feedwater from either the main or auxiliary supplies.

Loss of steam generator feedwater supply is a serious occurrence in pressurized water reactor operations. If secondary feedwater is not quickly restored and steam demand lowered, the steam generators can boil dry in minutes.

Apart from thermal damage to the generators, integrity of the reactor core may also be threatened. Steam generators serve as heat sinks to accept and transfer energy produced in the reactor core and circulated through the primary cooling system. (See Figure 15-2.) By pumping primary coolant water (circulating through and in contact with the reactor core) through tubes in the steam generators, heat is transferred across the tubes into the water reservoir in the secondary (steam generator side) of the system. As a result, uncontaminated steam is produced and piped to steam turbine-driven, electrical generators to produce electrical energy for the commercial power grid.

Though other means of core cooling exist, steam production in the secondary is the greatest single source of heat removal from the core. Without ample cooling, reactor fuel can reach temperatures exceeding 5000°F, resulting in severe core damage. Core, or reactor fuel rod damage, must be avoided at all costs since the fuel itself becomes highly radioactive during operation. The fuel is clad in thin metal jackets. If the jackets are breached due to mechanical, chemical, or thermal stresses, highly radioactive particles are released into the primary coolant and circulated throughout the primary system. As a result, radiation levels to which operating personnel are exposed increase beyond acceptable limits. Worse, any subsequent breach in the primary coolant system will then cause release of radioactive gases and particles into the reactor containment building or, if improperly controlled, into the atmosphere.

Operators desperately needed to restore feedwater to Unit 2's steam generators; but, unaware of the position of the auxiliary feedwater isolation valves, they didn't understand why steam generator water levels continued to decrease. Deprived of both main and auxiliary makeup feedwater, these steam generators boiled dry in a minute and forty-five seconds.

Stuck Open Primary Relief

On this morning of March 28, the control room at TMI-2 suddenly became a very busy and perplexing place. Just 6 seconds after losing steam generator feed, a primary pressure relief valve (Figure 15-1, Point 3) actuated by rising primary coolant pressure, the result of reduced ability to transfer heat to the steam generators. In another 6 seconds, high pressure in the primary coolant system caused automatic, rapid insertion of the reactor control rods to shut down the reactor.

Primary relief valve actuation was, by itself, not a great concern. In a pressurized water reactor such as Unit 2, an overpressure situation causes a primary relief valve to lift (open), routing a small portion of the primary coolant mass (as pressurized steam) to a drain tank where it is condensed and cooled. Pressure relief usually lasts only a few seconds, long enough to decrease pressure to an acceptable level, at which time the pressure relief resets (closes).

In this instance, however, the sequence of events was abnormally complicated. Rather than shutting as designed, the pressure relief valve stuck in the open position. A stuck-open primary relief valve is a dangerous situation since primary coolant is the medium—flowing over the hot fuel rods—to remove reactor core heat. If too much coolant is lost and not replaced, the reactor core can be uncovered, resulting in overheating and disastrous thermal damage to the fuel rods.

Operators in Unit 2's control room recognized that the primary relief valve had lifted but believed that the valve had already reset since its associated control panel light indicated it to be shut. This er-

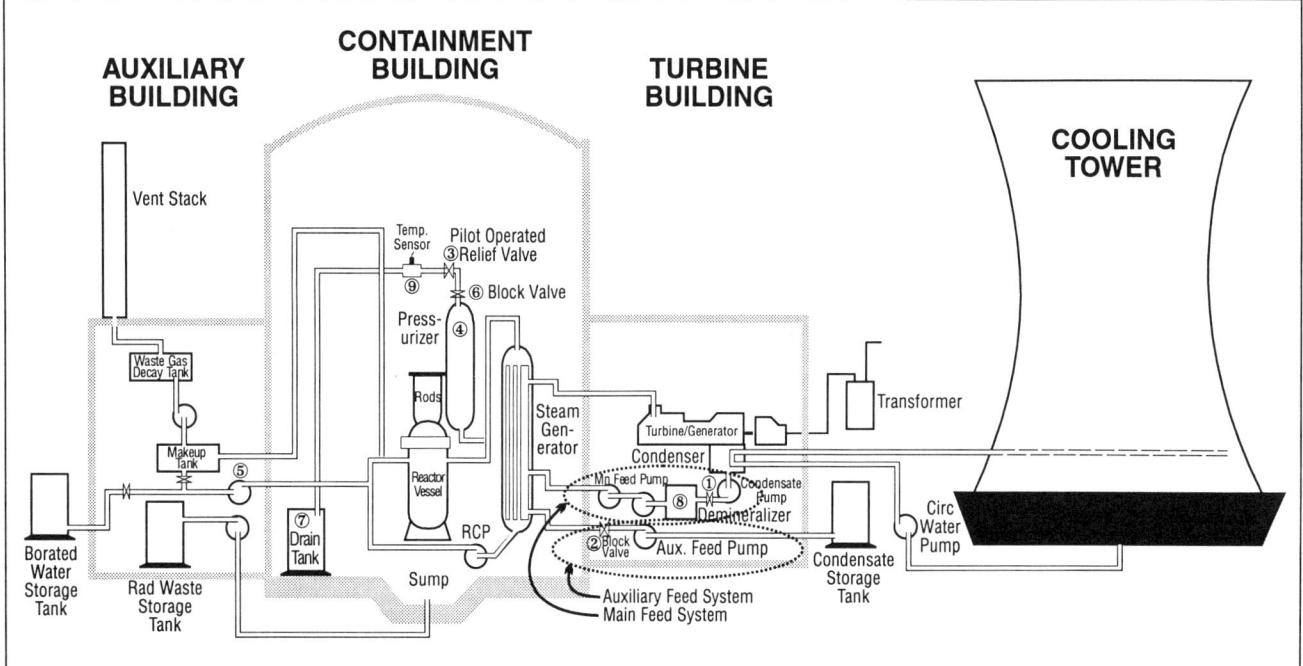

Figure 15-1: Simplified illustration of the Three-Mile Island TMI-2 nuclear generating plant.

ror was due to the design of the relief valve and its sensing/indicating circuitry. The valve and circuitry had been designed so that panel indicating lights would be actuated by the same electrical transducer signals that told the valve to open or shut. Though the signal had been sent to the relief to shut, it had stuck open. The control panel indicating light, however, had received the "shut" signal and, therefore, showed the valve to be shut.

Pressure in the primary coolant system continued to decrease. Control room operators knew that the primary system was losing coolant (and thus pressure) from some source. But, because the control panel indicated a shut relief, the possibility of an open relief valve was discounted.

"Rising" Pressurizer Level

Within 1 minute of the onset of the event, primary coolant pressurizer level indication began to rise rapidly, a confusing sign to the operators. The pressurizer is a surge tank used to maintain primary coolant volume and pressure. (Figure 15-1, Point 4.) Pressurizer coolant level is indirectly measured to provide the operators with an indication of water volume in the primary coolant system *under normal pressures and temperatures*. When primary coolant is lost, pressurizer level indication decreases if pressure and temperature are held relatively constant. Unfortunately, in this instance, conditions were not normal. Pressure was rapidly decreasing in Unit 2's primary system while, at the same time, temperature increased. Though the primary system was losing its mass of coolant, the volume of the remaining coolant was expanding rapidly, leading to an increasing (vice decreasing) pressurizer level. Operators did not recognize the correlation and believed the primary system had sufficient volume, even in the face of falling pressure.

Automatic Fill Initiation

In response to the low primary coolant system pressure (approximately 2 minutes from event onset), the reactor fill system automatically initiated to supply cooling water to the core. (Figure 15-1, Point 5.) The fill system used two high capacity, positive displacement injection pumps to force water into the primary system to keep the reactor core covered and cooled. Yet, control room operators, having already observed a rising pressurizer level, mistakenly believed that the primary system had an abundance of water. They soon shut off one of the reactor fill system pumps, isolated its outlet, and let the other reactor fill pump run with a throttled outlet valve. A few minutes later, they secured the second pump. All the while, the primary system continued to lose coolant.

Restoration of Feed

Approximately 8 minutes after the onset of the event, one control room operator, observing control panel indications, discovered that the auxiliary feed isolation valves were shut. Recognizing this as the cause of auxiliary feedwater supply failure, he rapidly opened them, causing a thermal shock to the generators. One of the generators sustained a severe leak while the other remained intact and could subsequently be used to remove residual heat from the primary coolant system.

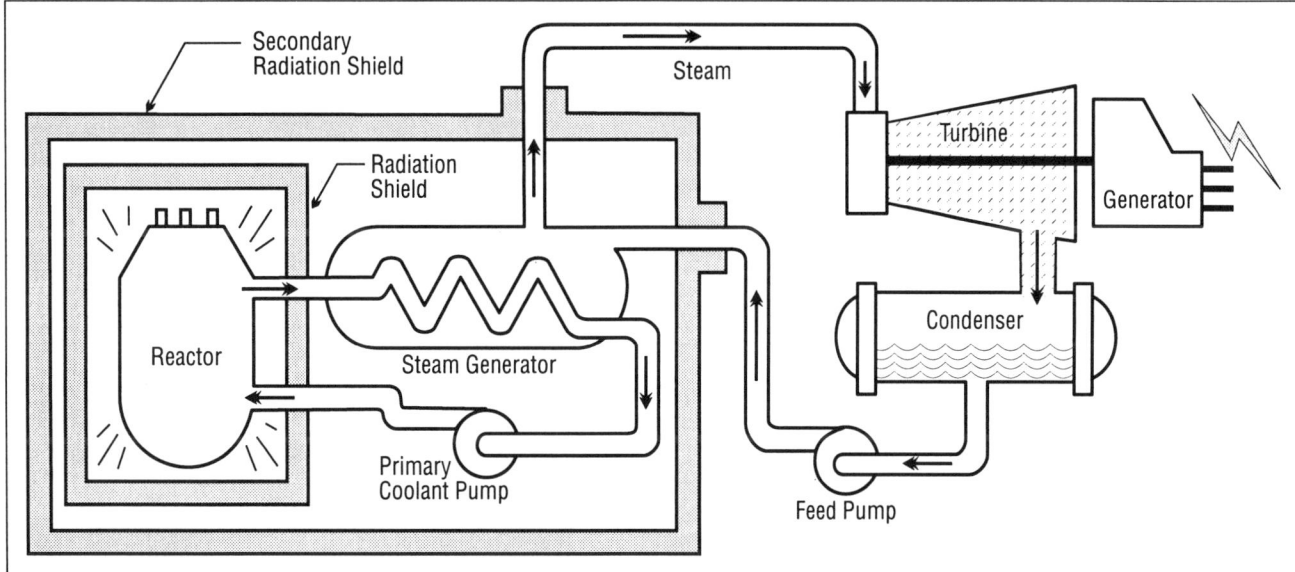

Figure 15-2: Typical Pressurized Water Reactor (PWR).

Combatting Emergencies and Casualties

Loss of Primary Flow

Seventeen minutes after automatic core shutdown, primary coolant pressure decreased to a point at which the water covering the reactor core began to boil. The loss of primary coolant continued and the tops of the hot reactor core fuel rods began to be uncovered, exposed to steam rather than pressurized coolant. A bubble was forming in the top region of the reactor core.

Up to this point, the reactor core remained intact even with reduced primary coolant volume and rapidly rising primary temperature. One reason was that operators continued to run the reactor coolant pumps, the normal driving force for circulating water through the primary system and around the hot fuel rods. Core cooling, though dramatically limited, continued.

As primary coolant pressure decreased, however, the reactor coolant pumps began to cavitate. (Cavitation is a process in which steam bubbles implode on the impeller vanes of a centrifugal pump. The steam bubbles form initially because the pressure of the coolant is too low for the temperature of the coolant. Cavitation can be very damaging to the impellers.) Responding to the pump cavitation, operators stopped the reactor coolant pumps at 1 hour, 15 minutes, and 1 hour, 40 minutes, respectively, from the accident sequence onset.

With no flow through the reactor core, a severe heat-up transient began. Within 3 hours of the event initiation, the reactor core would experience severe thermal damage from temperatures in excess of 4000°F. High core temperature readings, sensed by in-core thermocouples, warned operating personnel of the situation. Disbelieving their indications, they attached the thermocouple leads to auxiliary monitors which also read off-scale high. It didn't seem possible, so the temperature readings were judged to be in error.

Severe Core Damage

Meanwhile, the search for the source of loss of primary coolant continued. At about 6:20 A.M., an auxiliary operator was ordered to ensure that the primary relief isolation was shut (Figure 15-1, Point 6). Finding it open, he immediately shut it. Yet, rather than reporting that he had found it open and had then shut it, he simply reported it as "shut". This incomplete report would lead those combatting the casualty to believe for several more hours that the primary relief was not the source of the loss of coolant water from the primary system.

With the primary relief finally isolated, primary coolant pressure began to rise. Reactor coolant pumps were restarted to restore flow through the core. But because of the extraordinarily high core temperatures, the relatively cold water entering the core when the pumps started thermally shocked the embrittled fuel rods, causing them to slump to the bottom of the reactor vessel. Consequently, highly radioactive fission products were released from their fuel element cladding into the primary coolant system.

Release of Contaminants

As primary system pressure continued to rise, operators found it necessary to periodically re-open the primary relief. Unfortunately, with reactor core disintegration, each time the relief valve operated, radiological contaminants surged through the valve into the drain tank. (Figure 15-1, Point 7.) Normally, the drain tank would contain primary coolant routed from the primary relief valve; but, because of the continuous inflow of primary coolant from the stuck-open relief, the drain tank rupture disk failed 15 minutes into the event. Radioactive gases from the damaged fuel elements were released into the containment building. Eventually the tank overflowed, contaminating large areas of the reactor plant's containment building. Nevertheless, the containment building functioned as designed to check the spread of radiological contaminants into the surrounding atmosphere.

The Problems

Months would pass before analysts developed a thorough technical understanding of what occurred in the primary and secondary systems of TMI-2. Foremost among the problems to be addressed, however, was the issue of control. How had a relatively routine problem in a well-designed plant escalated into the worst commercial reactor accident in U.S. history?

Though serious, a loss of secondary feedwater supply should not have caused equipment damage if properly managed. Prompt restoration of steam generator feedwater from either the main or auxiliary source would have kept the steam generators filled. After discovering the cause of the event, the secondary system could then have been quickly restored. Even without steam generators to help cool the reactor, emergency water injection and core cooling systems were available to keep the core covered and to circulate cooling water through it. As long as the reactor core was shut down (control rods inserted), these systems had sufficient capacity to remove residual (decay) heat. What went wrong?

Extensive investigation of the accident determined that the foundation of the Three-Mile Island calamity, as in so many accidents, had been laid days and even months before the actual event. Poor maintenance practices, incomplete operator training, inadequate situational awareness, failure to follow procedures—all combined under non-conservative circumstances to create this disaster.

Incomplete System Restoration

On Monday, March 26, two days before the accident, a routine operating check was performed on Unit 2's auxiliary feed system. The test was designed to verify the automatic start feature of the auxiliary feedwater pumps in the event of loss of normal feedwater flow. In order to perform the check, the auxiliary feed system isolation valves had to be shut to prevent auxiliary feedwater from entering the steam generators when the auxiliary pumps started. (The steam generators were already being supplied by the main feedwater pumps). Warning tags were hung on the auxiliary feed system isolation valve actuating controls to indicate their closed position for the test.

Following a satisfactory check, the warning tags for the isolation valve controls were removed, but the valves were *not* repositioned to "open". The valves should have been opened at the same time as removal of the warning tags. The valve positions should then have been independently verified before completing this preventive maintenance check. With open auxiliary isolation valves, auxiliary feed water would automatically compensate for the lost normal feedwater supply on the morning of March 28, and the system could have been restored to normal operation in a controlled manner.

Poor Situational Awareness

Procedure required auxiliary feedwater isolation valves to remain open during normal operation. Indicating lights on the control panels for the supply line isolation valves both indicated that the valves were shut. One indicating light, however, was obscured by a warning tag on the control panel. Apparently, the status of neither of the lights was observed by any control room operator during numerous shift turnovers. Nor were they observed during routine operations even though operators were responsible to monitor the valve positions.

Faulty Maintenance

On Wednesday morning, in the hours prior to the accident, maintenance technicians elected to use compressed air in an attempt to clear an obstruction in a feedwater ion exchanger (the condensate polisher shown in Figure 15-1, Point 8). It is speculated that this technique forced water into the pneumatic air control system, causing a condensate pump to trip off line. The tripped condensate pump in turn sent a signal to the main feedwater control system, causing automatic closure of normal feedwater supply isolation valves. The auxiliary feedwater pumps started when their control circuitry told them that main feed had been lost. But, because the auxiliary isolation valves remained shut, auxiliary feedwater could not enter the steam generators.

Failure to Follow Procedure

Reactor core damage was, ultimately, the result of deprivation of cooling water to the core. The loss of primary coolant could have been stopped early in the event by isolating the stuck open primary relief. Immediate actions prescribed by procedure for such an event required performance of a sequential leak isolation process. One step in the procedure dictated isolation of the primary relief valve. As noted, that step was not accomplished until 2 hours and 20 minutes after the onset of the event. In the interim, the primary system lost a third of its coolant mass.

Masked Symptoms

Operators were misled by the control panel indication that the primary relief valve was shut. But the indicator light wasn't the only sign of an open relief. A stuck open relief valve could also be detected by a temperature sensor in the quench tank piping down-stream of the relief. (Figure 15-1, Point 9.) Steam entering the piping activates the sensor, causing an alarm in the control room. Unfortunately, this relief valve had a history of leakage, causing continuously high temperatures in the downstream piping. Further, operators thought, since the relief valve had just lifted, high temperature in the piping would remain higher for some time. The loss of primary coolant continued.

Combatting Emergencies and Casualties

Failure to Believe Indications

Downstream relief piping temperature wasn't the only indication that operators disregarded. As reactor core temperature increased, the in-core thermocouples provided operators with data indicating a severe temperature rise. Though thermocouple readings from two separate instruments confirmed very high temperatures in the reactor core, the instruments were considered to be in error. In retrospect, the instruments provided an accurate picture of the core temperature.

Incorrect Problem Diagnosis

Though the operators initially believed that the primary relief valve had closed, they still had indication of a primary coolant leak since pressure in the primary system continued to fall. Yet, since indicated pressurizer level was rising rather than falling, they believed incorrectly that the primary system had sufficient coolant. This faulty diagnosis led to a disastrously incorrect response.

Incorrect Response

One great irony of the TMI-2 accident is that, had the operating personnel done nothing, the primary system would still have been protected by the automatic features. Though the steam generators emptied quickly, the rise in reactor pressure shut down the reactor core by causing an automatic control rod insertion. Even with a stuck open primary relief valve, low coolant pressure caused the emergency core cooling system to start. The high pressure injection pumps actuated, injecting makeup cooling water into the primary system. Then, when the operators misinterpreted their indications, believing that the system was receiving too much cooling water, they secured the injection pumps. Had the pumps continued to run, there would have been little or no core damage.

Misplaced Priorities

Finally, core damage was guaranteed when operators secured the primary coolant circulating pumps because they began to cavitate. Not considered by the operators was the more significant damage that could occur if coolant circulation stopped. By securing circulation of the primary coolant, the core temperature was allowed to rise, creating a steam void in the upper portion of the fuel matrices. When the pumps were later restarted, the cold water shock essentially shattered the cladding of the embrittled fuel rods.

A Cautionary Note

At this point, you may be incredulous at the series of operating and maintenance mistakes that led to the TMI-2 event. A cautionary note, however, is in order. Though the response to this emergency-turned-casualty may seem straightforward in retrospect, you must consider the short and confusing timeframe in which these events occurred. Every step in the scenario—loss of feedwater, turbine-generator trip, primary coolant high pressure alarm, reactor shutdown, lifting of the primary relief valve, emergency injection initiation—occurred within the first 2 minutes of the event.

Far from being *user friendly*, the control panel indications and alarms at Unit 2 were more confusing than helpful. With the commencement of this event, over one hundred alarm indications were lit on the control panels at times. Silencing the audible alarms prevented the operators from determining what new alarms were actuating. In addition, a computer designed to assist the operators in analyzing reactor system functions and anomalies became so overloaded that key information and advice wasn't printed out for hours. The operating personnel were faced with a complex problem, the solution to which was not at all clear.

Responding to Industrial Crisis

The accident at TMI-2 stunned the commercial nuclear power industry. What began as a relatively routine feedwater emergency escalated within a few hours to a full-fledged core casualty. Incorrect diagnoses and erroneous operator actions changed a manageable situation to an unmanageable disaster. Though the general public and the environment had been protected, a $1 billion reactor plant was damaged beyond use, and another billion dollars would be spent in cleaning up after the accident.

Though already one of the most stringent and well-controlled of all U.S. industries, leaders in nuclear-powered energy production realized after the Three-Mile Island accident that even higher standards were necessary. Greater technical competence among operators, maintenance technicians, managers, and supervisors was needed. In particular, operating personnel had to improve their ability to respond to abnormal plant conditions.

Response to industrial crisis is a vital skill that can be taught to operating technicians and their leaders. It necessitates detailed initial training, thorough emergency planning, and rigorous practice. Therefore, the remainder of this chapter will explore the following topics:

- Emergency preparedness definitions,
- Planning for emergencies and casualties,
- Training for emergencies and casualties, and
- Learning from emergencies and casualties.

Emergency Preparedness

Emergency preparedness refers to the readiness of an individual or an organization to successfully combat or control abnormal conditions. It is the result of anticipating potential hazards (probable events capable of causing unacceptable damage), and then planning the means and the methods to counter them.

It requires the desire and ability to look beyond normal conditions in anticipation of future problems. It then demands the commitment to practice for events which might never transpire. Yet, it can never be achieved until the team members have first mastered the fundamental skills and knowledge necessary for routine operations.

Emergencies and Casualties

The term *emergency* is generally used to describe a crisis event which demands prompt action. The **American Heritage Dictionary**, Second College Edition, characterizes it as "an unexpected situation or sudden occurrence of a serious and urgent nature that demands immediate action".

Sometimes, though, this one term is too general to adequately characterize the severity or urgency of an event. In the field of emergency medicine, for example, first responders and care providers are confronted with a host of crisis situations. Some, if promptly and correctly treated, result in no significant, permanent damage to the patients. Other crises, however, are the result of trauma or illness in which damage has already been sustained. In the second case, regardless of what the care provider or first responder does, damage cannot be averted. It can only be limited.

Choking on food is a situation that illustrates the first category of crisis. If recognized and properly treated, choking will usually not result in serious or lasting injury to the victim. Choking, then, would well be characterized as an *emergency*.

In a serious automobile accident, on the other hand, the victims have already been injured. Even if the first responder recognizes and properly treats their injuries, they have already sustained physical damage. The automobile accident, then, would better be termed a *casualty* situation.

Emergency and casualty situations, then, can be differentiated by whether damage has already occurred or is likely to occur, regardless of the actions of the respondents. The immediate actions for emergencies are designed to avert damage. The immediate actions for casualties are designed to limit further damage.

Crises involving industrial equipment are often distinguished in the same manner. Some, if properly managed, will result in no damage to equipment, people, or environment. In this text, we define such a crisis in this category as an *equipment emergency*. Other mishaps, regardless of response, will result in significant, unavoidable damage. An event in this category shall be termed an *equipment casualty*.

In brief, then, an **equipment emergency** is an event that, if it occurs and is combatted by prompt immediate action, is unlikely to result in serious damage to operating personnel, the environment, or the general public. An **equipment casualty**, on the other hand, is an event that, if it occurs, will probably cause significant damage to the operating personnel, the environment, or the general public.

To illustrate the difference between an equipment emergency and an equipment casualty, consider for a moment your automobile. Were you to puncture the engine oil pan while driving on a difficult road, oil would begin to leak from the engine oil sump. The first symptom of trouble would probably be a low oil pressure light on the instrument panel. Without proper response, the sump would eventually drain, the oil pump would lose suction head, and lubricant would no longer be supplied to the rod and journal bearings. Significant damage would occur. Yet, with proper response (place

Combatting Emergencies and Casualties

the transmission in neutral, shut off the engine, and brake the vehicle slowly to a stop), no further damage would be likely. This situation would, therefore, be categorized as an equipment emergency.

On the other hand, if a connecting rod failed, you would be experiencing an equipment casualty. The symptoms would probably be a loud "bang" followed by a clatter of metal in the engine. The immediate actions for this event are the same as for a punctured oil pan. But major damage has already occurred.

It is apparent that, just as for human casualties and emergencies, the immediate actions for equipment casualties and emergencies also differ. Though both require prompt attention, the immediate actions for equipment emergencies are designed to *preclude significant damage* if promptly applied. Immediate actions for equipment casualties, however, cannot prevent serious damage. They are designed to *limit* the resulting damage.

Planning for Emergencies and Casualties

Investigation of the TMI-2 event disclosed flaws in design, operation, and maintenance; but the most serious concern focused on the performance of the control room operators on duty at the time of the accident. Experienced, intensively trained operators failed to recognize important equipment performance abnormalities. Not only did they fail to identify the emergency in progress, their subsequent actions allowed the situation to dramatically deteriorate until it became an equipment casualty. Primary system coolant volume was misdiagnosed because of a failure to recognize prevailing temperature and pressure conditions in the primary system. Reactor fill pumps and reactor coolant pumps were turned off or their output throttled when they should have been allowed to run without intervention. In summary, planning and training for plausible emergencies and casualties had been inadequate.

Determining the Hazards

The first step in the emergency planning process is identification of the hazards to which operating personnel, the environment, and the general public could potentially be subjected as a result of facility operation. The question to be answered is, "To what probable situations should we be prepared to respond?"

Fortunately, in well-designed facilities, the design process includes just such an analysis for the systems and components. By postulating design-basis accidents (and the mechanisms that could cause them), design engineers are able to establish the protective features and the operating/maintenance boundaries of the facility's safe operating envelope. These same plausible situations are an excellent starting point for emergency planning.

To the design-basis accidents must be added other casualties and emergencies which operating experience has demonstrated are likely. Operating and maintenance histories of plants with similar design and mission are the best source for identifying probable emergencies and casualties.

Further, the planning process must continue throughout the life of the facility. Ideally, each potential equipment crisis should be foreseen and analyzed at the beginning of the life of the complex. Yet, because of unforeseen factors and design modifications, continuous review of potential problems must be performed.

Planning the Response

Once significant hazards have been identified (and the resulting event mechanisms by which the events could occur), facility personnel must then plan the means to combat each event. Part of this process is to develop emergency procedures that provide response guidance to operating personnel. Perhaps the most important advice for developing such guidance is to ensure that the end users of the procedures are consulted during development.

Resource planning for emergency and casualty response is also a high priority in the planning process. Resources must be identified, budgeted, and procured. Further, the plan must provide for storing these resources so they are immediately available, maintaining them in a ready-for-use condition, and training personnel in how to use them.

Finally, on-site resources alone are usually insufficient. Mutual aid agreements with organizations such as police and fire departments, bomb squads, hazardous materials response units, and highway departments must be established.

Training for Emergencies and Casualties

Unpracticed procedures are almost worthless in an emergency. Events often develop so rapidly that, without prior mastery of the procedure steps, the procedure will be of little practical use. Emergency response training and testing is, therefore, necessary to evaluate the skills of the operating personnel, the resources staged for such events, and the procedures that have been developed to manage them.

Arguments Against Planning and Training

It is difficult, inconvenient, and expensive to plan for emergencies, acquire and stage the necessary emergency response equipment, and train team members in the tasks and skills necessary to avoid or mitigate damage. Emergency training must necessarily involve many, if not all, of the work groups within an industrial complex. Emergency drills affect production schedules, interrupt jobs, and disrupt normal activities. As a result, they are frequently avoided.

Here are some of the arguments against planning and training juxtaposed against their rebuttals:

1. One rationale for avoiding emergency planning and training is that they are too expensive. Why expend staffing and financial resources for something that might never occur? Such an argument fails for at least two reasons. The first is that most organizations are confronted with serious operational irregularities far more frequently than expected. The injury or death of a single employee often costs more than the preparedness process would require. The second is that emergency preparation and its attendant training, when properly performed, dramatically improve the awareness and ability of employees to intelligently analyze emergent conditions.

2. Another reason advanced for neglect of planning and training is that it is fruitless to train for contingencies since no organization is likely to identify all of the adverse situations with which it might reasonably be faced. It is certain that many unanticipated anomalies will arise and that emergencies usually don't follow the script. Through the emergency preparedness process, however, the most important factor, the ability of the operator to intelligently respond to anomalous occurrences, is sharpened. Also, though the actual emergency may be different from the planned emergency, it will usually be similar enough that the fundamental immediate corrective actions and the principles involved in those actions will be the same as those practiced.

3. A third argument often employed against emergency planning and training is that there isn't time to practice for emergencies. Yet, the best of organizations regard emergency preparedness as so important that to neglect it is unacceptable.

Certainly, the argument here is not for zero risk. Planning cannot be omniscient. Rather, the goal is *reasonable risk-minimization* through wise planning and intelligent use of resources. Mature organizations have learned, however, that the consequences and expenses of not planning and practicing for casualties and emergencies are, ultimately, greater than proper planning and training.

Guidelines for Emergency and Casualty Training

Emergency preparedness requires not only suitable initial training, but also ongoing classroom, seminar, and practical drill training to be effective. Most important are the practice sessions in which operating crews and their supporting elements are subjected to planned, realistic casualty drills followed by a thorough evaluation of performance.

Few organizations have developed casualty and emergency drill training to the level of the U.S. Naval Nuclear Power Program. Created by the temperamental genius, Admiral Hyman G. Rickover, this highly successful program is founded, in large part, upon rigorous drill training. After the Three-Mile Island accident in 1979, Admiral Rickover was called to testify before Congress regarding the principles upon which his program was based. He presented seven sound guidelines for developing a pragmatic emergency and casualty drill training program. His practical advice provides to every industry a standard for preparing operating teams for emergency and casualty response.

Combatting Emergencies and Casualties

1. Practice on sound theory. Emergency and casualty training won't work unless operating personnel thoroughly understand their equipment. Sound theory training must precede (and accompany) practical drill training. Rickover:

> *We train our people in theory because you can never postulate every accident that might happen.... [T]he only real safety you have is each operator having a theoretical and practical knowledge of the plant so he can react in any emergency.*

2. Practice often. Even if theory training is well-done, preparedness may fail in execution through inadequate practice. Emergency and casualty training drills must, therefore, be incorporated with regularity into the facility training plan. In fact, drill training should occur frequently, usually on a weekly basis. Rickover:

> *In addition to classroom type training, the recurring training program is also composed of practical evolutions and casualty drills. These form an important part of the shipboard [facility] training plan, allowing the...plant operator to build on his theoretical knowledge of the...plant and put into practice the principles of operating and casualty procedures he has studied.... The actual casualty drill may be pre-announced or may be a surprise to the watch section [operating crew]. The Engineer Officer [Operations Manager] will normally make this determination. Some combination of both methods is appropriate to ensure that the watchstanders [crew members] can properly handle unexpected plant casualties.*

3. Practice to perfection. When mistakes occur during drill practices, they must be corrected immediately. If necessary, the drill should be stopped, the mistakes discussed, and the proper responses delineated. Then, if possible, the drill should be reperformed on-the-spot. If time precludes repetition, the drill should be rescheduled for the near future.

In all cases, skilled drill coaches should be on the scene making immediate corrections. Corrections are more effective when they occur closer in time to the mistaken behavior. Rickover:

> *Poorly conducted casualty drill training, which allows improper actions to occur without identification and correction, simply reinforces the wrong way to do things in the...plant. In effect, we could train ourselves to operate the plant in an unsatisfactory fashion. During drills, monitors correct watchstander [crew member] errors on the spot, where failure to do so would reinforce improper actions.*

4. Practice by a plan. For casualty and emergency plans to be effectively, realistically, and safely run, they must be prepared in advance with great care. Rickover:

> *...I insist that casualty drills be carefully planned, closely monitored and thoroughly critiqued.... I will describe some of the considerations that are involved in the conduct of casualty drills on a nuclear ship [or an industrial facility]. First, a drill guide is prepared which describes the drill, how it will be initiated, what is to be accomplished.... [It] specifies safety monitors and observers.... Various... plant reference material[s]...are used. The Engineer Officer [Operations Manager] then submits this drill guide to the ship's [plant's] Commanding Officer [Plant Manager] for his approval. A file of these approved drill guides is maintained for recurring use. The Commanding Officer [Plant Manager] must approve the actual conduct of each drill even though he has previously approved the basic drill guide. Sometimes the watch section [operating crew] scheduled for a particular drill will be notified well in advance of the nature of the drill in order that specific training, such as a review of the appropriate casualty procedures, may be accomplished. This may be appropriate where the section [crew] will be doing a difficult drill for the first time or where the ship [plant] has just completed a lengthy period with the plant shutdown.*

5. Practice safely. During emergency and casualty drills, operating personnel are presented unusual situations which require prompt, decisive action. Without proper planning and control, drills can result in dangerous conditions. In order to be realistic, drills sometimes require unusual plant component and system configurations. Standby equipment (equipment that would normally start automatically upon loss of operating equipment) might be secured before the drill without informing on-duty operating personnel to establish realistic conditions. Therefore, qualified drill monitors and safety observers, must be briefed before the drill, must be controlled by a qualified senior drill observer, and must be on scene to stop unsafe actions and to ensure that equipment is not operated in an unsafe manner. Rickover:

Drill monitors and safety observers must be fully aware of what is expected of them and the limits to their responsibilities. This is accomplished at a briefing attended by all monitors and safety observers and normally led by the Engineering Officer [Operations Manager]. I consider it appropriate that the ship's [facility's] Commanding Officer [Plant Manager] or Executive Officer [Deputy Plant Manager or Operations Manager] be present at these briefings to the maximum extent possible. An important aspect of this session is to review in detail how the drill will be initiated and how the symptoms of the casualty will be made known to the watchstanders [crew members] in cases where the entire casualty cannot be allowed to occur because of reactor or ship [plant] safety.... Safety monitors are stationed to prevent incorrect watchstander [crew member] action which could hazard the...plant. Drills are allowed to progress long enough to evaluate the section's ability to restore the plant to its normal condition. Obviously there are practical limits to drill length and in some cases the first watch section [crew] will carry out the initial casualty actions and a second section will recover the plant back to a normal condition.

6. Practice realistically. Emergency and casualty drills lose effectiveness if they are not realistic. Crew members view them as a sham. Therefore, every effort must be made to incorporate realism within the bounds of safety. Imaginative simulation techniques developed by the team of drill and safety monitors will add immeasurable to the success of drills. Rickover:

Realism in the conduct of casualty drills is important, but safety considerations dictate that some casualties should not actually be done for training.... Within the constraints of reactor and ship [plant] safety, a conscious effort is made to carry out these casualty drills in a realistic manner.... Therefore, we use techniques for presenting the symptoms of these casualties in a manner that will, as nearly as practicable, appeal to the same senses that the watchstander [crew member] would normally use in the casualty situation. During this pre-drill briefing the applicable casualty procedures are also reviewed to ensure that all monitors and safety observers know the correct watchstander [crew member] actions.

7. Analyze the practice. Drill training is nearly worthless if operating team members and their leaders don't receive prompt feedback on their performance. So, besides on-the-spot coaching by monitors while drills are in progress, a post-drill performance analysis is necessary. The post drill analysis includes a critique of team members and the team as a whole for each drill. Also, the evaluation must provide recommendations (and assistance) for improvement.

A complete post-drill evaluation has four parts: (1) post-drill monitor meeting, (2) crew performance debriefing, (3) evaluation of drill monitor and safety observer performance, and (4) dissemination of lessons learned to other plant personnel.

The post-drill monitor meeting is designed to allow the drill administration team to form an overview of operating team performance. By the end of the drill session, it is unlikely that any one drill monitor or safety observer will have a thorough understanding of crew performance. Only by sharing observations can the drill team monitors and observers properly analyze performance and render a valid evaluation. Rickover:

Upon completion of the drill, a critique involving all drill monitors is immediately held to collect comments, determine where errors were made and evaluate the overall conduct of the drill. Appropriate reference material such as the operating manuals for the...plant are essential at this session to accurately assess all of the casualty actions taken.

Next, results of the evaluation must be provided to leaders and key members of the evaluated team. Therefore, a debriefing of the operating team must be conducted quickly following the drill session. Rickover:

After the Engineer Officer [Operations Manager] has assembled the significant comments from the monitor critique he conducts a critique of the drill for the watch section [operating team] after they come off watch.

If the whole operating team cannot be assembled, at least the crew leaders should be debriefed. Then, during another meeting (soon after the drill session), crew leaders can debrief the remaining team members. In any case, the verbal evaluation should be followed expeditiously with a written report of performance directed to the evaluated team as a guide for improvement.

Combatting Emergencies and Casualties

The third significant part of the post-drill evaluation process is for drill monitors and safety observers to analyze their own performance in planning and conducting the drill session. This should include an assessment of how the drill guides can be improved for future use.

Finally, some weaknesses observed during individual drill sessions may be applicable to others within the plant. Therefore, the lessons must be disseminated to other personnel by someone such as the Operations Manager. Rickover:

If training lessons are to be learned that would benefit other engineering department [plant] personnel, the Engineer Officer [Operations Manager] will cause this information to be disseminated. Finally, where drill deficiencies show weaknesses in the ship's [plant's] fundamental training program, corrective measures are taken to upgrade these areas.

It is clear from the foregoing steps that emergency planning and training cannot be a haphazard process. It requires the selection and training of a skilled drill team, extensive planning, and uncommon imagination. Yet, those organizations that invest the time and resources to develop a rigorous emergency and casualty training process improve far more than just the crisis preparedness of their team members. They also instill in them higher standards, greater awareness of the operating equipment, and a sense of teamwork—characteristics which will prove invaluable.

Responding to Emergencies and Casualties

No matter how well you train your team in emergency and casualty response, you will find that, when the events actually occur, they differ in some aspect than from the practice sessions. Remember, as Winston Churchill wrote:

In the problems which the Almighty sets his humble servants things hardly ever happen the same way twice over, or if they seem to do so there is some variant which stultifies undue generalisation.

Yet, though the crises differ in form, skilled operating personnel will probably manage each one successfully if they understand their own equipment, the nature of the crisis, the fundamental remedy for the emergency or casualty with which they are faced, and *the steps in the process by which any casualty or emergency should be combatted.*

There are nine fundamental steps that should guide the actions of any operator who is responding to an equipment emergency or casualty. They are:

1. Recognition of the symptoms of an emergency or casualty in progress
2. Initiation of the immediate actions prescribed to combat the event
3. Notification of plant leaders and support groups of the crisis in progress
4. Corroboration of the symptoms through use of other indications
5. Verification of system response to the immediate actions
6. Periodic reporting of the status of the crisis and of important equipment
7. Initiation of follow-up actions prescribed for the casualty or emergency
8. Determination of cause of the crisis
9. Recording the event and important steps taken to combat it

Symptom Recognition

Before an operator can successfully respond to an equipment emergency or casualty, he or she must first recognize the symptoms of the impending or ongoing crisis. In emergency medicine, first responders are taught symptom recognition as the initial step in effective, timely intervention. The same principle applies in industry. Untimely recognition yields untimely response. Operators must, therefore, be thoroughly educated in the symptoms associated with facility problems to which they are expected to respond.

Immediate Action Initiation

Having recognized the symptoms of an impending or ongoing equipment crisis, the operator must swiftly initiate the immediate actions prescribed to combat the emergency or casualty. **Immediate actions** are those steps necessary to place the equipment in a safe and stable condition. As you will recall, immediate actions for emergencies are designed to prevent damage. Immediate actions for casualties are designed to limit the damage.

Timely initiation of immediate actions for an equipment crisis can make a critical difference in the outcome. Just as at Three-Mile Island, a relatively routine emergency can soon become an irreversible casualty without proper response. Timely response is so important that, for some events, facility safety analyses often prescribe a maximum allowable time for an operator to observe the symptoms of a crisis and respond to it. If that time limit is exceeded, the safety analysis may be invalidated.

At the onset of an equipment crisis, there may be little or no time available to consult a procedure for immediate action steps. Therefore, the steps must be known from memory. Memorizing immediate action steps is often a point of contention. Remember, however, that you don't want an emergency medical technician to have to pull out the CPR manual if you're having a heart attack. At the earliest opportunity during the emergency or casualty, however, an operator should consult the procedure to ensure that no immediate action steps have been missed.

Procedures can't address *all* emergency or casualty situations that operating personnel are liable to confront. That's why thorough equipment understanding is so important. If no procedure is available, operating personnel must be taught to exercise their best engineering judgment to combat the crisis.

Finally, operating personnel must be taught to avoid delays in response. Immediate actions must be initiated based upon the *most conservative* indications. At Three-Mile Island, thermocouples indicated extremely high core temperatures. Operating personnel discounted them because they seemed implausible. In summary, operators must be taught to believe their indications.

Notification

Frequently, industrial casualties or emergencies require a team response. Notification, therefore, is an important step in the response process. Without timely notification, a crisis may unnecessarily escalate.

Notification of an emergency or casualty in progress may occur immediately upon symptom recognition or may have to wait until the initial immediate action steps have been completed. Nevertheless, early in the process, responding personnel must report the emergency or casualty to the central control room or team leader. Rendering a prompt report is vital if other resources are to be mustered in a timely fashion.

Once the initial report has been rendered, an announcement to the remainder of the equipment operating stations (usually via the facility public address system) should be made. A prompt report to the central control room allows team leaders to announce the existence of the emergency or casualty to other plant members. Others are, as a result, placed in an alerted status with heightened awareness toward plant conditions that might be affected by the crisis.

Though some crises may not require a plant-wide alert, most should be announced to all other stations. Events tend to expand in their effect. Early announcement prepares all stations to respond if necessary. Delay in announcing an event deprives others of advance information which may be of assistance in prompt response.

Symptom Corroboration

Once immediate actions have been initiated and others have been notified of the crisis, responding personnel should attempt to verify or corroborate the symptoms that triggered the response. For most problems, other instrumentation or indications can be checked to verify or refute the abnormality. If the indications conflict, a third source of information should be sought.

If indications confirm an equipment emergency or casualty in progress, the immediate action steps should be continued and completed. If corroborating indicators clearly show that the suspected casualty or emergency is not occurring, the immediate action steps can usually be terminated.

Sometimes, the nature of an emergency or casualty allows symptom confirmation *before* initiating immediate action steps. Responding operators must wisely determine whether it is safe to delay response steps until after indications are corroborated.

Verification of System Response

Throughout the emergency response process, operating personnel should perform periodic instrumentation checks to verify that key parameters are responding as expected and that immediate actions are having the desired effect. Important instrument readings should be recorded as required in the narrative logs or data record sheets for the affected equipment.

Combatting Emergencies and Casualties

Status Reporting Periodically during the crisis, operating personnel should inform the central control room or the team leader of the status of combatting the event. Likewise, the control room or team leader should inform other involved stations of the progress of combatting the emergency or casualty.

Follow-up Actions Once the immediate action steps for an on-going emergency or casualty have been completed, the follow-up actions must be initiated. In the case of an equipment casualty, the follow-up actions are designed to stabilize the processes and the equipment so that the cause of the casualty can be determined, the damage assessed, and an engineering judgment rendered regarding the steps for recovery. Recovery from an equipment casualty may be an extended affair, depending on the damage incurred. For equipment emergencies, the follow-up actions are normally designed to determine the cause of the emergency, correct the cause, and recover the equipment and processes to normal operation.

Determining the Cause Cause of an emergency or casualty must be determined to prevent unnecessary recurrence of the event. Cause determination may require additional follow-up actions, diagnostic equipment testing, operator interviews, or, in severe instances, an accident investigation.

Recording the Event Operating personnel should log the important events of an emergency or casualty chronologically. Log entries should include the observed symptoms of the event, the immediate and follow-up actions performed to combat the event, the discovered cause of the event, the results of execution of the immediate and follow-up actions, and the final status of the equipment.

Combatting an emergency or casualty may preclude timely recording of the symptoms and key response actions. If necessary, a log-keeping assistant may be stationed to record for the station operator. The assigned operator, however, should review the entries and make corrections if required before signing the logs. Another option is for the station operator to keep a set of scratch notes and then record log entries after the emergency or casualty has been brought under control.

Failure to log the cause of an event is a frequent problem. Once cause has been determined, it should be entered in the narrative logs for the station on which the equipment resides. If properly logged, other operators occupying the station will be apprised of the event, how it was caused, and how it was solved.

Learning from Emergencies and Casualties

One of the most important aspects of emergency preparedness is the process of analyzing abnormal events. Through the study of accidents and *close calls* (events which, under different circumstances, might have become accidents), operating teams are more likely to avoid similar problems.

But, for accident or close call studies to be successful, five important elements are necessary: (1) a motivation to study, (2) a source of relevant study material, (3) a time to study, (4) a method of study, and (5) an effective study leader. If any of the five elements is missing, the educational process is likely to fail.

Why Study? In 1977, two years before the Three-Mile Island accident, an electromatic primary relief valve (the same kind as at TMI-2) stuck open at the Davis-Besse plant. Fortunately, the reactor was at low power and recovery was not difficult. After the Davis-Besse event, engineers postulated the consequences of a similar occurrence with a reactor at high power—a scenario very much like the actual TMI-2 event. Unfortunately, the operators at Three-Mile Island Unit 2 didn't assimilate the lessons from the Davis-Besse event.

Abnormal event studies, when properly performed, force operating personnel to mentally assess the effect of equipment crisis on other operating systems and components for which they are responsible. Further, these "what if" scenarios compel operators to envision the actions necessary to combat such circumstances.

The benefits of abnormal event studies are threefold. First, by mentally analyzing potential crises, operators create a rational framework for responding when the postulated crisis (or a similar one) actually occurs. As a result, a confident, analytical approach is far more likely than a panic response.

Second, by thinking through potential equipment crises, operators often identify probable events for which no procedure exists and for which no training has been conducted. Consequently, abnormal event studies become an excellent source of emergency response planning and training. Third, by analyzing the effect of potential equipment crises, operators increase their understanding of the components and equipment that they control.

What to Study

The most obvious and accessible sources of abnormal event study information are the analyses of an operating team's own accidents and close calls. A properly investigated and well-reported event provides a made-to-order forum for abnormal event training for the operating crews of an entire facility. Since operating personnel are, ostensibly, already familiar with the technology and the equipment involved, the event scenario can be easily presented. The bulk of the training period can, therefore, be devoted to posing and answering questions regarding proper response.

Unfortunately, far too may organizations fail to thoroughly investigate events that lead to equipment damage or personnel injury. As a result, they squander valuable opportunities to learn lessons that can avert future crisis.

Event reports from similar industries provide another useful source of information for abnormal event studies. Usually, however, they are more difficult to acquire. While accounts of events from organizations within the same corporate structure are usually circulated for study, descriptions and analyses of events from competitors are shielded. No one likes to air dirty laundry. Some events, though, are so important that they are discussed in trade journals or are published by federal or state agencies after their investigations.

A third and vital source of abnormal event information comes from dissimilar industries. Industrial leaders sometimes err in believing that if a technology is not similar to their own, the accounts of accidents or near accidents hold no lessons for them. But anyone who studies accidents soon realizes that the technology under investigation is secondary to the mechanism of failure. A further benefit is that operators who are exposed to the mistakes of other industries acquire better sensitivity to the processes of risk analysis and risk avoidance.

Perhaps the best place to begin a search for abnormal event information is the aviation industry. Aviators, both general and commercial, have for decades shared stories of accidents and close calls within their own ranks. Though they are probably as a group the safest among equipment operators, they have learned that continuous "what if" study is an important element in improving their skills and, subsequently, their chances of survival. Aviators also seem to be the best at studying and publishing their errors. The process is evidenced by the Federal Aviation Administration's Aviation Safety Reporting System and its associated monthly publication, **Callback**. Equally as useful are publications such as **Aviation Safety** and the *Aftermath* section of **Flying** magazine.

One of the best sources, not only for aviation accidents but for transportation accidents of all kinds, are the accident analyses performed and published by the National Transportation Safety Board. These in-depth reports provide thorough accident descriptions, excellent analyses, and statements of probable cause.

The sources are almost limitless. The **Operating Experience Weekly Summaries** published by the Department of Energy outline abnormal events at DOE facilities across the nation. The **NUREG** publications of the Nuclear Regulatory Commission provide extensive information about abnormal events in commercial nuclear power generation. Finally, don't overlook publications from organizations such as OSHA and the National Fire Protection Association.

When to Study

Without time set aside for operating crews to study, all of the abnormal event research in the world will not pay off. Therefore, abnormal event study, just as with drill training, must be a planned, structured process. Approximately once every shift cycle (or once a month) each operating crew should have an opportunity to analyze an interesting event from their own or another industry and extract the relevant lessons. For some events, it will then be appropriate to devise and administer a drill based upon the event studied.

How to Study At times, the best method of disseminating abnormal event descriptions and lessons is through lecture. But, more often, small group seminars provide the best forum for study. Working in small groups, individuals can more easily participate. Further, it is through small group discussion that problems which might otherwise be overlooked are surfaced and addressed. Such small group seminars can also be used effectively to draft the drill scenarios which, when reviewed and approved, may be used to test the readiness of operating personnel.

Who Should Lead the Study The question of who should lead the abnormal event studies is critical. Normally, no one person will be designated to lead all such training for a facility or even for a single crew within the facility. Whoever is chosen, however, must have credibility, reasonable instructor skills, and in-depth knowledge of the technology to which the case studies apply.

Although some events can be instructed to each crew by someone on the crew, other events should be managed by a single plant instructor. In fact, some events are of such importance that leaders such as plant managers and operations managers should lead the instruction. But, regardless of who leads the instruction, it must be interesting and applicable or the process will quickly lose credibility and effectiveness.

Topic Summary and Questions to Consider

Napoleon Bonaparte, in speaking of combat, once said:

If I always appear prepared, it is because, before entering on an undertaking, I have meditated for long and have foreseen what may occur. It is not genius which reveals to me suddenly and secretly what I should do in circumstances unexpected by others; it is thought and meditation.

For Napoleon, emergency preparedness was a necessity for a seasoned fighting unit. It is equally important for an industrial team. Is your team prepared?

Ask Yourself Do you have a bank of approved emergency and casualty drill procedures? Are they used when you conduct drills? Do you devise new drills to challenge your teams?

Are emergency and casualty drill procedures regularly reviewed and updated?

Do you conduct drills at least once per month? Are your drills realistic? Are senior, technically knowledgeable drill observers in position and ready to stop unsafe actions?

Does your drill team debrief the operating team and prescribe or recommend corrective actions?

If your continuing training program doesn't include effective drill training, your team probably isn't ready to respond to actual emergencies and casualties.

Chapter 16

Overseeing Maintenance, Modification, and Testing

Controlling operating equipment is only one of the responsibilities of operating personnel. Another is to ensure that work on installed equipment is (or has been) correctly performed.

In Chapter 3, we learned that reliable equipment and facilities constitute one of the requisite elements of a successful operating strategy. We also learned that equipment can't remain reliable unless properly maintained and, when necessary, intelligently modified.

Since, in the final analysis, operators ultimately bear responsibility for the performance of their machinery, they fulfill a vital role in ensuring that equipment, once maintained or modified, is restored to safe and efficient service. Accordingly, operating personnel must understand how to monitor equipment maintenance, modification, and testing.

Maintenance, Modification, or Testing?

Sometimes, the definitions of maintenance, modification, and testing are confused. To understand the operator's role in each, we must distinguish between their meanings.

Maintenance Maintenance refers to the restoration of components and systems to original design specifications. If the deficiency is simple, maintenance can sometimes be performed by the assigned machinery operator. If complex, maintenance may require higher level assistance, perhaps even removal of the component or system from the plant, and shipment to the original vendor for repair.

Modification Modification, on the other hand, implies a change to original equipment design. As with maintenance, it may be simple or complex. But, in most instances, modification also requires an engineering study in advance of performance to ensure that personnel and equipment protection is maintained. Because of the potential changes to original design, equipment modification requires special controls.

Testing In contrast to maintenance and modification, equipment **tests** are controlled checks designed to diagnose problems or to verify operating functions. Equipment testing may be performed for a variety of reasons. Following most maintenance and modification, tests are performed to ensure that equipment has been restored to a safe operating configuration. Tests may also be performed to explore performance problems or to examine end-of-life characteristics.

Who is Responsible?

Whether the function is maintenance, modification, or testing, you can be sure that one debate will always arise: Who is responsible for the work? Operations, maintenance, or the test group?

The Danger of Dilution The answer is a resounding "Yes!" Everyone involved bears responsibility. But, in a stricter sense, final responsibility for safe equipment operation always resides, like it or not, with the operating team and its leader.

Responsibility and its inseparable counterpart, accountability, motivate people to achieve high standards of performance. Without them, "teams" become individuals and performance degrades. Somebody has to "stop the buck".

In a team environment, especially where there are many players and many leaders, responsibility can be dangerously diluted and accountability lost. Admiral Rickover often advised his subordinates that:

> *The practice of having shared responsibility really means that no one is really responsible. Unless you can point your finger at the one person who is responsible when something goes wrong, then you have never had anyone really responsible. That is the crucial test of responsibility.*

As in every team operation, whether sports, commercial business, or military endeavor, the question of who is responsible and who is in charge must be settled before the team can progress.

"Command Responsibility"

For most situations in an industrial environment, the plant or activity manager is *the* one person who, in the end, is responsible. The plant manager is the chief operator, chief engineer, chief maintenance technician, and the chief trainer.

In military organizations, this concept is called *command responsibility*. Every commander is held responsible for the performance of his or her subordinates, even in the absence of direct control.

Such responsibility emphasizes the need to train subordinates so that, if lines of control are slackened or cut, performance continues unabated. In the early years of the War Between the States, Lee's Virginians, though overmatched in numbers and equipment, frequently bested the Army of the Potomac under generals such as McClellan. One of the primary reasons was that McClellan's officers were afforded little flexibility. They always had to wait for the next order. In contrast, Lee's officers were taught to act even without direct orders when necessary, so long as the actions taken were within the scope of the battle plan. The Confederate Army rewarded flexibility, whereas, the Union Army did not. As a result, Confederate leaders were better able to mass their troops at critical junctures on the battlefield, one of Napoleon's maxims of war. They actually achieved superior strength at critical times, despite their inferior numbers.

The practice of assigning overall responsibility to the chief line leader is age-old. In maritime operations, a ship's master (captain) bears ultimate responsibility for the safety of the ship, its crew, and its cargo. The performance of the *Exxon-Valdez* master is a prominent example. Though the third mate was the deck watch officer on duty at the time of the accident, the master bears final responsibility for whatever happened (or failed to happen) aboard his ship, even though he wasn't on the bridge when the vessel grounded. The same is true for Captain E.J. Smith and the *Titanic*.

Aircraft operations are no different. The Federal Aviation Administration assigns final responsibility for safe flight to the chief operator in the cockpit, whether he or she has the controls or not. And, though the aircraft owner bears final responsible for maintenance, the chief aircraft operator must make reasonable checks of maintenance and operability to ensure that the craft is fit to fly.

The same is true for industrial operations. Though supporting organizations are held accountable for the special skills that they provide, line leaders are ultimately responsible for the safety of the plant, its operating and support staffs, the general public, and the environment.

A Team "Sport"

Maintenance, modification, and testing of plant equipment are conspicuous examples of the need for industrial teamwork. They are seldom individual efforts. Operators rely upon skilled engineers and maintenance technicians to repair systems and components which are not functioning as designed. Maintenance and engineering personnel require accurate communication of symptoms from the operating staff before proper diagnoses of equipment problems can be rendered. Testing technicians must engage the skills of operators, maintenance technicians, and engineers to contrive and perform their tests. Then, all elements must work together, once the task has been successfully performed, to safely restore the repaired component or system to its original operating configuration.

Yet, we have already seen many examples of the failure of teams to coordinate important maintenance and testing tasks. At the Bhopal pesticide production facility, several safety-related systems and components—storage tank chiller, caustic scrubber, flare tower, and nitrogen pressurization system—were out of service while the MIC plant was running, a violation of the plant's safety analysis and plant operating procedures. At Three-Mile Island, maintenance and operations failed to coordinate the repositioning of the auxiliary feed pump stop valves following a pump start test. As a result, auxiliary feed pumps could not supply water to the steam generators when the main feed supply was lost.

Obviously, maintenance, modification, and testing are important team processes shared by operating and supporting teams—all of whom must cooperate closely to ensure that equipment is installed, repaired, restored, and operated within design specifications. Their efforts must be coordinated by the leaders of the cooperating agencies and, ultimately, by one leader responsible for the entire organization.

Tiers of Maintenance

To fully participate in the maintenance process, operators must understand what resources are available for equipment repair. They range in scope from operator maintenance to depot maintenance.

Operator Maintenance The first level (and one of the most important) is operator maintenance. Operators should be taught to perform routine, minor maintenance (both corrective and preventive) on equipment that they operate. Even in motor vehicle operations, every driver should know how to change a flat, check the oil, replace washer fluid, check radiator coolant level and condition, and check and adjust tire pressure. Unfortunately, many industries have created environments and structures which actually prevent operators from performing minor maintenance. Yet, properly performed first-line maintenance is the first element in a sound maintenance program.

Facility Maintenance The second tier of maintenance is performed by the facility maintenance staff (a maintenance group assigned full-time to an operating plant or activity). They perform preventive and corrective maintenance that is either too time consuming or too complex for the operating staff but which doesn't require a site maintenance team. Most preventive maintenance tasks in a plant are performed by the facility maintenance staff and the operating staff.

Site Maintenance Site maintenance is the third maintenance tier for most organizations. Site maintenance teams usually have the tools and expertise necessary for major repair jobs that are beyond the capability of facility maintenance teams. Site teams typically rotate among the facilities on a site, providing service during scheduled outages or shutdowns. However, they may also be called upon, as a result of major equipment breakage, to repair major components and systems during unplanned or emergency shutdowns.

Depot Maintenance Finally, when the scope of the task exceeds the capability of the site, depot or vendor maintenance is usually appropriate. Either the vendor will send support teams to the equipment, or the equipment will be removed, packaged, and shipped to the vendor for repair.

Operations/Maintenance Interface

Operating teams and maintenance teams (properly aligned and properly led) actually comprise a single, larger team directed toward the same mission. In fact, maintenance is little more than a specialized outgrowth of operations. One need only look at the relationship between driver and pit crew in a world class NASCAR race team to see how maintenance and operations are *supposed* to work. The driver provides vital information to members of the pit crew, allowing them to isolate and correct problems. The pit crew works diligently and professionally to give the driver the greatest machinery advantage for mission accomplishment, and they also advise the driver on machinery performance and limitations. Neither can do the other's job; yet, without each other, neither can succeed. Accordingly, the interface between operations and maintenance (or *any* supporting team) must be clearly understood and masterfully managed if an industrial team is to grow beyond mediocrity.

The Operator's Role

Regardless of the tier at which maintenance is performed, equipment operators play a vital role in repairing or preparing the equipment. In fact, maintenance and modification almost always require the assistance and oversight of operating personnel. When installed equipment must be worked on, operators usually establish necessary plant conditions, perform the valve and switch line-ups to isolate people and equipment from high energy sources (through installation of locks, tags, flanges, etc.), restore plant conditions to allow testing of repaired or modified equipment, concur with post-work tests to be performed for verifying restoration of components and systems to safe operating configurations, and perform (or monitor) the post-work tests.

Though operators cannot oversee every aspect of maintenance, through the process of *post-maintenance testing*, they can, for the most part, ensure that the equipment is functioning as designed before they accept it back for use. Further, operators should always have the opportunity to review the steps of maintenance, modification, or testing processes to satisfy themselves that the proposed work is reasonable.

The Team in Action

To illustrate the importance of these team interrelationships, consider the ordinary process of finding and fixing equipment deficiencies. When an equipment deficiency is discovered, operating personnel (often advised by maintenance technicians) must determine whether the deficiency will lead to hazardous conditions if equipment operation continues. If the deficiency is minor and does not pose a threat, operations may continue (perhaps necessitating a reduced operating status) until repair can be effected. Sometimes, however, the deficiency may be serious enough to require complete shutdown of the component or system and, perhaps, electrical or mechanical isolation from other equipment.

If the deficiency can be (and *is*) corrected immediately by the operator or maintenance technician, no other action may be necessary. But, if the defect cannot be immediately fixed, the deficient condition should be recorded and identified via a *formal* deficiency tag system. Only then can the problem be properly assessed and prioritized for repair.

Once a deficiency is identified, a plan for repairing the component must be developed. That plan may be as simple as scheduling and assigning a technician to the job. On the other hand, special plant conditions may be needed to conduct repairs. In those cases, repairs (for minor deficiencies) may be deferred until the next scheduled maintenance outage; or (for serious deficiencies), an unplanned maintenance shutdown may be required. In either case, a formal deficiency and maintenance tracking system *must* be utilized to ensure that *all* deficiencies are corrected in an appropriate and timely manner.

Once the maintenance plan is established, preparation for maintenance is apt to require reconfiguration of other plant components and systems, disconnection of the defective component from power and fluid supplies, and creation of physical boundaries around the component to protect others during the repair process.

Reconfiguration is often a team effort involving both operations and maintenance. It may require valve repositioning, installation of blank flanges in piping systems, electrical isolation, or mechanical disconnection. The operating team must control this process until the equipment is effectively isolated from other operating systems. If care is not exercised, maintenance activities can adversely affect other operating equipment.

Once the component or system has been prepared for maintenance, it can be released to a maintenance team for repair. Though the maintenance team will exercise local control of the component or system during the maintenance period, operations must still remain responsible and knowledgeable in the overall sense for the equipment being repaired.

After completing maintenance, the component or system must be restored to an operating state. For this to occur, the component or system must be realigned with other operating systems and prepared for *post-maintenance testing*. **Post-maintenance testing** is a verification process designed to validate the effectiveness of repairs and ensure that the component or system has been restored to its design configuration.

Post-maintenance tests are often specified by component or system technical manuals. They should, however, be critically examined by both the operating technicians and the maintenance technicians.

Sometimes, testing will require a component test while the component is on the repair bench or otherwise isolated from the system. A second stage of testing will then be required to *operationally* test the component after it is reinstalled in the system.

Specification of post-maintenance tests requires special care. Failure to appropriately specify and execute post-maintenance testing can pose a serious threat to operators, equipment, and the environment if the machinery has not been successfully repaired or correctly reinstalled. When post-maintenance testing has been properly completed, the repaired equipment should meet or exceed its original design specifications.

Guidelines for Maintenance Control

Maintenance performed on installed equipment within an operating facility obviously requires close coordination between operations and maintenance teams. Though conflicts are bound to arise, team leaders must ensure that the relationship between these teams is not antagonistic. Again, neither can succeed unless both succeed.

Here are some simple guidelines to ensure that the maintenance process is properly controlled:

Operations Review — Work procedures for performing maintenance on equipment installed in an industrial facility should be reviewed and approved by members of the operating staff. The coordinated review of operations and maintenance (along with any other necessary cognizant work groups) is the basis for conducting the work in a manner that is unlikely to endanger members of either team.

Pre-Task Component Identification — Identification and marking of the component or system to be repaired should be accomplished by an operating technician who is currently qualified or certified to operate (or supervise operation of) the subject component or system. If possible, the repair technician should accompany the operating technician to visually identify the component or system. The purpose of this guideline is to avoid working on the wrong component or system.

Component Isolation — Isolation of the component or system from energy sources by repositioning breakers, switches, valves, or other devices should be done by equipment operators using the facility's lock and tag program. Maintenance technicians may then each install their own locking device as prescribed in OSHA Regulations and the facility lock and tag procedure. A well-coordinated lock and tag program provides independent forms of protection to both the operating staff and the maintenance team.

Operations Work Approval — Approval to commence maintenance of equipment should be granted by a cognizant operations manager or supervisor to the maintenance team. Such approval should normally be a standard part of the work document prescribing the steps of preparation, repair, and restoration.

Operations Overview — Operating teams should maintain overall control of the equipment during most maintenance tasks. Decisions to disconnect or connect energy sources and fluid paths, energizing equipment, and testing must be made only with operations concurrence and coordination. Again, the coordinated review by maintenance and operations protects both work groups.

Post-Maintenance Tests — Post-maintenance testing requirements for installed components and systems which have undergone maintenance should be reviewed, approved, and performed only with operations concurrence and coordination. The role of the maintenance team is to provide expert advice in accordance with current technical instructions regarding proper testing. The role of operations is to ensure to their own satisfaction that the recommended tests are sufficient to guarantee that the equipment will function as designed when returned to service. Operations must also ensure that the tests can be safely performed in the context of the entire operating environment.

Overseeing Maintenance, Modification, and Testing

Types of Testing

As we just learned, operating and maintenance teams routinely conduct post-maintenance testing to ensure that equipment is restored to design configuration following repair.

Like maintenance and modification, testing is a continuous process that transpires throughout the life of an industrial complex. It begins early in the design phase as newly designed components and systems are tested, modified, and retested as a part of research and development. The testing process intensifies during the construction and startup phase as acceptance testing is conducted to validate the performance of components and systems following installation. Then, as modifications are made and new equipment is installed, testing must again be conducted to validate design.

Acceptance Tests

When new (or modified) equipment is received and installed in a plant, it must be tested before acceptance by the operating organization; hence, the name *acceptance testing*. Often, the equipment is accompanied by vendor representatives and startup engineers—often not qualified on the plant in which the equipment is being installed—who must coordinate with plant operators to put new components and systems into operation.

Vendors and startup engineers serve as special assistants to the operating staff. They provide expert guidance on how the testing process should be performed and they oversee the tests. Nevertheless, the operating staff must always remain in charge and in control of the tests.

Diagnostic Tests

Testing is sometimes required as a part of "troubleshooting" activities. When equipment isn't performing as designed and the cause is not readily apparent, *diagnostic tests* may be necessary to evaluate the problem.

As with medical problems, diagnostic testing provides information which helps to develop a clear picture of the problem. For machinery, diagnostic testing might include vibration analysis, sound signature evaluation, spectral analysis of lubricating oil, electrical circuit analysis, or simply an ongoing visual observation check of performance.

There are as many diagnostic tests as there are problems. But, for each type, if the test has the potential to alter or affect component and system operation for which operators are responsible, the operating staff must be involved.

Functional Tests

Routine equipment checks, often called *functional tests*, must be conducted on a daily basis in most operating plants. Functional tests are usually performed by equipment operators using built-in instrument test circuitry. By placing an instrument in its "TEST" mode, the test circuit inputs a calibrated signal to the instrument. If the meter or gauge output falls within a range specified by the component technical manual, the operator is assured that the instrument is operable and functioning correctly.

Operations/Testing Interface

All that has been said about the relationship between operating and maintenance groups also applies to operating and testing groups. Neither group succeeds unless both succeed. It is a team effort.

Not all testing, however, is within the technological or resource capabilities of the industrial complex in which it must be performed. Some testing is so extensive or specialized that it requires the services of a distinct test engineer or test group. Therefore, as with maintenance activities, testing organizations may need to be established at the site or vendor level. In such instances, support agreements and policies should be established beforehand to ensure that the lines of authority and communication are clearly understood.

The Operator's Role

Plant equipment tests, to be safely and successfully performed, usually require involvement of qualified operators who must understand their role in testing. Otherwise, the lines of responsibility and accountability for completing the tests become obscure—and the outcome may be equipment which won't function as designed when expected or needed.

Tests must be thoroughly reviewed and approved by qualified operators to ensure that the operating staff, the environment, and the general public will not be jeopardized. At the Chernobyl Atomic Power Station, testing control was not in place. Unqualified electrical engineers wrote an engineering test with serious safety implications; but, the test was never reviewed or approved by qualified operations personnel prior to performance. Operators were then placed under duress to accomplish the test by a supervisor who had little understanding of nuclear power operations.

Since equipment under test may affect performance of other plant machinery, operating personnel must maintain a keen awareness of component and system status. Further, they must preserve a clear relationship to the testing technicians so that testing can occur safely and efficiently.

Guidelines for Testing Control

Test control is very similar to maintenance control. Here are some simple guidelines to ensure that the testing process is properly controlled:

Operations Review — Procedures for testing installed equipment should be reviewed and approved by members of the operating staff. The operational review must ensure that proposed equipment testing can be safely conducted in the context of the overall operating environment.

Operations Control — Tests conducted on installed facility equipment should normally be controlled by operators who are qualified or certified on the equipment and in the testing procedure. Test engineers should serve as expert advisors to operations to ensure that testing is conducted in the prescribed manner.

Qualified Testers — Test engineers should be qualified or certified in the general process of testing and, specifically, in the conduct of the proposed test. If possible, they should also undergo rigorous training in the systems that their tests potentially affect. Such training helps to avert conceptual mistakes such as those that occurred at Chernobyl.

Pre-Test Training — Pre-test training (using pre-test briefings, mockups, and walk-throughs) for the integrated operations/testing team should be conducted before complex tests with associated high risk. NASA presents the quintessential example of mockup and walk-through training. Since astronauts usually have a limited window of opportunity to conduct experiments and accomplish tasks, they practice extensively in realistic simulated environments on earth. They leave as little to chance as possible.

Approved Deviations — Deviations from normal operating procedures required by tests should receive critical engineering and operational review prior to approval. This review must ensure that the test procedure does not violate equipment design specifications.

Component Isolation — As with all maintenance and testing activities, isolation of components or systems from energy sources or fluid flows required by the test procedure should be accomplished by equipment operators using the facility's lock and tag program. Decisions regarding connection (and disconnection) of energy sources, valve and switch realignments, or other significant changes must be made only with operations concurrence and coordination.

Approval to Commence — Approval to commence testing of equipment installed in a facility should be granted to the test team by a cognizant operations manager. Such approval should normally be a standard part of the test document which prescribes the testing steps.

Post-Test Tests — Upon completion of testing activities, validation tests should be conducted to ensure that involved equipment has been restored to design configuration. Validation testing requirements should be reviewed, approved, and performed only with operations concurrence and coordination.

Topic Summary and Questions to Consider

Most industrial operations have dozens of support organizations and hundreds of team members, all necessary to sustain the core of operators around whom the activity mission is based. No single part can function successfully without the other. They are inextricably linked to the same mission. To succeed, they must cooperate, for, if one fails, eventually, they all fail.

Maintenance, modification, and testing are among the most important of the support activities. Without close coordination with the operations group, they can dangerously erode the margins and boundaries of the safety envelope. Unless they remain closely attuned to the needs, desires, and limits of operators, they may lay the foundation for a Chernobyl or Bhopal.

Ask Yourself Do your mantenance and testing teams work closely with your operating team?

Does the operating team feel responsible for maintenance and testing?

Does the operating team review maintenance and testing procedures for installed equipment?

Do maintenance and operations personnel coordinate in prescribing and performing post-work tests?

Chapter 17

Isolating Energy Hazards

In an industrial facility, people may be exposed to a host of hazards—fires, falls, chemicals, ionizing and non-ionizing radiation, and biological agents—all of which can jeopardize health and safety.

As we have already seen, design and operating safeguards are usually employed to protect against such hazards for routine operations. Not all operations are routine, however. During maintenance and testing, unusual or dangerous equipment configurations can result in harmful energy release. Therefore, special constraints and warnings are necessary to protect personnel and machinery. Useful controls include protective valve, switch, and electrical breaker line-ups, installation of blank flanges, locks on isolation devices, and using warning signs or tags to restrict use of equipment.

We have already learned that operators are an integral part of maintenance and testing—they have the expertise and knowledge of component and system interactions, of the potential hazards introduced by maintenance or testing, and of coordinating and regulating the operations/maintenance interface. Accordingly, one principal responsibility (and vital operating skill) of operators is to correctly isolate components and systems to preclude inadvertent release of (and exposure to) hazardous energy before commencing maintenance or testing.

Sources of Hazardous Energy

In general terms, **hazardous energy** is any type and amount of energy that will cause injury to a person or damage to equipment. The definition is vague because the levels of energy that will injure vary from situation to situation. Electrical energy, for example, can stop or fibrillate a human heart at potentials as low as 50 volts if conditions are favorable. Fortunately, hazard levels are usually addressed by standardized codes such as the National Electrical Safety Code.

OSHA The OSHA Standard for "Control of Hazardous Energy—Lockout/Tagout" (29 CFR 1910.147) provides guidance for:

> ...the servicing and maintenance of machines and equipment in which the unexpected energization or start up of the machines or equipment, or release of stored energy could cause injury to employees.

It requires employees to:

> ...establish a program and utilize procedures for affixing appropriate lockout devices or tagout devices to energy isolating devices, and to otherwise disable machines or equipment to prevent unexpected energization, start up or release of stored energy in order to prevent injury to employees.

OSHA Standard 1910.147 identifies six specific types of energy to be considered by industrial teams in developing protective procedures and methods: electrical, mechanical, hydraulic, pneumatic, chemical, and thermal energy.

Electrical Energy Electrical energy is a fundamental resource for almost all industrial operations. Accordingly, it is also a fundamental hazard. Since it is not easily seen (unless released uncontrollably), team members sometimes become complacent in its presence.

Physical protection against electrical hazards relies primarily upon disconnecting the source of electrical energy to the machinery under repair. But, one should never forget that electrical energy can also be stored in batteries or capacitors. Just because a component or system is disconnected does not mean that it is de-energized.

Mechanical Energy Mechanical energy is usually stored energy. Its force derives from its mass and height above a positive stop (such as the ground), from stored spring energy, or from rotating equipment (such as fans, flywheels, turbines, etc.).

Protection against uncontrolled release of mechanical energy relies primarily upon physical blocks that prevent falling, rotation, movement, or release of spring energy.

Hydraulic Energy Hydraulic energy is, of course, force transmitted through liquid in a closed system by the build-up or application of pressure. It may result from applying a compressive force (e.g., action of a hydraulic motor or positive displacement pump) or through the introduction of heat to the system.

Protection against release of hydraulic force usually relies upon opening or venting the system, or, in some manner, physically blocking exposure to pressure release points.

Pneumatic Energy Pneumatic energy is force transmitted through gas, usually as a result of compression or heating. Remember, however, that for both liquids and gases (liquids in particular), cooling lowers pressure in a closed system, resulting in a vacuum. Venting an evacuated (or cooled) system or pressure vessel presents all of the problems associated with implosions.

Protection against pneumatic forces is, in general, the same as for hydraulic forces.

Chemical Energy Molecular combination and dissociation often are accompanied by energy release or absorption. A reaction, for example, that gives off heat energy is called exothermic. One that absorbs heat energy is termed endothermic.

Depending upon how rapidly energy transfer occurs, the reaction may be mild or explosive. The runaway reaction of methyl isocyanate at Bhopal was a severe exothermic reaction, releasing extraordinary amounts of both heat and toxic gas.

Protection against uncontrolled releases of chemical energy depend primarily upon preventing the combination of reactive compounds. At Bhopal, the installation of blank flanges before initiating a water flush was required to prevent water from entering methyl isocyanate water tanks.

Thermal Energy Thermal or heat energy may be a product of chemical reaction, electrical current flow, or nuclear reaction. Thermal energy hazards in industry can result in burns to the skin or eyes.

Protection against thermal hazards depends upon the ability to either prevent heat generation or to shield humans from heat which is generated.

Other Hazards

Although not specifically mentioned in OSHA 1910.147 (but discussed as hazards in other OSHA standards), other forms of energy should also be considered for isolation via lockout or tagout.

Ionizing Radiation Ionizing radiation includes high energy electromagnetic radiation (such as gamma and X-ray) and high energy subatomic particles (such as alpha, beta, and neutrons). OSHA (1910.96) defines ionizing radiation to include "alpha rays, beta rays, gamma rays, X-rays, neutrons, high speed electrons, high speed protons, and other atomic particles." These high energy waves and particles can result from spontaneous disintegration of radioisotopes (radioactive sources), from nuclear reactions (nuclear reactors), or from artificial sources such as X-ray tubes, particle accelerators, and cathode ray tubes.

When biological organisms are exposed to ionizing radiation, cell damage occurs. Accordingly, lockouts or tagouts may be needed to protect personnel from uncontrolled radiation sources.

Non-Ionizing Radiation Non-ionizing radiation includes the remainder of the electromagnetic radiation spectrum—radio and microwave, infrared, visible light, and ultraviolet. Specific biological hazards associated with non-ionizing radiation are highly dependent on the frequency (wavelength), intensity, and duration of exposure. For example, although the hazard of visible light is restricted primarily to eye damage, ultraviolet light at high intensity (or prolonged exposure at lower intensities) poses an additional risk of skin cancer. Although OSHA's discussion (1910.97) is limited to radio and microwave frequencies, with supplementary guidance for laser radiation (1926.54); lockout or tagout protection may be required to prevent inadvertent exposure to any of these non-ionizing radiation sources.

Sound Energy OSHA's discussion of sound or noise exposure (1910.95) is limited to the effects on hearing and requirements for ear protection. But it should also be noted that intense concussive forces can have other adverse biological effects. For example, the concussion from a blast can cause lung damage as well as hearing loss.

Industry-Specific Hazards Although we've discussed many types and sources of energy hazards in general terms, we obviously can't provide an exhaustive list in this text. Our purpose here is to provide an overview of the kinds of hazards that need to be considered. Each industry—and facility within an industry—has specific hazards unique to its particular technology and operation, and has a responsibility to identify those hazards and to implement the necessary means of isolation to provide personnel protection.

Means of Isolation

The means of isolating hazardous energy sources are based upon the type of energy and the design of equipment. The OSHA standard applies one general term to physical devices that block or disconnect energy sources—*energy isolating devices*.

Energy Isolating Devices An **energy isolating device** is defined by OSHA as:

A mechanical device that physically prevents the transmission or release of energy, including but not limited to the following: a manually operated electrical circuit breaker, a disconnect switch, a manually operated switch by which the conductors of a circuit can be disconnected from all ungrounded supply conductors and, in addition, no pole can be operated independently; a line valve; a block; and any similar device used to block or isolate energy.

But warns that:

Push buttons, selector switches and other control circuit type devices are not energy isolating devices.

Double Valve Protection For fluid systems, valves which block the flow of fluid (and which will hold pressure) are energy isolation devices. Valves, however, tend to leak. Therefore, when it is necessary to use valves as pressure boundaries, *double valve isolation* is a good practice.

Double valve isolation simply means that two valves are shut to prevent the escape of fluid from a leak path. The purpose is to put two valve barriers between a source of fluid energy and the point of repair.

Blank Flanges Blank flanges are considered by the OSHA standard to be *lockout devices*, a term which will be explained momentarily. **Blank flanges** (also called pancake flanges or slip blinds) are plates which are inserted into bolted piping connections to physically block fluid flow. When inserted, they must be marked (or in some way annotated) to ensure that they are removed before the system is again pressurized.

Circuit Breakers and Disconnects Electrical isolation is most often accomplished by opening circuit breakers or electrical disconnects. It is a good practice to use double breaker (or disconnect) isolation when working on electrical gear. Isolation at the source and at the local control breaker (using locks or tags) provides an extra margin of protection against error.

Lockout Isolation

Locks may be used in conjunction with energy isolation devices to prevent re-energizing or release of stored energy until the equipment is repaired and appropriately reconfigured. They provide a physical barrier against repositioning or removing an isolation device.

Lockout Definition

Lockouts are the preferred method of isolating energy sources to protect both workers and equipment. Each worker (or work group) may have an individual lock placed on the energy isolating devices to afford positive, individual protection.

OSHA guidance for use of locks and tags defines **lockout** as:

The placement of a lockout device on an energy isolating device, in accordance with an established procedure, ensuring that the energy isolating device and the equipment being controlled cannot be operated until the lockout device is removed.

The standard further defines a **lockout device** as:

A device that utilizes a positive means such as a lock, either key or combination type, to hold an energy isolating device in the safe position and prevent the energizing of a machine or equipment. Included are blank flanges and bolted slip blinds.

"Capable of Being Locked Out"

Lockouts should be used whenever possible to isolate equipment from energy sources if work on the equipment would subject a worker to dangerous energy hazards.

To use a lock, the energy isolation device must be capable of being locked out. Standard 1910.147 states that:

An energy isolating device is capable of being locked out if it has a hasp or other means of attachment to which, or through which, a lock can be affixed, or it has a locking mechanism built into it. Other energy isolating devices are capable of being locked out, if lockout can be achieved without the need to dismantle, rebuild, or replace the energy isolating device or permanently alter its energy control capability.

Choosing Lockout Locations

When a situation requires that locks be installed, the determination of where they are placed should be made by leaders, technicians, and workers qualified or certified to operate or work on the equipment.

Isolation locations should be chosen based upon review of approved procedures and drawings of components and systems. If drawings are suspect, reviewers should verify the locations by physically tracing systems.

The review should consider not only external energy supplies, but also internal sources such as compressed springs, batteries, capacitors, or pressurized fluids. The result of this review should be a map or list of devices that must be positioned correctly to isolate the subject equipment.

Independently Verifying Locations

Since energy isolation is always related to personnel or equipment safety, the map or list created by the first person should be independently verified by a second qualified worker. In order for the review to be "independent", it should be done separately and without benefit of the first person's map or list. If differences arise, they must be reconciled, perhaps with the aid of a third person.

Up-to-Date Drawings

It is clear that without up-to-date drawings and procedures that present an accurate picture of how components and systems are installed or connected, isolation of energy sources becomes tenuous. As just stated, isolation should never be left to chance. If necessary, the affected systems should be physically traced to guarantee correct isolation.

Obviously, if components, piping, and cable runs aren't well-labeled, accurate determination of isolation points is apt to be ineffective. It pays to label correctly during plant construction and to keep labeling up-to-date.

Installing Locks

Once the points of interdiction have been selected (and the desired positions of the isolation devices determined), individual or group locks should be installed and independently checked. An index or record of the where locks have been installed should also be created, filed in a lockout record log, and stored in a central location to allow for immediate review and retrieval.

Removing Locks

Before locks may be removed and energy restored to equipment, it must be determined that workers are clear of the equipment and that re-energizing will not present a hazard to personnel or equipment. As discussed in the previous chapter, operators bear primary responsibility to ensure that equipment is properly restored to operating configuration.

Tagout Isolation

When locking is not possible, warning tags may be used to alert team members to abnormal equipment conditions. OSHA defines the **tagout** process as:

The placement of a tagout device on an energy isolating device in accordance with an established procedure, to indicate that the energy isolating device and the equipment being controlled may not be operated until the tagout device is removed.

Tagout Device   A **tagout device** is:

A prominent warning device, such as a tag and a means of attachment, which can be securely fastened to an energy isolating device in accordance with an established procedure, to indicate that the energy isolating device and the equipment being controlled may not be operated until the tagout device is removed.

Since tags don't provide a positive means of protection, it is particularly important that all employees know what the tags look like and what they mean. Standard 1910.147 states:

Tags may evoke a false sense of security, and their meaning needs to be understood as part of the overall energy control program.

Using Danger and Caution Tags

Two general types of tags are commonly used in industrial complexes to warn of abnormal or dangerous conditions. The first is a *danger tag*. **Danger tags** are warning tags, often bright red in color, installed on the operating controls and power sources of equipment which, if operated, could result in injury to personnel or serious damage to the machinery. The type of tagout device referenced by OSHA is a danger tag. (See Figure 17-1.)

Caution tags, on the other hand, are warning tags, frequently yellow in color, indicating unusual or abnormal operating precautions that must be observed when operating the subject equipment.

Figure 17-1: Typical Danger (red) and Caution (yellow) Tags as referenced by OSHA.

Isolating Energy Hazards

OSHA states that "Caution tags shall be used in minor hazard situations where a non-immediate or potential hazard or unsafe practice presents a lesser threat of employee injury." Caution tags require *careful* operation of equipment while danger tags *forbid* operation.

Tag Identification Tags must be clearly and individually identifiable. Each tag should have a pre-printed identification number or an information block for assigning an individual identification number. Individual tag identification is used in conjunction with an index of installed tags to document tag locations and positions or conditions of the devices on which they are installed.

When prepared for installation, each tag should correspond to a single isolation device. As such, each tag should be traceable and easily located by another person qualified on the components or systems involved.

Condition or Position In addition to specifying the devices upon which they are placed, tags should also identify the positions or conditions in which the devices are to remain until the tags are removed.

When delineating positions or conditions of valves, switches, and breakers, terminology conventions should be observed. Policy should specify terms such as "open", "shut", "off", and "on" to minimize confusion that could occur during tag installation or subsequent tag audits.

Signature Verification Finally, tags usually must include three signature blocks: one for the individual who authorizes tag installation, one for the person who positions the energy isolation device and places the tag, and one for the person who independently verifies proper positioning of the energy isolation device and tag placement. The authorizing official can sign any time before the tags are hung. But, tag installers and checkers should sign only at the time of tag placement and verification, respectively.

Determining Tag Locations

When danger tags must be prepared and installed, the determination of where they should be positioned should—as for lockouts—be made by qualified personnel based on approved procedures and accurate drawings. A map or list of devices and tags should be prepared (including the conditions necessary for reconnection and clearing the tags).

As with lockout devices, the use of danger tags is always related to personnel or equipment safety. Thus, the same independent verification requirements apply.

As with lockouts, up-to-date drawings and procedures that accurately represent "as-built" configurations are essential; and again, affected systems should be physically traced if necessary to ensure accurate and complete selection of isolation points.

Conservatism is a good guideline. For example, if electrical leads are disconnected, it is desirable to tag both the lead terminations and the associated upstream breakers. Similarly, when blank or blind flanges are installed in piping runs, it is wise to install a danger tag on each blank flange to ensure that it is removed after completion of the job.

Positioning and Installing Persons chosen for isolation device positioning and tag installation should be qualified on the components or systems involved. It makes little sense for team members who are unfamiliar with systems and components to be making critical adjustments to their status.

When the tags have been prepared and authorized for installation, the installer must reposition (or verify the position of) each isolation device, install the tag, and sign the tag in the installer's signature verification block. The installer should sign the tag record sheet at the same time. This process should normally be performed one tag at a time.

Then, a person designated to verify or *second check* the isolation device position should *independently* recheck the device positions and sign the tag and record sheet verification signature blocks. As discussed in Chapter 11, the second-checker should (in order to maintain independence) not accompany the installer.

Tag Record Index A record of each tag (and the information contained on the tag) should be made in a tag index. This *tag record sheet* should be a pre-printed form with information blocks to record the same information as on each individual tag. When completed, the tag index will serve as a register of all of the danger tags installed within the facility.

After the tags have been installed and all of the information on the record sheet has been completed, the tag index should be filed in a central and accessible location, usually in a tag record log. The log must remain available for review at all times.

An accurate tag index is an important tool for conducting periodic audits to verify the effectiveness of component condition and tag installation. If the index isn't both accurate and used with discipline, the entire tagout process is jeopardized. Therefore, the index should be closely controlled and frequently reviewed.

Tag Removal When the task for which the tags were installed is complete, removal of the tags should be authorized only after ensuring that workers are clear of the equipment and re-energizing does not present a hazard to personnel or equipment.

Just as tag installation is controlled and recorded, their removal should also be closely controlled. Directions for tag removal should clearly specify the position in which the energy isolating devices are to be left upon tag removal. Remember the problem at TMI-2: The auxiliary feed pump stop valves were not repositioned to "open" when the tags were removed.

Once the tags have been removed, they should be returned to the authorizing official for destruction—tags should *never* be reused—and the tag index should be removed from the tag record log.

Topic Summary and Questions to Consider

Successful configuration control requires that every operator acquire the skill of protecting people and equipment by isolating energy hazards. The accidents at Three-Mile Island and Bhopal were greatly influenced by deficient application of this vital operating skill.

Ask Yourself Does your team perform work on equipment which exposes team members to energy hazards?

Do you have a lock and tag program in place for such situations? Does it conform to OSHA standards? Do you regularly audit the lock and tag program—including physical tag and record sheet inspections?

Have there been injuries or near misses as a result of improper lockout or tagout at your facility?

Where hazardous energy isolation is being improperly performed, someone will eventually be seriously (perhaps fatally) injured. Energy isolation is a common sense method for minimizing your risk.

Chapter 18

Training On-the-Job

The Big Bayou Canot Bridge accident (Chapter 6) illustrates that the selection and training of equipment operators and their leaders is a critical factor to ensure that equipment is consistently controlled within the boundaries of its safe operating envelope. The best procedures and equipment cannot compensate for an operating deficit incurred by poor training.

Yet, as we have seen, excellent training is neither simple nor inexpensive. It requires skilled analysis, design, development, implementation, and evaluation. Successful training must progress through a process of selecting personnel in accordance with strict selection criteria, performing initial training in accordance with written qualification standards, and improving personnel through continuing training.

Proper training pays invaluable dividends over the long term. Less supervision is required for technicians who are well-trained. Fewer mistakes occur when the training of team members is in-depth. Time and resources are more efficiently used by personnel who know where they are going and how to get there.

Perhaps the most important and effective training technique employed in an industrial facility is *on-the-job training* (OJT). Studies indicate that, in a training environment, an average student retains nearly ninety percent of what he or she does and explains at the same time; whereas, only ten to thirty percent of what a person reads or hears is retained.

Since the acquisition of operating skills constitutes such a large part of industrial operator training, the ability to provide excellent on-the-job training is a critical expertise. In that light, this chapter explores what it takes to run a good on-the-job training program.

The Purpose of OJT

The purpose of on-the-job training is to transform education into skill. It requires that a student's mind be shepherded from theory to practice.

Most people learn by forming pictures in their minds. The process of on-the-job training capitalizes on that trait by explaining a process to a student, showing the student how to perform the task at the actual equipment operating site, letting the student perform and explain the task under the watchful eye of an instructor, and then evaluating the student's performance. This process is conducted until the student achieves the minimum accepted level of performance and knowledge as verified by instructor evaluation.

Elements of a Good OJT Program

As discussed in Chapter 6, a successful training program has several key elements:
- Qualified Candidates
- Explicit Objectives
- Thorough Qualification Standards
- Skilled Instructors
- Valid Evaluation
- Coordinated Administration
- Clear Training Policy

The same elements are necessary for on-the-job skills training.

Qualified Candidates   The first element—qualified candidates—requires that candidates be carefully selected. It is senseless to begin training someone incapable of achieving the training goals, especially when aptitude can be evaluated through interview and testing. Hence, the need for clearly established selection criteria.

Explicit Objectives   Next, the program must provide explicit training objectives. Both instructor and student must know the goals of the program. Training policies will probably not be carried out if objectives aren't clear.

Overall objectives should be established with this question in mind: "What must the student know or be able to do upon completion of this training program?"

Thorough Qualification Standards   With objectives clearly established and communicated, qualification standards can be created to delineate training tasks. As we learned earlier, consistency in operator training (as well as the training of all other supporting activities) is fundamental to dependable industrial team performance. Consistency is achieved, in part, through the use of written qualification standards.

Recall that a qualification standard is a written set of requirements that state succinctly what a candidate must know and be able to do to be considered qualified in a position. A qualification stantard is usually segmented into distinct tasks required for the job position. For each task, the knowledge and performance requirements are delineated. Signature blocks are usually provided for the instructor to validate successful completion of each required task—when satisfied that the candidate is fully capable of performing a task without aid, the evaluator signs (validates) the student's qualification standard for that task.

We learned also that the development of good qualification standards can be achieved only by involving experienced and capable team members. Through round table discussions with competent technicians and through the administration of questionnaires, the individual tasks, knowledge requirements, and performance requirements can be accurately determined.

Skilled Instructors   The on-the-job training process will only be as good as the instructors who administer it. Skilled OJT instructors, therefore, constitute the next element of a good on-the-job training program. Potential trainers should be chosen from the ranks of those operating and maintenance technicians who have performed well in their jobs. The practice of removing incompetent performers and placing them in training positions is absurd. Never should we make coaches out of incompetents just to remove them from the operating and maintenance ranks.

Excellent job performance, a desire to coach, and demonstrated teaching skills are among the criteria necessary for selecting trainers. If an instructor has excellent performance skills, but is unable to teach, students will be frustrated. Worse, an instructor with excellent skills but a poor attitude will transform enthusiastic new candidates into sullen, cynical workers.

The skills necessary to be an excellent on-the-job trainer include an ability to communicate well, an abundance of patience, reasonable reading and writing skills to provide written evaluation, and competence in the technology for which the training is provided. Attaining these skills is not an easy task for many people. Therefore, it is important that on-the-job trainer candidates receive instruction and evaluation in the process. Many organizations require a certification examination rather than simple qualification of their trainers. Periodic re-qualification or proficiency demonstrations are often included as a part of the job.

Valid Evaluation   Good training is wasted if the evaluation process does not actually test whether the student has achieved the desired standards. Therefore, valid evaluation is the next element of the OJT process.

One-on-one instruction carries the advantage of personal attention, detailed questioning, and careful control. But, with poorly trained instructors, it may carry the disadvantage of incomplete testing. As an instructor and student develop a good relationship, the instructor may be tempted to "shortcut" the evaluation process for the sake of friendship.

Since the goal of on-the-job training is to bring operator candidates to the point of proficiency at which they can perform specified tasks without assistance, qualification standards for on-the-job training often require OJT instructors to sign a student's qualification progress book for each task completed. Instructors should be fully aware of what their signature means. Normally, it is an affirmation by a recognized authority that the student has demonstrated satisfactory proficiency in the performance of a task *without the instructor's assistance*.

OJT programs are very susceptible to unauthorized coaching during practical performance exams since the instruction is frequently one-on-one. Each signature is actually a part of the qualification process. If there is no signature integrity—if an instructor signs a student's progress card even though the student has not demonstrated satisfactory, independent performance—both the student and the organization eventually suffer.

Incomplete evaluation creates an operating liability for the one who was slighted and for the facility in which he or she works. Maintaining friendship is a poor reason for inadequate evaluation.

Accordingly, one principle of on-the-job training is that **an instructor should never validate a knowledge or performance requirement unless the student has demonstrated—without instructor assistance—the minimum knowledge and skill necessary to perform the task alone**. To ensure compliance, wise instructors carefully read the requirements of qualification standards so that complete evaluations are rendered.

Coordinated Administration

On-the-job training usually must be performed within the context of on-going operations and maintenance. Consequently, training coordination constitutes a critical element in successful OJT.

When plant activities are predictable, on-the-job training activities should be formally scheduled and published far enough in advance to allow both students and instructors time to prepare. Accordingly, establishing a weekly training schedule is an important function of a facility's training department.

The training manager or plant training coordinator should liaison frequently with the operations and maintenance managers. Attending daily activity planning meetings assures an opportunity to coordinate schedules and determine equipment and system status. Armed with such knowledge, the training coordinator can orchestrate training needs with plant activities.

Clear Training Policy

The training plan for an industrial facility should contain a clear text of the policies which guide the training process. That portion which deals with on-the-job training should include direction for OJT instructors regarding how OJT is to be performed. In particular, it should specify controls for on-the-job training. Such OJT controls include:

- Guiding Philosophy
- Instructor-Student Relationship
- Talking Before Performing
- Suspension of Training
- Span of Control

On-the-Job Training Controls

The process of acquiring skills is largely experiential. It dictates hands-on training. Just as an aspiring trapeze artist cannot acquire aerial skills *only* by studying a book, neither can a potential industrial operator become skilled *only* by sitting in a classroom.

Certainly, abstract education is important to set a mental foundation upon which skill can be built. But, education alone is insufficient. It must be followed by hands-on instruction and practice. Accordingly, on-the-job instruction forms an essential part of the training for any industrial operator.

Yet, on-the-job training in an industrial facility presents a dilemma. If manipulation of equipment introduces high risk, or if the system and component status at one station can affect operations at other stations, there is usually little room for operating error. How, then, can skill training be successfully conducted in an operating facility, especially when students must have opportunity to try and, sometimes, fail?

For the trapeze student—and for potential industrial operators—the answer lies in additional protection. Neither the trapeze student nor the operator candidate should be allowed to attempt risky exercises without excellent instructor supervision and a safety net.

A good on-the-job training instructor creates a *safety net* by exercising close control—being in a position to arrest or mitigate actions which would otherwise have an adverse effect. The rigor of that student control, obviously, must be based upon the risk associated with failure. Control may be looser if the risk is low; but, if failure would incur unacceptable cost, control must be strict.

Clearly, guidance for both instructors and students must delineate how the on-the-job training process is performed and controlled. Some basic precepts for controlling on-the-job training:

Guiding Philosophy The guiding philosophy of on-the-job training during industrial operations should be this: **All actions related to the operation and control of plant equipment must be observed and verified at the time of performance by a person who is qualified (or certified) to operate the equipment.** In other words, though the student is physically performing the steps of a task, another operator, currently qualified (or certified) on the systems and components of the station, must observe and confirm each action.

Since the instructor is usually also the assigned station operator, the instructor retains responsibility for all that happens or fails to happen at the operating station. Normally, to fulfil this responsibility, the instructor must be in a position to stop the student with a voice command or physical restraint should the student attempt to perform an action which could cause personal injury or equipment damage. Usually, such a level of control requires the instructor to be no more than arm's length from the student.

Instructor-Student Relationship For on-the-job training controls to be effective, an understanding of responsibilities must be established between instructor and student. For example, though the instructor never relinquishes responsibility for safe operation of the equipment, the student must understand that his or her actions must allow time for the instructor to intervene if necessary. Discussing actions before performing them, pointing to switches prior to operating them, making an effort to remain no more than arm's length from the instructor—student actions such as these help to ensure safe and efficient on-the-job training.

Talking Before Performing The process of talking through a task prior to performing it provides far more than a measure of operating control. It also serves as a pre-task briefing in which expected equipment response, potential hazards, and contingency actions can be discussed. Talking before performing also provides an excellent opportunity for the instructor to question the student regarding the reasons for each step. In so doing, the instructor can thoroughly evaluate the student's comprehension.

Suspension of Training The instructor must at all times ensure that the training process never interferes with safe operation of the equipment. If an emergency arises during the performance of training, the instructor must make a judgment as to whether the student currently has the capability to respond to the emergency situation (observed and controlled by the instructor). If, in the instructor's judgment, the student can't handle the situation, the instructor should personally take control of the equipment until the situation has been resolved. Only then should the training process resume.

Span of Control As already discussed, one important advantage of on-the-job training is that it is usually conducted one-on-one. Sometimes, however, an instructor needs to teach more than one student at a time. For example, students may be required during qualification to perform a task which seldom occurs during normal operations. Repetitive task performance during a window of opportunity may be the only solution. How many students should an instructor be allowed to oversee during on-the-job training?

The instructor's span of control (the number of students trained at one time) depends upon a host of factors. They include situational risk, instructor experience, and importance of the task being performed. If the task is not complex, carries little risk, and if the instructor can safely oversee performance with several students on station, increased span of control may be warranted.

Two concerns, however, must govern the judgment of expanded span of control. The overriding consideration must be personnel and equipment safety. No training action should unreasonably jeopardize operating safety.

The second consideration is whether an instructor can ensure that appropriate learning has occurred with an increased span of control. Validation through signature of a student's competence is a serious (and perhaps legal) matter. Therefore, the instructor should be personally satisfied that each student on station has met the qualification standard requirements for the training received.

Topic Summary and Questions to Consider

In this brief chapter, we studied the elements of a good on-the-job training program along with many of the controls necessary for well-conducted skill training. Few training processes are as effective and long-lasting as on-the-job training. Confucius summarized it well:

Tell me, and I forget; show me, and I remember; involve me, and I understand.

We should also remember that training is not an isolated function. In reality, each member of an industrial team plays a vital role in the training process. In a general sense, every person who interacts with others in the course of work performance is actually training (and being trained by) someone in some way each minute of the day.

Ask Yourself Does your team have a formal, documented on-the-job training program for high risk positions?

Have you developed written qualification standards that delineate training requirements? Are your qualification standards formally reviewed and approved? Do changes to the standards undergo similar review and approval?

Do you frequently monitor OJT to ensure that it is performed conscientiously and consistently?

Do your OJT instructors know what their signature validation means?

Remember, someone is always watching you. Are you setting a good example?

Chapter 19

Performing Shift Turnovers

A large percentage of operating, maintenance, and testing errors in the industrial environment are the result of poor communication between off-going and on-coming crews or work groups. In fact, improperly executed *turnovers*, as they are sometimes called, probably account, as a single cause, for more accidents in industrial operations employing shifts of personnel than any other factor.

Continental Express Accident, September 1991

On September 11, 1991, Continental Express (Jetlink) Flight 2574, a 30 passenger twin turboprop commuter aircraft en route to Houston International Airport (IAH) from Laredo (LRD), crashed three miles south-southwest of Eagle Lake, Texas, when the upper tail section (including the left and right horizontal stabilizers) separated from the aircraft at an altitude of approximately 11,500 feet.

The Accident Continental Express Flight 2574, an Embraer 120 Brasilia (identification number N33701), departed LRD at 9:09 A.M. central daylight time (CDT) on the second leg of a scheduled round trip flight. [The flight originated in Houston that morning at approximately 7:00 A.M., arrived in Laredo without event, and was returning to Houston.] This return trip proceeded routinely until approximately 10:03 A.M., when, after providing instructions to the flight crew, the Eagle Lake Sector route air traffic controllers lost both radio and radar contact with the aircraft. Before losing contact, the cockpit crew did not indicate, nor did the air traffic controllers detect, any difficulty or evidence of aircraft malfunction.

A search was immediately initiated and the wreckage site located within minutes. Two fire trucks arrived on scene by 10:20 A.M.; but, the aircraft had been destroyed by impact and fire. None of the fourteen aboard—two pilots, a flight attendant, and eleven passengers—had survived.

The National Transportation Safety Board (NTSB) was notified of the accident at 12:30 P.M. eastern daylight time and dispatched an investigating team to the scene. The team arrived in Houston at 7:30 P.M. CDT and soon learned from eye witnesses that the aircraft had been engulfed in flame and appeared to have broken up in flight, plummeting to earth.

The Investigation NTSB's investigating team began evaluations of the wreckage site, flight data recorder (FDR) information, the cockpit voice recorder (CVR) tape, structural performance of the aircraft, and performance of the flight crew and the ground maintenance crews. The cockpit voice recording tape substantiated the possibility of in-flight structural break-up. The CVR analysis team found that:

At [09:59:57 CDT], the last radio communication between Houston Center and flight 2574 is recorded.... At [10:03:07 CDT], the sound of objects being upset in the cockpit can be heard. These sounds are consistent with what might be heard during an extremely abrupt flight maneuver and are followed closely by a sound similar to that of a human grunt. There are no further human voices (or sounds) detected through the end of the recording. The remaining sounds are aural warnings produced by the aircraft along with mechanical sounds consistent with what might be heard during the breakup of an aircraft in flight. A noise, characterized as wind noise is evident on the area microphone at [10:03:13 CDT].

Whatever happened to Flight 2574 occurred so quickly that the flight crew couldn't communicate to the ground controllers or even to one another.

Evaluation of FDR data corroborated the possibility of structural disintegration. The FDR analysis team discovered that:

...the airplane was descending through 11,500 feet (pressure altitude) at 260 knots indicated airspeed (KIAS) when it abruptly pitched down and entered a steep dive. The airplane was 12 knots below the upper limit (272 KIAS) of the EMB-120 airspeed envelope when the upset occurred.

Though well within the limits of its airspeed operating envelope, the aircraft had experienced an event that sent it into a severe nose-down pitch from which it was unable to recover.

Wreckage analysis and structural evaluation provided additional indications of an in-flight structural break-up. The entire T-tail assembly (consisting of the one piece horizontal stabilizer assembly and the upper section of the vertical stabilizer) was found [Figure 19-2] 650 feet west-southwest of the main wreckage location, having separated from the aircraft while in flight. Missing from the left side of the horizontal stabilizer was the leading edge assembly containing the de-ice boot.

The Analysis

What caused such a sudden and unexpected failure of the airframe? Why did the horizontal stabilizer and upper section of the vertical stabilizer separate from the aircraft? Where was the leading edge assembly from the left horizontal stabilizer?

The leading edge and attached deice boot were found three-fourths of a mile from the main crash site, much farther away than any other debris. They had clearly detached just before or just after the abrupt flight upset. Why did the edge detach? Did this separation initiate structural disintegration?

A metallurgical study of the horizontal stabilizer provided the answers. It disclosed that:

...the black composite leading edge (LE) was missing from the left horizontal stabilizer of the airplane.... A total of 47 screws were missing from the top side (forward spar) of this stabilizer—43 screws along a straight row on the top side of the LE and 4 screws from the inboard and outboard top curved portion of the LE.... The LE attaching screws on the left horizontal stabilizer forward spar lower cap appeared to be firmly fastened to their respective nut plates.... On the Embraer 120..., the horizontal stabilizers are installed above the vertical stabilizer. The vertical stabilizer separated chordwise between approximately six to twelve inches below the horizontal stabilizers, resulting in the separation of the horizontal stabilizers from the airplane. [See Figure 19-3.]

The leading edge separated from the aircraft during flight, resulting in a cascading sequence of structurally-related stress failures. The final moments of the flight probably occurred as follows:

The airplane was descending at 260 KIAS [knots indicated airspeed], which was well within its operating envelope, the wings were level, both engines were operating normally, and the pitch attitude was 10° nose down. As the airplane descended through 11,500 feet, the leading edge of the left horizontal stabilizer separated from the airframe. The left horizontal stabilizer leading edge was the first piece of wreckage found along the wreckage path, preceding the next piece by almost ½ mile. This indicates that it was the first piece to separate from the airplane. The loss of the leading edge exposed the front spar of the left side of the horizontal stabilizer to the airstream, and an aerodynamic stall occurred that greatly reduced the downforce produced by the horizontal stabilizer. The reduction in downforce created a large nose-down pitching moment, and the airplane pitched down immediately. A peak load factor of approximately -5 g was reached at the end of only 1 second.

The airframe remained intact (minus the leading edge), and the load factor fluctuated around -2 g, for approximately 6½ seconds. The airplane pitch attitude decreased to 68° nose down, airplane heading moved 20° nose left, and a 15° right roll attitude was reached at the end of this period. The airplane's

Figure 19-1: An Embraer 120 Brasilia (EMB-120).

Figure 19-2: Tail section as found. [NTSB Photo, edited]

altitude was 9,500 feet, and it was flying at an airspeed of 280 KIAS. A second peak in negative load factor was then experienced, and the Safety Board believes that the left wing failed and the right wing tip detached at this point.

The airplane then rolled to the right at a roll rate exceeding 160° per second. The Safety Board believes that the lift produced by the intact right wing produced the extreme roll. The high airspeed and roll rate created large airloads on the airplane's structure. The Safety Board believes that excessive airloads induced by the high airspeeds and/or roll rate caused the horizontal stabilizer and left engine to separate from the airframe. The airplane then entered a spin to the right, fell uncontrollably toward impact, its pitch attitude oscillating between approximately -40° and +40°.

The Cause Evidence of leading edge separation and the missing screws led the investigating team to examine the recent maintenance history of Flight 2574. Since the screw holes in the bottom joint of the left hand horizontal stabilizer leading edge joint were distended and torn while those in the top were undamaged, it appeared that the forty-seven top screws had not been installed. What the team found was an incompletely performed maintenance task conducted the night before the accident. The task deficiencies had gone undiscovered because of a tangled web of defective and nonexistent shift turnover communications.

At about 9:30 P.M. on September 10, the night before the accident, N33701 was delivered to the Continental Express maintenance facility at IAH to replace both horizontal stabilizer de-ice boots. The need for replacement had been identified on August 26 during fleet-wide inspection of de-ice boots in preparation for winter operations. A quality control inspector had designated the horizontal stabilizer boots on N33701 as "watch list" items because of dry rot observed on both boots. De-ice boot replacement for N33701 was subsequently scheduled for the night of September 10.

De-ice boots are an integral part of the leading edge surfaces of the horizontal stabilizer unit of the Embraer 120. (See Figure 19-3.) The boots are a rubber and fiberglass composite material, bonded to the leading edge airfoil surfaces to preclude ice accumulation. Periodically, they must be replaced as a part of the aircraft's preventive maintenance program.

Boot replacement requires removal of the leading edge assembly from the horizontal stabilizer, removal of the old boot from the leading edge, disconnection of fluid lines, installation of a new boot, reconnection of fluid lines, and reinstallation of the leading edge assembly. The leading edge assemblies are rejoined to the stabilizers with screws and lock plates at top, bottom, and sides.

At the Houston Continental Express maintenance hangar, watch list items and routine preventive maintenance tasks were normally performed by the third shift (10:30 P.M. to 6:30 A.M.) maintenance crew. However, on September 10, the *second* shift (2:30 P.M. to 10:30 P.M.) started the boot replacement task. The task was to be performed in accordance with the aircraft maintenance technical manual and recorded on preventive maintenance record documents (M-602 cards), one card for each boot.

Each maintenance crew consisted of two shift supervisors, several maintenance mechanics, and three quality control inspectors, all having current Airframe and Powerplant Certificates. According to the Continental Express General Maintenance Manual (GMM), both shift supervisors were to serve under a single leader having the title of "lead mechanic". But, after the merger of Britt Airlines with Continental to form the Houston branch of Continental Express, the lead mechanic was no longer part of the on-shift maintenance organization. The two crew supervisors subsequently shared responsibility for direction of the maintenance crew: one directed flight line maintenance operations and the other directed hangar maintenance operations.

When N33701 was delivered to the maintenance bay at approximately 9:30 P.M., one hour before the end of second shift, one of the second shift crew supervisors (designated herein as Supervisor 1) received the aircraft and passed the job to the other supervisor (Supervisor 2) who was in charge of maintenance for N33701. Supervisor 1 designated two second shift mechanics (Mechanic 1 and Mechanic 2) to assist Supervisor 2 with the de-ice boot replacement job. One of the second shift quality control inspectors (Inspector 1) volunteered to assist the two mechanics.

Near the time for shift change, about 10:15 P.M. on the night of September 10, Supervisor 1 walked back to the aircraft to review the boot replacement job status. He was accompanied by a third shift crew supervisor, Supervisor 3, who had arrived early for the midnight shift (about 9:45 P.M.). Supervisor 1 asked the two mechanics about the status of the de-ice boot replacement. The mechanics stated that they had removed most of the screws for the right hand horizontal stabilizer leading edge, but that some of the screws would have to be drilled out. Supervisor 3 asked Supervisor 1 if any work had been performed on the left hand boot. He recalls that Supervisor 1 looked up at the tail assembly and replied, "No". Supervisor 3 then told Supervisor 1 that he didn't believe the third shift would have time to complete both boot replacements. Supervisor 1 recalls that he subsequently told the two mechanics not to remove the left hand leading edge.

Just after this event, a third shift mechanic (Mechanic 3) asked Supervisor 1 about the status of boot replacement. Supervisor 1 told him that the screws for the right hand leading edge were almost all removed, but that some would have to be drilled. Mechanic 3 then joined Mechanics 1 and 2 to receive a job status report.

During testimony, Supervisor 2 (in charge of the N33701 boot replacement job) said that he checked on the task status about 10:00 P.M. He noted that, in addition to Mechanics 1 and 2 working on the stabilizer, Inspector 1 was kneeling on the left hand stabilizer. Supervisor 2 stated, however, that he was unaware of what the inspector was doing. (The horizontal stabilizer assembly of the Embraer 120 is approximately 20 feet off the ground and must be accessed by a lift or platform. Neither Supervisor 1 nor Supervisor 2 inspected the job in close proximity.) Supervisor 2 left the job site that night without receiving a verbal turnover of the status of boot replacement.

When questioned by the investigating board, Mechanics 1 and 2 stated that Inspector 1 had offered assistance. He had removed the top joint screws for both the left and right leading edge assemblies. Mechanic 1 stated that, when questioned at the end of shift by Supervisor 1, he advised the supervisor of the status of the work. He could not recall, however, specifically informing Supervisor 1 that the top screws had been removed from the left hand horizontal stabilizer leading edge. He said, however, that he "felt sure" that he had. He also stated that Supervisor 1 directed him to turn over the status of the job to another third shift mechanic (Mechanic 4). He recalls specifically informing

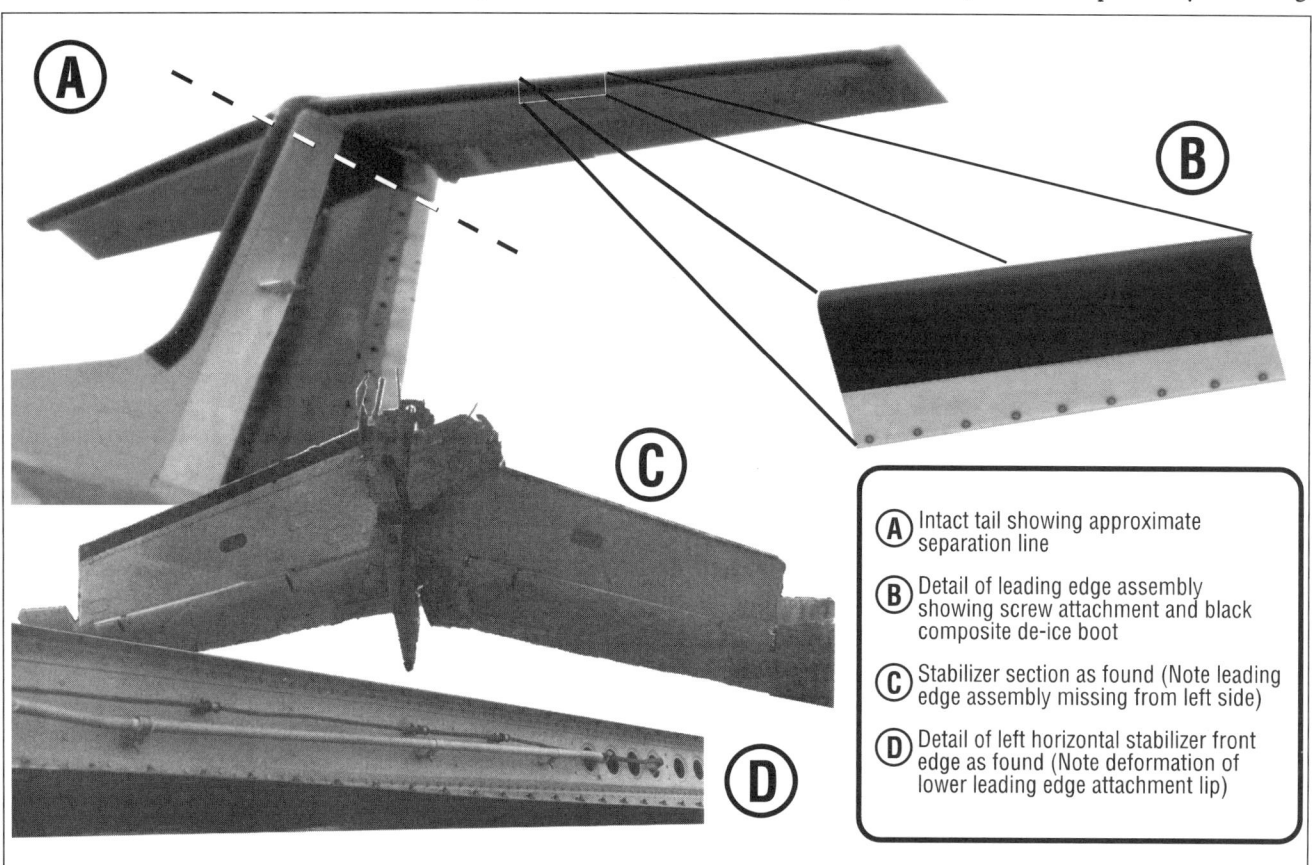

Figure 19-3: EMB-120 tail section components relevant to the September 11, 1991 Eagle Lake, Texas crash.

Mechanic 4 that only the top screws for the left hand leading edge had been removed. He stated that he gave Mechanic 4 the bag of screws collected from the removal job.

Neither Mechanic 1 nor Mechanic 2 filled out the back of the M-602 cards to record the status of the boot replacement job. Mechanic 2 did not record any information, he stated, because he had been working on another aircraft when assigned the N33701 job. After assisting Mechanic 1 in removal of the right hand leading edge, he returned to his original job. He also testified that he did not give a final verbal turnover to either Supervisor 1 or 2 on N33701 before leaving for the night.

Mechanic 1 did not log his work, he said, because Supervisor 2 had not given him the M-602 cards. The NTSB report records that, when asked why he had not issued the M-602 cards for the job, Supervisor 2 replied that:

> ...given the limited time left on his (evening) shift to work on N33701, he did not think it was important for the mechanics to have the M-602 cards.

The report also records that Supervisor 2:

> ...said that normally all routine and watch list maintenance was performed on the third shift.

When asked by the investigation board, Inspector 1 stated that during the time that Mechanics 1 and 2 were removing the right hand leading edge assembly, he had removed all of the top screws for both sides and (except for about five unserviceable screws) placed them in a bag which he left on the lifting platform. He stated that he and the two mechanics then gathered their tools and prepared to leave. He testified that he did not conduct a verbal turnover with any third shift mechanics or inspectors, but he did log in his inspector shift logs that he "Assisted mechanics with removal of deice boots." Unfortunately, Inspector 1's log entries were reviewed by his oncoming counterpart *before* that final entry was made. The relieving inspector arrived early for the shift (about 10:05 P.M.) and conducted the log review at that time.

Mechanic 4 acknowledged during testimony that he had received a verbal turnover from Mechanic 1, but couldn't remember whether he had been told that the screws for the top of the left hand leading edge had been removed. After assuming responsibilities for the midnight shift, Supervisor 3 assigned Mechanics 3 and 5 to the boot replacement job. Mechanic 4 prepared to continue the boot replacement job, but was soon reassigned to another task and, therefore, turned boot replacement over to Mechanic 5. Mechanic 4 says that he told Mechanic 5 that the screws were removed and that the boots needed to be replaced, and also stated that he gave the bag of screws found on the lifting mechanism to Mechanic 5.

The NTSB report states that Mechanic 5:

> ...did not recall receiving a verbal shift turnover from [Mechanic 4], and was not aware of any previous work performed on the leading edge of the right side of the horizontal stabilizer.

Sometime after the shift change, N33701 was moved out of the maintenance bay to accommodate higher priority maintenance on another aircraft. The boot replacement job on N33701, therefore, continued outside in the dark.

Mechanic 5 completed the right side boot replacement with help from another inspector on his shift, Inspector 2. Mechanic 5 stated that he had some screws left over from the job but didn't think it significant since several screws had been replaced with new ones.

Sometime during the process, a third inspector, Inspector 3, examined the top of the horizontal stabilizer assembly and "assisted with the installation and inspection of the deice lines on the right side of the horizontal stabilizer." The NTSB report records that:

> He stated that he was not aware of the removal of the screws from the top of the left leading edge assembly of the horizontal stabilizer. In the dark outside the hangar, he did not see that the screws were missing from the top of the left side leading edge assembly for the horizontal stabilizer.

When the job was finished, Supervisor 3 gave Mechanic 3 the M-602 cards to record completion of the right hand leading edge boot replacement. Mechanics 3 and 6 along with Inspector 2 (all from third shift), completed the paperwork for the job while Mechanic 4 cleaned up the work site. Mechanic 4 stated he was told that the left hand boot job would be deferred to a later time. The NTSB report also records that:

...since he had received no information that work on the deice boot on the left side of the horizontal stabilizer had been started, he and [Mechanic 3] had agreed that they would not begin to remove the deice boot on the left side of the horizontal stabilizer until they completed the work on the right side deice boot.

Supervisor 3 checked the reverse side of the M-602 card for the left hand leading edge boot replacement, saw no indication that work had been done on the boot, and therefore released the aircraft for a 7:00 A.M. departure to Laredo.

The Problems

Clearly, the failed maintenance process that occurred in Houston on the night of September 10 was fraught with error. A quality control inspector who should have been providing independent oversight was, instead, assisting with the work. The maintenance organization lacked the lead mechanic and lead inspector required by the Continental Express GMM. The local FAA inspector was operating under such a heavy workload that FAA oversight was inadequate.

Of greatest impact, however, was the failure of the second and third maintenance shifts to maintain continuity of work—to conduct proper shift turnovers. As with so many accidents, the Eagle Lake misfortune could have been averted if the maintenance crews involved had understood and exercised the basic principles and elements of proper shift turnover. A well-conducted turnover would have addressed the current status of work, the steps necessary to continue and complete the work, and the hazards attendant to work performance. Clearly, in this case, the process failed.

Procedural Violations

Shift turnovers were considered by Continental Express to be an important factor in successful maintenance. The NTSB investigating team found that:

GMM 1, Section 3, Paragraph 10, specifies that it is imperative for maintenance/inspection forms to be completed to ensure that no work item is overlooked. Such work includes the completion of maintenance/inspection shift turnover forms, so that oncoming supervisory personnel can be made aware of complete/incomplete work, and the documentation of incomplete work that the mechanic can note on the reverse side of the M-602 work cards. GMM 1, Section 5, Paragraph 7, specifically addresses several methods to ensure proper turnover during shift changes. These methods include briefings by mechanics to supervisors and briefings by outgoing supervisors to incoming supervisors.

Yet, on the night of September 10, M-602 cards were not issued or used by the second shift crew. Shift turnover checklists appeared unused by either crew, and verbal turnovers were, at best, incomplete and, at worst, not performed.

Further, shift turnover procedural errors were not isolated to one or two team members. The investigating board reported that:

The failure to follow proper turnover procedures—the most dramatic failure in the accident—involved mechanics, supervisors, and inspectors from two shifts and noncompliance with GMM procedures.

These teams had procedures that, if heeded, would have ensured successful work continuity. They chose, however, not to follow them.

Confused Task Accountability

One primary function of shift turnover is to clearly establish accountability for tasks that will be assumed by the on-coming crew. At the Continental Express maintenance facility, however, the second and third shift maintenance crews failed to establish distinct accountability for the performance and overview of boot replacement on N33701. Responsibility for the task passed between so many supervisors and mechanics that accountability was diluted and task status information was clouded.

This dilution of responsibility was exacerbated by a dual chain of command. Mechanics seemed to work interchangeably between the hangar supervisor and the flight line supervisor. As a result, status reports were sometimes provided to the wrong leader. NTSB investigators commented:

The other second shift supervisor, who was not responsible for N33701 [Supervisor 1] but was in charge of a C check on another airplane, assigned two mechanics to the second shift supervisor responsible for N33701 [Supervisor 2]. He received a verbal shift turnover from one of the mechanics he had assigned to the other supervisor. However, this turnover came after he had already given a verbal shift turnover to the oncoming third shift supervisor [Supervisor 3], informing him that no work had been done on the left stabilizer. The Safety Board found that when he did receive the verbal turnover from the mechanic, he failed to fill out a maintenance shift turnover form and failed to inform the oncoming third shift supervisor. Also, he did not direct the mechanic to give his verbal shift turnover to the second shift supervisor who was responsible for N33701 or to the oncoming third shift supervisor. Instead, he instructed the mechanic to seek out a third shift mechanic and report to him the work that had been accomplished.

Though multiple, disconnected verbal turnovers occurred, no one person remained truly responsible and cognizant of work on N33701. A dissenting opinion from the investigating board highlights this lapse:

Still another factor that I believe to be highly relevant here was the absence of a Lead Mechanic and a Lead Inspector as specified in the GMM. Senior management's failure to fill these positions in effect diffused and diluted the chain of authority and accountability among maintenance and inspection personnel at Continental Express. A detailed examination of the organizational aspects of the maintenance activities the night before the accident reveals a melange of crossed lines of supervision, communications and control. This situation, more than any other single factor, was directly causal to this accident.

Deficient Work Log Entries

A fundamental tenet of proper shift turnover is that the previous shifts' work logs or operating log entries be reviewed by the on-coming relief to determine status of components, systems, processes, and tasks. It follows that accurate work logs and operating logs are a critical part of successful shift turnovers. Yet, when poorly written, inadequately reviewed, or when entries are not completed in a timely fashion, they represent a liability rather than an asset.

In Houston on the night of September 10, work logs of the maintenance crews were inadequate. In at least one case, an important entry was seriously deficient. Inspector 1's last log entry before he left his station stated, "Assisted mechanics with removal of deice boots". The entry fails to clearly communicate the current status of the task, a fatal flaw.

Inadequate Work Log Review

Work log and operating log reviews constitute an important part of independent oversight in an industrial operation. Review of logs by supervisory personnel provides a check to ensure that entries are proper, accurate, and representative of the history of the work or operations conducted on that shift. Reviews also assist leaders to determine if their subordinates are controlling maintenance and operations in accordance with prescribed practices and procedures. It is clear that, in Houston, log reviews by maintenance supervisors had become deficient.

Definition, Purpose, and Applicability

Definition of Turnover

Turnover, or shift change, refers to the exchange of information between work units, one of whom is relieving the other. It can occur between operating units, maintenance units, design teams, or construction teams.

Purpose of Turnover

Though turnovers differ in shape and form depending on the task and the nature of the work team involved, all share a common purpose: *Information which is critical to the successful continuation of a process must be passed from one group to another in a fashion that limits interruption of work and ensures safe and efficient work completion.*

Applicability of Turnover

The concept of turnover is not new or unique. Anyone who has worked in a team environment knows that turnover is as relevant on a friendly sporting field or a hostile battlefield as it is in the industrial environment. Track and field relay teams, for example, expend great effort in perfecting the baton pass. They have learned that the most likely source of failure in the relay is a poor handoff.

Within the industrial arena, shift turnover is the *baton pass* between an on-coming work group and an off-going work group. It is appropriate for nearly every type and size of operating and supporting team. Whether operators, maintenance technicians, engineers, health physicists, or craftsmen—all teams engaged in activities which require passing responsibility for a task to an on-coming work group should acquire the skill of proper shift turnover.

Eight Principles of Shift Turnover

Your industrial team needn't be plagued by the turnover problems experienced by the maintenance crews just studied. Eight simple principles govern the process of shift turnover. They are easily remembered because they are the same principles that regulate the baton pass on a successful relay team.

Principle 1   **Make sure you have a trained relay team.** No matter how hard the members of an industrial team try, they are unlikely to conduct successful turnovers unless they are skilled in their technology. Team members must be capable of knowledgeably controlling the components, systems, and processes with which they are charged. That implies, obviously, that they must also understand the language of their vocation.

Principle 2   **Practice the baton pass often.** The skill of shift turnover is critical to the success of the industrial team. As such, it must be practiced to perfection. *Practice* occurs every time a shift turnover is conducted—*so long as competent, caring leaders are there to observe and correct the process!* Without leaders on scene to advise, correct, and encourage, shift turnovers will soon degrade to the level observed in the Eagle Lake event.

Principle 3   **Make sure everyone shows up for the race on time.** When crew members are late, turnovers are usually rushed or omitted. Punctuality, therefore, must be highly valued among industrial team members and their leaders.

Principle 4   **Make sure all the team members know the race strategy.** Almost every shift has a different set of priorities and, necessarily, a different strategy. Therefore, early in the shift, either right before or im-

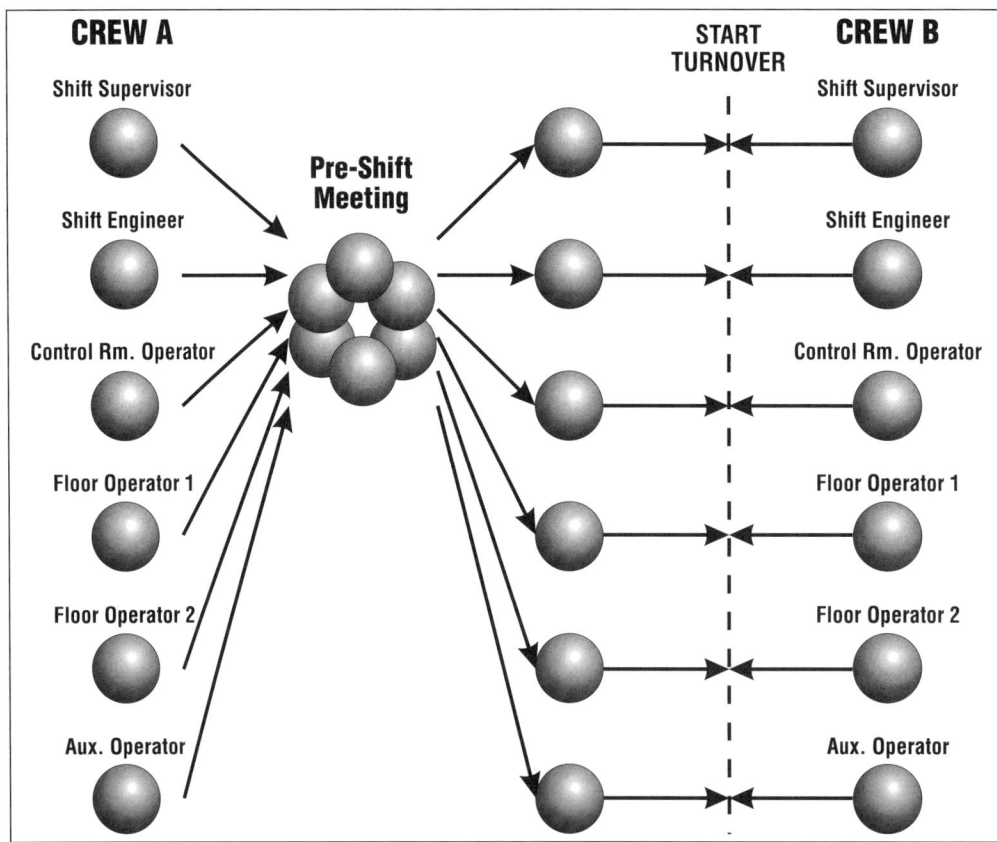

Figure 19-4: Sequence of events in a typical shift turnover for two hypothetical crews.

mediately after shift change, a strategy meeting must occur that results in all of the team members knowing the *game plan* for the shift. Included should be a discussion of the status of major operating systems, maintenance or testing work in progress, and the principal tasks that must be accomplished during the shift. Without this "big picture" the ability of team members to assist, especially in time of crisis, is severely limited.

Principle 5 **Make sure there is a baton to pass.** In an industrial environment, the baton is the bundle of vital information passed to the next runner for the succeeding leg of the relay. The information is packaged in a variety of media including verbal interview and response, visual observation of the work area (with its related systems, components, and instrumentation), and review of written logs, data record sheets, and status boards. As such, the off-going crew members have a vital responsibility to ensure that logs, turnover checklists, and status boards are up-to-date and complete.

Principle 6 **Minimize the number of times you pass the baton.** After the Three-Mile Island accident (March 28, 1979), the nuclear-powered electric generating industry was forced to re-evaluate operating and maintenance team practices. Deficient shift turnovers played a vital role in the accident. Many of the utilities elected to operate on a 12-hour shift schedule rather than an 8-hour rotation in order to eliminate one of the turnovers.

Every time a crew (or individual) turnover occurs, there is danger that vital information will be omitted. People filter information, no matter how skilled they are. Every turnover results in added filtering. Therefore, wise industrial team leaders devise shift schedules and policies for on-shift reliefs that minimize the number of turnovers needed.

Principle 7 **Every time the baton must be passed, make sure that a skilled runner is in position to accept it.** Whenever a crew or individual turnover must occur, crew leaders and team members are responsible to ensure that competent, fit-for-duty team members are in place to accept the turnover. Only then can clear accountability be established for task performance and reporting.

Principle 8 **When you pass the baton, don't drop it!** No matter how good the policies of an industrial team, if they are not well-executed, they are ineffectual. Team leaders must, therefore, ensure that team members concentrate on the process *every time it is performed* so vital information is not omitted. Shift turnover checklists should *always* be used to ensure that critical information is not forgotten.

The Process of Shift Turnover

Now that the principles of shift turnover are clear, the fundamental elements of the shift turnover process and how they should be executed must be outlined.

Figure 19-4 depicts the hypothetical membership of two industrial operating crews. The objective of the turnover is to successfully replace each crew member with his or her counterpart. For that task to be effectively accomplished, each crew member and leader must know the status of his or her particular work station and the general status of related functions. Therefore, every on-coming team member must coordinate with his or her counterpart, normally at their respective work stations.

Before that can occur, each team member must learn the overall plant, process, or task status. Without overall plant status knowledge as well as station-specific knowledge, work team members will operate at a deficit. (Team response to a problem that affects more than one work station will be limited.) Accordingly, a central or *pre-shift meeting* is necessary to initially inform all team members of the status of plant systems and components, newly discovered abnormalities, and the progress of current work. This meeting prepares team members to perform as a team rather than separate work group members.

The changing of the shift usually entails four steps which, if well-executed, minimize or eliminate problems. These four steps are (1) work or station preparation by the off-going crew, (2) pre-shift meeting of the on-coming crew, (3) pre-shift tour of individual work stations by on-coming crew members, and (4) station turnover conducted jointly by both off-going and on-coming crews.

Though the elements may differ slightly in complexity or form, depending upon the nature of the task and the team, this four-step process provides the basic structure for successful turnovers. You should consider each element carefully before discarding any one as inapplicable to your situation.

Preparing the Station for Turnover

In the industrial environment, successful turnover is a coordinated effort. Though much of the responsibility for station turnover lies with the on-coming station operator, the off-going crew member can expedite the process through skillful advance preparation. Such preparation and organization will pay great dividends through improvement of station turnover effectiveness.

Actions of the off-going watchstander prior to station relief should include:
- Log and Data Record Completion
- Checklist and Status Board Completion
- Storage of Tools and Equipment

Log and Data Record Completion

The off-going crew member should ensure that the last set of required equipment operating data (e.g., temperatures, pressures, tank levels) has been recorded on the data record sheets, and also that the narrative logs for the station are updated. An off-going crew member who has failed to complete data recordings or log entries in a timely manner is likely to be hurriedly making rounds and recording data rather than conducting a turnover discussion with the oncoming crew member. The result is often a poorly conducted turnover with incomplete information transfer. (The oncoming crew may be robbed of critical station information and the historical record becomes thin.)

Included in the narrative log should be the current status of alarms and equipment anomalies. Though the oncoming crew member must tour the equipment station to determine the status of equipment and alarms, the off-going crew member can improve the turnover by properly recording the status of alarm conditions or equipment abnormalities beforehand. This record also serves as a reminder to both the oncoming and off-going crew members to discuss these important items.

Finally, the off-going crew member should ensure that a sufficient number of blank data record sheets and narrative log pages are available. Since one of the first tasks of the oncoming crew member will be to complete a set of equipment data record sheets, if a new set of record sheets is not on station, recording may be delayed. With blank record sheets immediately available, the oncoming crew member has a better opportunity to begin the shift in an orderly and controlled manner.

Checklist and Status Board Completion

Prior to station turnover, the off-going crew member should complete the turnover checklist for the station. As indicated earlier, the turnover checklist should summarize equipment status and anomalies, tasks in progress, and tasks that must be completed.

Also, throughout the shift, the off-going crew member should record on the checklist reminder notes of informational items that should be communicated to the oncoming watchstander but that might otherwise be forgotten. If committed simply to memory, many will probably be overlooked.

In addition, status boards or status sheets (besides the logs or turnover checklist) should be updated prior to turnover. Status boards are designed to reflect the current status of equipment. When they are not kept up to date, they at best fall into disuse. At worst, they can mislead the oncoming station operator regarding the actual status of equipment. They should be updated at the earliest opportunity following changes in the status of equipment illustrated on the board.

Storage of Tools and Equipment

The off-going crew member should ensure that the station is clean and uncluttered. Tools and equipment used during the shift should be replaced and the area cleaned.

Besides being a safety hazard at the station, loose tools and equipment lend to a picture and an attitude of undisciplined operations. A well-organized station, on the other hand, helps to set an attitude of control and consistency. Tools and equipment which are cleaned and returned to appropriate storage locations will be available when next needed, especially in cases of emergency.

Pre-Shift Meeting

The pre-shift meeting represents the start of the shift change—the beginning of the baton pass. It consists of a meeting of approximately 15 to 20 minutes duration prior to the commencement of shift in which pertinent facility conditions and plans for the upcoming shift are conveyed to the shift team members.

Figure 19-4 illustrates the ideal shift turnover sequence. A pre-shift meeting is conducted to provide each team member with a thorough, overall picture of equipment and plant conditions before individual station reliefs.

Unfortunately, pre-shift meetings are frequently neglected because they require the oncoming team members to arrive earlier than their assigned shift. Rather, team members go directly to their shift counterparts without attending a pre-shift meeting. In such cases, many team members enter the shift with isolated understanding, a potentially dangerous situation.

Here are some guidelines for properly executing the pre-shift meeting:

Arrival in Advance of Shift — The oncoming crew should arrive approximately 30 minutes before the scheduled shift change. Provision must be made for crew members to arrive in time to receive a proper pre-shift briefing. If there is no overlap between crews, there cannot be a proper baton pass.

Pre-Shift Briefing — The oncoming crew should gather in a quiet location for the pre-shift briefing. The location chosen for this meeting is very important. It must be quiet and free from interruption so that everyone can hear the briefing and ask questions

Meeting Constituency — The meeting constituency must include all crew members if possible. It is particularly important for the crew leaders to attend the shift change meeting. When crew officials such as the shift supervisor do not attend shift change, the meeting loses significance and effectiveness.

Briefing Conduct and Content — The first order of business of the pre-shift meeting is to determine if all crew members are present. Though duty assignments should have been allocated at least one day prior, absences may require reassignments.

Next, an off-going crew official such as the shift supervisor or shift engineer should brief the oncoming crew regarding current plant conditions, out-of-service systems and equipment, work in progress, and schedule for the upcoming shift. The off-going official should then field questions from the oncoming crew.

The oncoming shift supervisor should then brief his or her crew on plans for the shift (usually described in the daily work plan). Recent plant or crew problems should also be reviewed and lessons learned from the problems disseminated. In addition, the oncoming shift supervisor should read and explain to the crew any written instructions, directions, or orders from the plant manager or plant staff directed to the crew since the crew was last on duty.

After answering questions, the shift supervisor should release the crew members to report to their assigned duty stations.

Pre-Shift Tour

Upon completion of the pre-shift meeting, crew members with assignments requiring on-station relief should conduct pre-shift tours of their stations. A pre-shift tour is a "walk-through" of a duty station of approximately 10 minutes duration to personally determine the status of equipment and facility conditions. During the tour, the oncoming station operator should identify station deficiencies or discrepancies.

The length and depth of the pre-shift tour is dependent upon the amount of equipment over which the station attendant is cognizant, the abnormalities of current plant conditions, and the skill and alacrity of the oncoming station attendant. For operators who are responsible for physically large areas, a control room tour may be adequate initially.

Performing Shift Turnovers

Equipment Review One of the most important elements of the pre-shift tour is to determine the status of equipment controlled at that station. Therefore, the duty station equipment review should usually include:

- A visual inspection of the equipment controlled by that station to determine its status (e.g., operating, standby, out-of-service). The visual inspection should include a review of key indications and investigation of any abnormal sounds, smells, or vibrations.
- A review of danger tags, caution tags, deficiency tags, alarm conditions, and any other warning indicators relevant to the watchstation.

To ensure complete reviews, checklists should be developed and used as a part of the review to assist operators in consistently observing the appropriate items. Turnover checklists should include a summary of the status of equipment and alarms, equipment anomalies, tasks in progress, and tasks to be completed.

The use of checklists may seem to some to be nothing more than additional paperwork. Yet, if left to memory, some equipment checks will eventually be overlooked. Industries with high risks use checklists even when staffed with the most rigorously selected and highly trained professionals. In flight operations, for example, pilots and ground crews use checklists in preparation for most significant events. Well-designed checklists can assist operators in the same way to effectively conduct turnovers.

Status Board Review Another important part of the pre-shift tour is a review of pertinent status boards. The status board review should determine:

- Whether the representation of equipment as indicated on the duty station status boards coincides with the visual inspection results.
- Whether any abnormalities or anomalies exist which were not detected during the visual inspection tour.
- Whether the status board has been initialled or signed and dated if required by facility policy to indicate its currency.

Data and Log Review Next, the oncoming crew member should review the narrative log entries and data record sheets for the last 24-hour period (or since last operating the station, whichever is shorter). If the oncoming operator has not controlled the station for several days, a longer review may be appropriate. The purpose of the review is to determine equipment or station abnormalities, trends in equipment performance from the review of equipment parameters, recorded equipment status, alarms, out-of-service equipment or instrumentation, and any plant casualties or emergencies which have occurred since the operator was last on station.

Station Turnover

The oncoming crew member should now be prepared to competently accept the baton pass from the off-going crew member. Station turnover should then proceed in a controlled, sequential manner.

Permission to Relieve Permission to relieve the off-going crew member may seem an unnecessary step in shift turnover. Sometimes, however, a delicate equipment operation may be occurring which should be interrupted for station relief only at a time prescribed by a central controlling station. Also, some operations may routinely be so sensitive that permission granted by a central control room to relieve must always occur.

In cases that require such control for a whole plant, the central controlling station should announce permission for station reliefs to commence over the public address system. For single stations, permission should be requested over a phone line for the station in question.

The Relief Process Once permission to conduct the relief has been received, the on-coming crew member should indicate to the off-going crew member that he or she is prepared to exchange responsibility for the station. For formal reliefs, plant policy should specify words such as "I am ready to relieve you". To some, such words may seem unnecessary; but, to ensure that the process is seriously performed and properly controlled, adoption of such a practice will pay well over time through elimination of confusion regarding who is actually in control.

The off-going crew member should then brief the oncoming crew member regarding station equipment status and abnormalities. The off-going crew member should answer any questions regarding station status using appropriate checklists and status boards to ensure consistency and completeness in transmitting information.

(Note: This process can be combined with the oncoming crew member's visual inspection tour. A joint tour can serve the purpose of acquainting the oncoming crew member with station status while, at the same time, accomplishing the turnover briefing.)

The oncoming crew member should relieve the off-going crew member with words such as "I relieve you". Again, performing this ritual, though apparently tedious, leaves no question as to who controls the station.

The off-going crew member should then acknowledge relief with words such as "I stand relieved".

The off-going crew member should then sign out of the station narrative log with words such as "Properly relieved as X-Line Process Operator by Ms. Smith" followed by a signature.

The oncoming crew member should log onto the station with words such as "Properly relieved the X-Line Process Operator" followed by his or her signature.

For situations requiring permission to relieve, the relieving crew member should then call the central control room and announce that he or she has properly relieved the X-Line Process Operator.

Two cautions should be included here regarding station relief. First, if an impropriety in the status of the station is discovered during the turnover process, both crew members, off-going and oncoming, must make a judgment regarding whether the impropriety needs to be corrected prior to station relief. Second, high risk tasks should not normally be started if a station relief will intervene. If a crucial task is being performed at the time of a scheduled turnover, crew leaders and station operators must make a sound judgment regarding whether the off-going crew member should remain on duty until the task is completed.

Post-Turnover Meeting

Shortly after station reliefs are conducted and initial shift tasks are begun, a meeting of key crew leaders may need to be conducted to confirm or alter shift work plans based upon station conditions and new information. Lead technicians and principal crew leaders should meet with the shift supervisor to ensure that personnel resources are being utilized to best advantage.

In addition, actual equipment status and configuration should be confirmed. Anticipated equipment status and configuration changes should be analyzed for impact to safety and shift operations. Current or potential problems anticipated for the shift should be discussed to preclude unnecessary errors.

Guidance for Successful Turnovers

Shift turnovers are not difficult. Here are six simple questions and a checklist of guidelines and reminders that will assist any industrial team in conducting successful turnovers:

Six General Questions

Six general questions are always appropriate for turnovers, whether in operations, maintenance, or testing. In fact, the effectiveness of most turnovers in *any* organization can be judged by these six questions:

1. Where are we at? (What is our location in the task or process?)
2. What is our direction? (What is the objective of the task or process?)
3. How fast are we going? (At what speed is the task or process progressing?)
4. What is the status of the equipment? (Is it operating? In standby? Under repair? Is everything working?)
5. What is the current and future environment? (What hazards, conditions, or obstacles confront us?)
6. What is the plan? (How are we going to get to where we are going?)

The first three questions establish a starting point, a direction, and a speed. In geometric terms, these are the three elements of a *vector*. The next three questions serve to establish the environment in which the work is to be done and the path by which the obstacles will be met and overcome.

Guidelines and Reminders

A few important guidelines should be considered in developing and executing a turnover process:

- ❏ Conduct a turnover! It is disturbing how often in industry one operator departs from a work station before the relieving operator has arrived. Off-going crew members are so anxious to leave their stations that "turnovers" are sometimes performed in the parking lot as crew members pass one another. In such instances, mutual equipment observation, log review, and status overview are absent from the process. The chances for error are increased dramatically.

- ❏ Conduct a pre-shift meeting. Failure to conduct a pre-shift meeting is akin to attempting to run a football team without a huddle. It can be done, but it is *very* difficult. If a pre-shift meeting is not possible, conduct a strategy meeting immediately after the change of shift. Remember, though, that substituting a post-turnover meeting for a pre-shift meeting is less than ideal. Many team members probably will not be able to attend because play has already started!

- ❏ Enforce the requirement that all crew members be present for the pre-shift meeting. Every team member must be aware of important equipment status, major maintenance or testing in progress, and the critical tasks that must be accomplished during the shift. If a team member misses the pre-shift meeting, ensure that he or she is individually briefed, and, if necessary, don't allow the team member to perform until such a briefing has occurred.

- ❏ Make sure the team leaders attend the pre-shift meeting and set a good example in conducting their own individual turnovers. Breakdown in the discipline of shift turnover is usually attributable to inadequate leadership. When leaders don't demonstrate concern for the process, neither will subordinates. It is imperative, then, that crew or team leaders attend pre-shift meetings and preserve the process integrity. Not only is their input and perspective important, but their presence emphasizes the importance of the process. When leaders cease to attend pre-shift meetings, the meetings will assuredly lose their effectiveness.

- ❏ Make sure that there is a clear, singular line of authority and accountability for each critical function. One person should be responsible for each job. Don't fall prey to the Eagle Lake event mistakes.

- ❏ Use a turnover checklist. Even though you have conducted turnovers many times before, don't trust your memory. Remember, airline cockpit crews, though exceptionally well-chosen and trained, rely upon checklists to supplement their skills and knowledge. Though they have conducted thousands of takeoffs and landings, they still use checklists. If the task is non-routine or, for some other reason, does not have an established turnover checklist, write a simple checklist to ensure that you don't forget something. Remember, the period of turnover is usually crowded with many other tasks and distractions. Regardless of how good your memory is, you'll forget a vital piece of information.

- ❏ Avoid, if possible, turning over work that is at a critical juncture. If the current work group can successfully complete the task in a reasonable time (or reach a phase of the work at which turnover can be more safely conducted), delay the turnover.

- ❏ Don't forget to conduct thorough equipment "walk-downs" during individual turnovers. A successful turnover process is easily derailed by the failure of the oncoming party to perform a thorough observation tour of the equipment station prior to receiving the verbal turnover. The equipment walk-down, when properly performed, provides a visual review of the status of important components, the location and meaning of warning tags, and the condition and hazards associated with the equipment station.

- ❏ Beware midnight shift turnovers! The weakest link in the shift change "relay" normally occurs between the second, or swing, shift (typically 4:00 P.M. to 12:00 A.M.) and the third, or midnight/graveyard, shift (typically 12:00 A.M. to 8:00 A.M.). Though fatigue is a part of the reason, the lack of observant leaders is probably the most important factor. Well-disciplined industrial teams, however, have learned that good turnovers are critical to their own safety and efficiency. Therefore, whether observed or not, skilled team members excel in turnover to enhance their performance.

And a final note: Many operating problems result from incomplete communication of plant status from shift to shift. Because of this, some organizations have chosen to operate in twelve-hour shifts in order to eliminate one turnover during the operating day and thus ensure better operating continuity.

Topic Summary and Questions to Consider

The tragedy at Eagle Lake is yet another costly demonstration of the need to establish a systematic approach to industrial operations. In this event, fourteen people unnecessarily lost their lives and a $7.75 million dollar aircraft was destroyed—an event which probably would have been prevented by a reasonably well-conducted shift turnover.

In the dissenting opinion to the NTSB report on N33701, the author states:

> *By permitting, whether implicitly or explicitly, such deviations to occur on a continuing basis, senior management created a work environment in which a string of failures, such as occurred the night before the accident, became probable.*

The responsibility for establishing and maintaining a viable shift turnover process falls directly to the leaders of industrial teams. Policies and practices for turnovers must be clearly delineated and rigorously enforced. The *baton pass* must be taught, practiced, and monitored to maintain high standards. The decay of that process is a harbinger of impending failure. Are *you* willing to take the risk?

Ask Yourself Do your work teams conduct formal turnovers? Are turnover checklists used?

Do you routinely conduct pre-shift meetings before individuals relieve their counterparts? Do team leaders attend the pre-shift meetings?

Do facility or activity leaders observe turnovers frequently to ensure that they are correctly and consistently performed?

Part IV

Implementing the Systematic Approach

- ○ Investigating Abnormal Events
- ○ Conducting Continuing Training
- ○ Evaluating Operating Performance
- ○ A Case Study in Implementation
- ○ Your Challenge

The first three parts of this text have provided the basics for understanding the need for an operating strategy, establishing an overall strategy, and identifying and applying key elements and skills of the strategy.

Part IV continues in the vein of implementing the strategy by discussing how to put all the elements together, going beyond simple failure avoidance toward process and product *improvement*—true operating success.

Chapter 20

Investigating Abnormal Events

The first three parts of this text focused on the dangers of industrial operations, a strategy to minimize those dangers through systematic equipment and process control, and the skills necessary to accomplish that strategy. But talking about the strategy and implementing it are worlds apart. Therefore, Part IV is devoted to three essential implementation processes, the first of which is investigating abnormal events.

Despite the best efforts of even the best industrial teams, people make mistakes and machines aren't perfect. In 1979, after the Three-Mile Island accident, Admiral Rickover told Congress:

> ...[M]istakes must be taken into account.... [I]t is important to recognize that mistakes will be made, because we are dealing with machines and they cannot be made perfect. The human body is God's finest creation and yet we get sick. If we cannot have perfect human beings then why should we expect, philosophically, that machines designed by human beings will be more perfect than their creators? ...That is the basis on which I have conducted all my work in this field and I believe it true just as strongly today as I ever have.

Yet, few engineers have been more devoted than Hyman Rickover to eliminating error in design, construction, operation, and maintenance. In fact, Rickover's work predates today's popular quality movement by several decades. He realized early in his labors that, through investigating mistakes, great strides could be achieved in personal and organizational improvement.

The ability (and commitment) to investigate abnormal events is an important ingredient in implementing the strategy for operating success. Through aggressive investigation, the causes of things that go wrong (or things that almost go wrong) can be determined and others within the industrial team inoculated against similar errors.

The Challenger Accident, January 1986

On the morning of January 28, 1986, the Space Shuttle Challenger (Figure 20-1), carrying seven astronauts (including Christa McCauliffe, America's first "teacher-in-space"), was launched from Cape Canaveral amidst intense national media attention. This was to be the twenty-fifth successful launching of the space shuttle in a record-paced schedule.

But, seventy-three seconds after lift-off, disaster struck. At an altitude of 46,000 feet (nine nautical miles), the Challenger throttled up to one hundred percent of power on its main engine (fueled by a hydrogen and oxygen mixture from the vehicle's external tank) and two solid rocket boosters to propel it into orbit. A fiery explosion lit the sky, breaking the vehicle apart. The capsule compartment separated from the vehicle carrying the astronauts, and at least three emergency breathing packs were activated. But all of its occupants died approximately two-and-one-half minutes later when the capsule compartment ditched in the Atlantic at a speed of nearly two hundred miles per hour. The nation watched in horror.

A Presidential Inquiry On February 3, 1986, President Reagan appointed a presidential commission, chaired by former Secretary of State William P. Rogers and astronaut Neil Armstrong, to investigate the Challenger accident. Their assignment was, within 120 days, to:

(1) Review the circumstances surrounding the accident to establish the probable causes of the accident; and (2) Develop recommendations for corrective or other action based upon the Commission's findings and determinations.

The commission was sworn in on February 6 and began interviewing NASA officials regarding the agency's own internal inquiry. They would not finish their work until June.

Background The Space Shuttle Program was the latest U.S. aerospace endeavor. The shuttle's anticipated missions ranged from commercial satellite deployment to classified military projects including the Strategic Defense Initiative (SDI). Originally, NASA committed to an optimistic schedule of one shuttle launch per week. It soon became clear, however, that such a schedule could not be supported by the limited spare parts inventory and the logistical nightmare of recovering and refurbishing the boosters as well as the shuttles themselves. Turn-around time was further extended by a problem with the shuttle's brakes which necessitated landing at the huge dry lake at Edwards Air Force Base in California rather than the original landing site in Florida. Recovery at Edwards required piggy-backing the shuttle on a transport aircraft and ferrying it cross country to its home base in Florida.

By 1985, shuttle flights had become commonplace. Networks were reluctant to interrupt normal programming to broadcast launches because of viewer complaints. Senator Jake Garn of Utah had ridden aboard one flight and plans were developing to include a school teacher in an upcoming flight to spark the interest of students around the nation in spaceflight.

Shuttle flights, however, were not as routine as most Americans believed. The shuttle had been plagued with problems, not only in its braking system, but also in a host of other components and systems. Cannibalization of shuttles to meet launch schedules had become an accepted practice.

One worrisome problem was that solid rocket booster joints didn't consistently seal as designed. The solid rocket booster joint design—patterned after the Titan missile, a military rocket whose casings were seldom recovered for study—relied on two O-rings to prevent flame leakage at the joints. Joint performance wasn't a major concern until O-ring erosion was observed as early as September, 1977, during static testing of the solid rocket boosters. The joints were not performing as expected.

Figure 20-1: Challenger launch (STS 61-A 22) with detail showing field joint ring.

O-Ring Erosion

By February 10, the commission began to assemble accident investigation experts and administrative support personnel from other agencies and from the private sector (soon to total 42 permanent staff as well as 100 other contract personnel) to assist in the commission's work. They included, among others, Dr. William Graham, Acting Administrator of NASA; Air Force Major General Donald Kutyna, serving as chairman of the Accident Analysis Panel; Nobel Laureate Physicist, Dr. Richard P. Feynman; and astronaut, Dr. Sally Ride.

On February 10, in a closed committee session, commissioners learned about the solid rocket motor joint sealing problem. The motors were constructed of stacked segments, the joints of which were closed using a putty and O-ring seal. (See Figure 20-2.) Primary and secondary O-rings were the last line of defense in blocking 5000 degree flame from exiting the joints, should other barriers fail.

Severe O-ring erosion had occurred on the second shuttle flight, flight STS-2, in November, 1981. By April, 1984, Marshall Space Flight Center and Morton-Thiokol, maker of the solid rocket boosters, established a joint task force to study the problem. Their results were inconclusive and no major redesign resulted. Again in January, 1985, O-ring erosion, the worst yet, occurred on a flight launched at the coldest temperature up until that time. The O-rings were recovered and examined in June of that year and a Morton-Thiokol O-ring engineer warned his manager of impending disaster if the design wasn't changed. Plans to redesign the seals were initiated, but the shuttle flights continued.

At the outset of the shuttle program, the joint seals on the solid rocket booster segments were categorized as "Criticality 1R". The "R" categorization meant that redundancy existed in the component; if the primary seal failed, the secondary O-ring would also have to be breached to cause a burn-through of the solid rocket motor. But the high criticality level of "1" was assigned because, if both seals failed, a high probability existed that the burn-through would ignite the shuttle's external tank, a three-section vessel containing both hydrogen and oxygen used to help propel the shuttle vehicle into orbit. Detonation of the external tank and destruction of the entire vehicle would result.

Based upon a history of O-ring erosion involving both the primary and secondary rings, the joint seals had been upgraded by 1986 to "Criticality 1" meaning that no real redundancy existed. Any burn-through could create a disaster.

In August, 1985, NASA headquarters personnel in Washington, D.C., were briefed on the history of

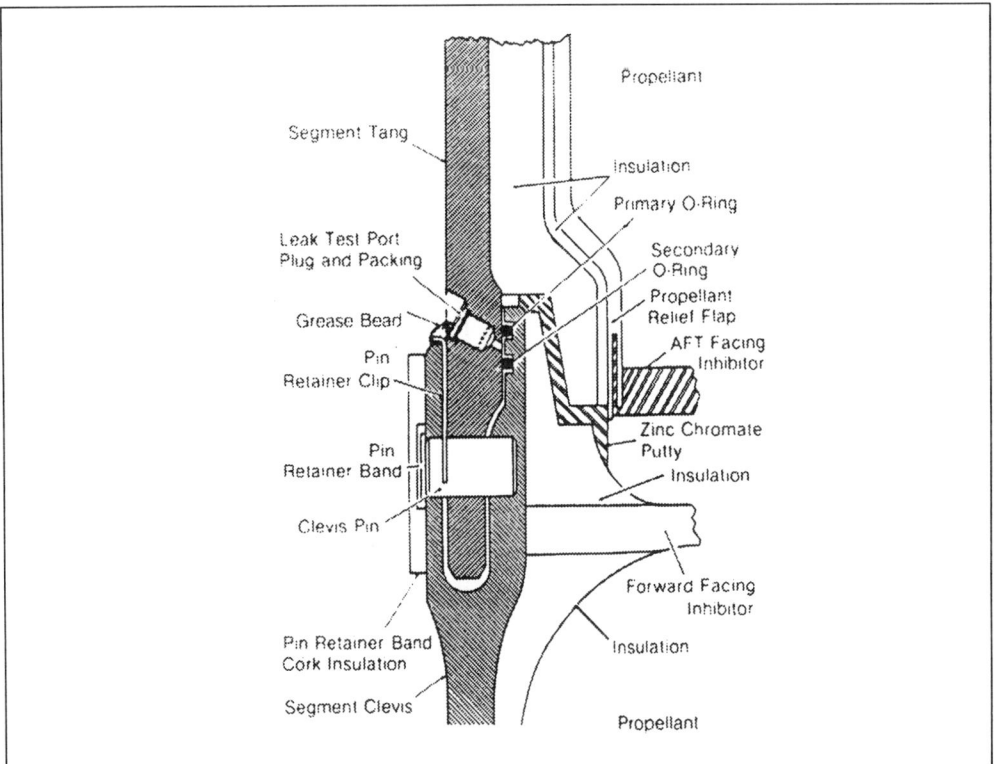

Figure 20-2: Booster-rocket field joint detail. [From Congressional Report]

O-ring problems; but, the problem was not elevated to a high level of concern. The O-rings continued to be treated as Criticality 1R items rather than the new Criticality 1 classification.

Late Night Decision

As the record of the O-rings became known to the commission, a pall was cast upon NASA's organizational soundness. By February 14, commissioners heard testimony from members of NASA and shuttle component contractors which spread doubt upon the decision to launch Challenger. The commission reported:

In recent days, the Commission has been investigating all aspects of the decision making process leading up to the launch of the Challenger and has found that the process may have been flawed. The President has been so advised.

Late in the evening of January 27, 1986, the night before the scheduled launch of the twenty-fifth space shuttle flight, a tele-conference transpired between managers and engineers at Morton-Thiokol and NASA managers at both Marshall Space Flight Center in Huntsville, Alabama, and Cape Kennedy, Florida. The subject of the call was flight readiness for the following day's launch. During the conference, Morton-Thiokol engineers recommended that the shuttle launch be delayed because of expected low temperatures at the launch site. The predicted temperature for the launch was 18°F, the coldest in shuttle history. Morton-Thiokol engineers were concerned about the ability of the primary and secondary O-rings to expand rapidly enough to protect against flame erosion to the outside of the rocket casing. Previous data, though incomplete, indicated that the O-rings could experience severe erosion at reduced temperatures. (See Figure 20-3, Correlation Between O-Ring Erosion and Temperature.) Following a heated debate in which NASA managers questioned the responsibility of Morton-Thiokol's engineers to raise such an issue the night before a scheduled launch, the general manager of Morton-Thiokol told the vice president of engineering to take off his "engineer's hat" and put on his "manager's hat". The engineering recommendation was overruled and Morton-Thiokol management, reversing their original decision, recommended a launch for the following day.

Accident Sequence

Challenger was cleared for launch and lifted off on the morning of January 28, 1986. In the 73 seconds after the shuttle cleared the pad, a sequence of events occurred which culminated in the explosion of the shuttle's external tank. A detailed technical analysis of film, video, telemetry data, and recovered wreckage was performed by NASA, the National Transportation Safety Board, the United States Air Force, and a host of personnel from the shuttle component contracting agencies. They reconstructed for the commission this probable scenario:

- *Liftoff began with ignition of the Solid Rocket Boosters (6.6 seconds after ignition of the Space Shuttle Main Engines).*
- *At .678 seconds after liftoff, the first puff of smoke was observed emanating from the right Solid Rocket Booster in the vicinity of the aft field joint between the booster and the External Tank, near the*

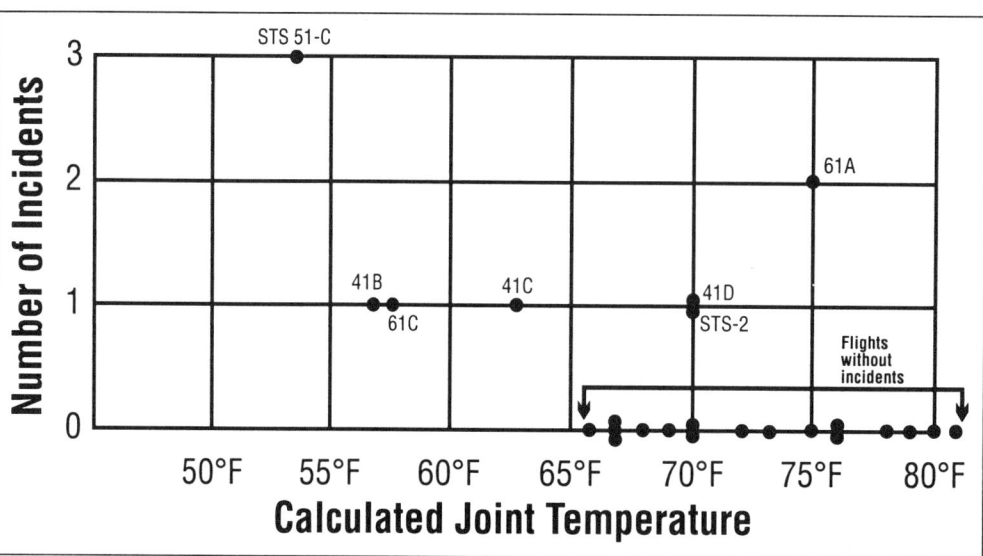

Figure 20-3: Correlation between temperature and O-ring incidents.

External Tank attack strut.

- *By 2.5 seconds after liftoff the generation of the smoke stopped. About nine puffs of smoke had been generated.*
- *During the ascent—beginning at about 37 seconds and lasting until about 64 seconds—heavy wind shears were encountered that, although not producing excessive loads, did provide a "bumpy ride" that could have had an effect on an already damaged system; namely, the seal in the aft field joint.*
- *Everything looked normal until about 59 seconds after liftoff. At this time flame started coming out of the right booster in the area where smoke had been seen before.*
- *The flame and hot gas plume grew in size during the next 14 to 15 seconds. It was impinging on the aft (hydrogen tank portion) of the External Tank close to where the tank is connected to the Solid Rocket Booster.*
- *At about 64 to 65 seconds the structural integrity of the External Tank was breached and hydrogen began leaking from the aft region near a welded seam.*
- *Beginning at about 72 seconds, a rapid sequence of events began. The heat and flame weakened connection (strut) to the lower part of the External Tank failed. At about 73 seconds, the bottom portion of the External Tank (hydrogen tank) failed.*
- *Failure of the bottom of the External Tank caused the pressurized liquid hydrogen to be released rapidly, which in turn propelled the hydrogen tank, with about 2.8 million pounds of force, into the intertank area (between hydrogen and oxygen tanks), and probably into the bottom of the oxygen tank (upper portion of External Tank).*
- *About the same time, the forward part of the booster (frustum) impacted with the forward part of the External Tank, which contained the oxygen tank. Failure of the aft booster attachment strut had allowed the bottom part of the booster to move away from the External Tank, rotating about its forward attachment point.*
- *This nearly instantaneous mixing of hydrogen and oxygen, in an environment of sparks produced when the hydrogen tank was propelled into the intertank area, caused a fire, or nearly explosive burning of these propellants. The Orbiter, under severe aerodynamic loads, broke into pieces within fractions of a second.*

Cause of the Accident

Evidence of direct cause was clear. The joint seals had indeed failed. The commission reported that:

> *The consensus of the Commission and participating investigative agencies is that the loss of the Space Shuttle Challenger was caused by a failure in the joint between the two lower segments of the right Solid Rocket Motor. The specific failure was the destruction of the seals that are intended to prevent hot gases from leaking through the joint during the propellant burn of the rocket motor. The evidence assembled by the Commission indicates that no other elements of the Space Shuttle system contributed to this failure.*

Ambient temperature at the time of launch was 36°F, 15 degrees colder than any previous launch. Calculated O-ring temperature was 28°F, 25 degrees colder than any previous launch. The commission concluded:

> *A compressed O-ring at 75°F is five times more responsive in returning to its uncompressed shape than a cold O-ring at 30°F. As a result, it is probable that the O-rings in the right solid rocket booster aft field joint were not following the opening gap between the tang and clevis at the time of ignition.*

Contributing Problems

But O-ring erosion was not NASA's only problem. In the course of the investigation, the commission discovered other troubles within the agency. Additional warning signs of impending danger—accident precursors—were present before the event. Leadership, communications, independent safety checks—all had fallen prey to the pressures of meeting launch schedules.

Overextended Resources By 1985, schedular pressures were taking a subtle toll on personnel and equipment. Long hours and late nights were the rule. On the shuttle mission prior to the accident, just 31 seconds before launch, a delay was invoked because instrumentation showed that 18,000 pounds of fuel had been inadvertently drained from the shuttle's external tank. The lapse was attributed to personnel error, in part because of fatigue.

Early in the shuttle program, optimistic planners had projected one mission each week. But, as the reality of resources set in, projections were revised. By 1985, even one mission a month had become a struggle. The commission would write:

> One effect of NASA's accelerated flight rate and the agency's determination to meet it was the dilution of the human and material resources that could be applied to any particular flight. The part of the system responsible for turning the mission requirements and objectives into flight software, flight trajectory information and crew training materials was struggling to keep up with the flight rate in late 1985, and forecasts showed it would be unable to meet its milestones for 1986.

Equipment readiness fared little better as budgeted funds for spare parts dwindled. To meet launch schedules, one shuttle was cannibalized for parts to equip another. The commission reported that:

> With respect to the flight rate pressures the Commission found...[t]he Shuttle program made a conscious decision to postpone spare parts procurements in favor of budget items of perceived higher priority. Lack of spare parts would likely have limited flight operations in 1986.

Failed Independent Verification

Clearly, the system had been stretched to (and perhaps beyond) the limits of safety. Production and scheduling pressures had led to short-cuts in a system that previously had been a model for safety. The commission concluded that:

> The unrelenting pressure to meet the demands of an accelerating flight schedule might have been adequately handled by NASA if it had insisted upon the exactingly thorough procedures that were its hallmark during the Apollo program. An extensive and redundant safety program comprising interdependent safety, reliability and quality assurance functions existed during and after the lunar program to discover any potential safety problems. Between that period and 1986, however, the program became ineffective. This loss of effectiveness seriously degraded the checks and balances essential for maintaining flight safety.

Strict, independent verification had been the mark of NASA during the Apollo program. Pre-flight readiness normally revolved around the concept of "prove that you are ready to fly". By the time the Challenger was launched, however, the concept seemed to have changed to "prove that you are *not* ready to fly". Though independent verification in the Shuttle Program continued to be forceful in components such as the main engine and control computer hardware and software, the less complex boosters seemed to get less independent confirmation of readiness.

Deficient Communications

Failed independence was, in part, the result of degraded communication. Free flow of information within NASA had become a victim of the cancer of bureaucracy. What once had been a well-knit team had deteriorated into a competition between subordinate groups.

The commission wrote:

> Commission testimony reveals failures in communication that resulted in a decision to launch 51-L [Challenger] based on incomplete and sometimes misleading information, a conflict between engineering data and management judgments, and a NASA management structure that permitted internal flight safety problems to bypass key Shuttle managers.

Somewhere, the critical thinking that marked Apollo was lost in the pressure to meet schedules.

Flawed Leadership

Each contributing cause pointed to a more serious defect—flawed leadership. Organizational command and control had failed. Decisions in the flight readiness review process had been made at a third tier of management, precluding the escalation of serious problems to the top. Not only had Morton-Thiokol originally recommended against launch, but Rockwell, the prime contractor responsible for the launch pad and facilities, had also recommended that the launch not proceed because of ice on the pad. Neither concern had been elevated to a higher level of authority.

The commission's report states:

> The Commission is troubled by what appears to be a propensity of management at Marshall [Marshall Space Flight Center in Huntsville, AL] to contain potentially serious problems and to attempt to resolve

them internally rather than communicate them forward. This tendency is altogether at odds with the need for Marshall to function as part of a system working toward successful flight missions, interfacing and communicating with the other parts of the system that work to the same end.

An Investigative Model

The Challenger tragedy contains many lessons for improved industrial operations—lessons of scheduling, resource utilization, communication, verification, and leadership. But, above all, it provides extraordinary insight into the process of abnormal event investigation.

The science of investigation is age-old. It begins with reasonable gathering and preservation of the evidence available at the scene of the event, interrogation of witnesses to the event, analysis (and synthesis) of the evidence, and determination of the causes of the event. Based upon the analysis, corrective measures are formulated and implemented to prevent the event from recurring. Finally, the event analysis is documented and publicized so that others have an opportunity to learn from it.

As with the process of complex problem-solving, there are a few rudimentary steps that govern investigation. They are:

- Decide to investigate.
- Designate investigators.
- Gather and preserve evidence.
- Conduct a critique.
- Analyze the evidence.
- Determine the causes of the event.
- Correct the causes of the event.
- Document the investigation.
- Disseminate the lessons learned from the event.

In the remainder of this chapter, we shall explore each step and provide guidance for performance within the industrial environment.

Deciding to Investigate

Every abnormal event is worthy of investigation. Yet, some require an in-depth, formal investigation while others can be adequately addressed with a simple, informal analysis conducted by an on-duty operator. When, then, is a formal investigation required? There is no perfectly objective answer. A good rule of thumb, however, is that **a formal investigation should be conducted if an occurrence led to (or potentially could have caused) serious equipment damage or personnel injury**.

Ultimately, team leaders must make the decision whether a formal inquiry is warranted. Yet, without policy and training, no consistency in reporting and investigation can be achieved. Here are some general guidelines for determining when a formal investigation should be conducted:

- A design limit or safety specification for operation of facility equipment has been violated.
- An equipment malfunction, whether related to material failure or personnel error, has resulted or could have resulted in serious injury to personnel or significant damage to the equipment.
- A component or system has functioned in an unexpected manner with no clear cause indicated.

Note that, though an actual accident may not have occurred, an investigation may still be warranted. Most accidents are not first-time events. They have usually been preceded by a number of *near misses*—events which, under less conservative circumstances, may have resulted in damage to people or equipment. Recall that, though the grounding of the *Exxon-Valdez* was the first such event in over 8800 tanker passages past Bligh Reef, a tanker came within one ship length of striking the reef just the day before. By investigating near misses, facility personnel are made aware of dangerous conditions, allowing an opportunity to adjust behavior or correct conditions.

Clearly, not every abnormal event can or should be investigated formally. Bear in mind, however, that most industrial organizations err in the direction of too little formal investigation. The result is often repeated events from the same cause.

Investigating Abnormal Events

Designating Investigators

The practice of investigating and reporting abnormalities should be taught and expected as a job responsibility for all equipment operators and their leaders. Investigation is everyone's responsibility. It should begin as soon as an anomalous situation is discovered.

The first investigator—the person who discovers an abnormal condition—is usually an operator or maintenance technician. If the situation is serious, the operator should promptly report the event so that the investigation can be escalated to the level of on-duty team leaders.

In a medical emergency, the first actions of care are taken by the first responders to the scene. In industrial operations, operators and their team leaders are the usually the first responders, or *primary investigators*, at the site of an event. Not only must they stabilize the situation, they must also initiate the first actions of investigation. Hence, their skills must include protecting people and equipment from further damage, gathering and preserving evidence, and performing interviews (and responding to interview questions). They should know from their training when to begin an investigative process without being told.

Secondary investigators are more experienced investigators appointed to expand an inquiry. They are analogous to trained medical personnel—such as paramedics and doctors—who accept a patient as provided by the first responder, and then further the care. Since secondary investigative response is not always required for abnormal events, a leader such as the plant manager or operations manager normally should determine if an investigation is to be escalated and who is to lead the task.

Qualified Investigators

Because many events are extremely complex, secondary investigators must usually have superior investigative skills, especially for event analysis, cause determination, witness interviews, and event synthesis (event reconstruction). Such skills require special training, so it is wise to designate selected personnel to become formally trained as investigators so they will be ready when circumstances warrant.

Investigators may not have all the technical knowledge necessary to conduct the investigation alone, and thus may require assistance from *subject matter experts*—leaders and technicians currently qualified on the equipment and processes under investigation. For example, in the Challenger investigation, Dr. Neil Armstrong and Dr. Sally Ride were chosen to assist in the investigation, not so much for their formal investigative skills, but rather for the practical point of view and expert knowledge that they could bring to the inquiry. The combination of skilled investigator and subject matter expert, when properly employed, is vital to successful investigation.

Gathering and Preserving Evidence

An investigation can only be as good as the evidence that supports it. Therefore, if evidence is incompletely gathered or inadequately preserved, a successful inquiry is unlikely.

What is Evidence?

Anything that assists in determining what happened should be considered *evidence*. Operating logs, instrument recordings, photographs or videotapes of the scene, location identifiers such as charts, floor plans, or drawings, debris from the event, failed parts or components, and pertinent tools and materials—all may be important elements in determining the sequence of events from an accident or near accident.

Witness Statements

Often, the most important (and sometimes only) sources of information regarding what occurred are witness statements obtained through written deposition and oral interview. Brief oral interviews immediately after an event has occurred are necessary to determine the extent and seriousness of the event. But as soon as possible, all team members who may be able to provide information regarding the event should be asked to provide written statements of their observations and actions.

Written statements should be composed without consulting other team members. The purpose of obtaining separate, written statements is not to test the honesty of team members. Rather, it is a technique that recognizes that two people who see the same event will report it in two different ways. One will see what another has not. By comparing and contrasting the statements, a better picture of the actual event is more likely. Also, if two witnesses discuss the event before writing their statements, they tend naturally to report their observations in light of what the other has observed.

For oral and written testimony to be effective, the respondents must be truthful. If witnesses choose not to be forthcoming, the cause of an event may never be determined, a very dangerous situation. For, without accurate cause determination and dissemination of lessons learned from the event, it is likely to be repeated. Accordingly, leaders of an industrial complex must establish high value on the quality of integrity, both by proclaiming and exemplifying it. "Do as I say, not as I do" leaders inevitably breed untrustworthy subordinates.

Leaders must also strive to use the investigatory process, not as a "witch hunt", but as an objective fact-finding process. Though disciplinary measures may ultimately be necessary, if such measures are perceived as the *reason* for investigation, witnesses will soon become very unresponsive.

Evidence Preservation

Often, when an event occurs, it will be necessary to preserve the scene in an undisturbed condition. Preservation of the scene must be accomplished with reason as well as caution. For example, isolation of the scene must be balanced with stabilization of the event itself. Safety of personnel must take priority over evidence preservation.

If the scene has been stabilized and evidence must be gathered, it is usually wise to blockade the area with signs, boundaries, or guards. Materials and components should be left in their post-event state if possible until technical experts observe and gather the evidence. If the scene can't be preserved, photographs, videotapes, sketches, and maps will provide substantial assistance in the investigation.

Evidence gathered from the scene also must be sensibly preserved. Techniques for preservation may include unique handling techniques to preclude contamination, bagging, and special preservation measures. In some instances, a *chain of custody* may be indicated if it is probable that the evidence will be used in a criminal proceeding. If criminal activity is evident or suspected, appropriate law enforcement and security personnel should be promptly notified and the scene carefully preserved. Again, however, preservation must take second priority to personnel and equipment safety.

Event Critique

Another crucial step in the investigatory process is an oral review of the event often called a *critique*. A **critique** is a formal proceeding similar to a grand jury, preliminary hearing, or congressional hearing. Fact-finding is its purpose—to recount and clarify the details of the event in order to learn and prevent recurrence. (See also Figure 20-4.)

Within the context of critique, several actions should occur. The critique should:

- Organize and preserve the evidence,
- Establish the facts of the occurrence,
- Establish a chronology of events,
- Document the occurrence and its supporting evidence,
- Establish the "probable cause" of the occurrence to extent possible,
- Verify the initial actions taken (and the adequacy of the actions),
- Identify additional actions or investigations necessary to resolve the occurrence,
- Categorize the occurrence for tracking and trend analysis, and
- Establish (if possible) lessons that should be learned from the event.

As part of evidence gathering, the critique should establish the conditions at the time of the event, information which provides a baseline from which to proceed. Also, the signs and symptoms observed prior to, during, and after the event should be determined. They will provide clues to the cause of the event.

The critique should also establish a chronology of the event sequence. Chronology will become an important tool in the analysis phase.

Before the Critique:

- ❑ Ensure immediate actions to place equipment in a safe condition are complete.
- ❑ Relieve involved personnel.
- ❑ Preserve the scene/evidence to maximum possible extent.
- ❑ Have all involved individuals prepare written statements (separately and without counsel) as soon as possible after the event. Provide guidelines for written statements if necessary.
- ❑ Gather, identify, and preserve evidence including:
  - ✔ Operating logs
  - ✔ Instrument recordings
  - ✔ Photographs, videotapes of scene, charts, floor plans, location identifications, or drawings
  - ✔ Accident debris or failed items
  - ✔ Tools and materials as applicable
  - ✔ Procedures used (or applicable to event)
- ❑ If criminal activity is evident or suspected, notify appropriate law enforcement and security personnel and preserve scene but do not disturb evidence.
- ❑ Make all required notifications.
- ❑ Identify critique leader(s) and participants including:
  - ✔ Cognizant management (who should lead the critique)
  - ✔ Note-taker (recorder)
  - ✔ Involved personnel
  - ✔ Technical experts (if required)
- ❑ Keep the number of attendees to a minimum. Unnecessary "interested" people shouldn't be there. It's not a circus.
- ❑ Leader(s) review the evidence and written statements and:
  - ✔ Develop agenda for the critique
  - ✔ Schedule and notify participants
- ❑ Schedule the critique as soon as possible after occurrence, generally within 24 hours.
- ❑ Have all required evidence/supporting materials organized and ready for critique.
- ❑ Write notes and questions that should be highlighted or asked during the course of the critique.

During the Critique:

- ❑ Establish the purpose of the critique with the attendees.
- ❑ Recount the overview of the event as you know it.
- ❑ Ask the questions that you prepared prior to the critique, allowing others to participate, as the critique progresses. You needn't limit questioning only to prepared questions since new information will arise during the critique.
- ❑ Solicit questions and pertinent information from participants.
- ❑ Maintain order and decorum.
- ❑ Keep personal notes.

After the Critique:

- ❑ Review the results of the critique including the recorder's notes.
- ❑ Consult with other experienced personnel if necessary regarding the information now known.
- ❑ Determine if there are other questions that need to be asked.
- ❑ Determine if further critique meetings will be necessary.
- ❑ Determine the causes of the event, if possible.
- ❑ Issue a preliminary report of the event based upon what was learned during the critique.

Figure 20-4: Guidelines for conducting a critique.

The event critique should be led by a skilled critique leader. Participants in the critique meeting should include all team members who can contribute significantly to the review or who have a need to learn the details of the event.

Critique leaders should exercise care to keep the critique process objective. If it is allowed to become a "witch hunt", future critiques will suffer.

The critique should also serve as a check to ensure that proper immediate corrective actions have been completed and documented. Follow-up actions can then be prescribed or verified.

What a Critique is Not We have emphasized that the critique must be (and must be perceived as) an objective, fact-finding process. It is instructive at this point to consider some things that a critique is not. A critique:

- Is not a trial.
- Is not a "fault-finding" or "finger-pointing" exercise.
- Is not a disciplinary proceeding.
- Is not a time or forum for speculation or theorizing.
- Is not a shouting match.

Analyzing the Evidence

Critique can be a very important part of analyzing the evidence surrounding an abnormal event. Yet, if the event is complex, evidence analysis may require much more than a single critique. The Challenger investigation, for example, spanned a period of several months, engaging thousands of people to retrieve, preserve, examine, and analyze the evidence. One critique was not possible. An extended complex problem-solving process was necessary.

Resolving Complexity As we learned in the first two chapters, abnormal events are usually complex in nature, consisting of a number of contributing factors. Complex problems may be rooted in technological difficulties, inadequate care and control of equipment, or in the human controllers themselves.

Evidence gathering and event analysis is nothing less than an exercise in unraveling the contributors, determining how they interacted to cause the problem, and developing methods to preclude recurrence. In short, it is an exercise in complex problem-solving.

There are many methods of solving complex problems. The process can be as comprehensive or abbreviated as the situation warrants; but, it should almost always follow the logical five-step problem-solving sequence: *(1) define the problem, (2) identify possible alternative solutions, (3) analyze and select a solution alternative, (4) implement the selected solution, and (5) evaluate the effectiveness of implementation.* Formal investigation modifies the steps somewhat since the process of solution now involves an alliance between investigators and operating personnel. Investigators *(1) define the problem* and *(2) identify causes* with the assistance of facility personnel. Operators and their leaders then must *(3) select corrective actions, (4) implement the corrective actions*, and *(5) evaluate the effectiveness of implementation.*

Without a thorough understanding of the event contributors, successful resolution is unlikely. Individuals alone, however, are limited in perspective and, therefore, limited in their abilities to thoroughly identify contributors. Multiple perspectives are necessary.

Perspective is set by background, education, and experience. Therefore, multiple perspective is achieved through including in the investigation and resolution phases people from different backgrounds and disciplines.

Maintaining Objectivity While analyzing evidence (and throughout the investigatory process), investigators must labor to maintain objectivity. Especially, if the investigators have a vested interest in the outcome of the investigation, objectivity may be difficult. Rickover stated it in this manner:

Another principle for successful application of a sophisticated technology is to resist the human inclination to hope that things will work out, despite evidence or suspicions to the contrary. This may seem obvious, but it is a human factor you must be conscious of and actively guard against. It can affect you in subtle ways, particularly when you have spent a lot of time and energy on a project and feel personally responsible for it, and thus somewhat possessive. It is a common human problem and it is not easy to admit what you thought was correct did not turn out that way.

The question naturally arises as to whether investigators from within an organization can ever be independent. They can be, but they must be of high character and integrity and committed to long-term outcomes rather than short-term expediency.

Determining Causes

Within the medical community, competent physicians search beyond the symptoms until the causes of an illness are clear. Only then can a satisfactory solution be advanced. While investigating an abnormal event, the investigators must similarly avoid focusing on symptoms rather than determining the underlying causes—the roots—that have given growth to the problem.

Cause Nomenclature

Accident analysts classify causes in a bewildering variety of ways. There are primary causes, secondary causes, root causes, probable causes, direct causes, proximate causes, principle causes, and contributing causes. The meanings of these terms often overlap and the application of terms seems to depend upon who is teaching the concept.

The National Transportation Safety Board chooses to use the term *probable cause* to identify the most important cause(s) of an accident. The commercial nuclear power generating industry elects to speak of *root causes* and *contributing causes*. The definition of "root cause" is often stated as the cause or causes of a problem which, if corrected, will prevent the problem from recurring.

In general, causes are delineated based upon *how near in time the cause was to the event* and *how important in priority the cause was to the event*. In this text, for the sake of simplicity, our greatest concern is the importance of the cause to the event. Accordingly, we shall employ the terms *primary cause* and *secondary cause*.

Primary and Secondary Causes

Primary causes are the central or principle causes of a problem. If these can be isolated, effective problem resolution is far more likely. **Secondary causes** are causes which contributed to the event but in a far less significant role. Secondary causes must be addressed for complete long-term resolution but do not carry the same importance of primary causes.

Some organizations believe that any complex problem has only one primary cause. Accident studies, however, show that premise seldom to be the case. Investigators should expect to find multiple problem components and, consequently, multiple causes. The importance of the component problem to its complex parent determines whether the cause is primary or secondary. If more than one component of the complex parent is very important, more than one primary cause exists.

Declaring Causes

The information obtained during the critique meeting may be sufficient to establish cause. Often, however, further investigation must be conducted to determine cause. Based upon the information obtained during the critique, decide whether further event analysis is necessary. If the causes are manifest and if no further benefit will accrue by continuing the investigation, the causes may be declared, corrective actions established, and the lessons promulgated. On the other hand, if there remains a question of cause, further formal analysis is probably warranted.

Correcting Causes

Since each component of a complex problem is likely to have a central cause, it is also likely to have a central solution. Consequently, application of a number of different remedies is usually necessary to resolve a complex problem.

Sometimes, a complex problem can be solved by separately addressing each of its components. More often, however, the individual problem components are so intertwined that their solutions are apt to be similarly interdependent. Other times, the resolution of one problem component may also correct other components. In fact, some complex problems are formed from a central simple problem that has been surrounding by a group of attendant smaller problems. Solution of the central simple problem may cause the entire problem to dissolve.

Declaration of cause and subsequent corrective actions is greatly dependent upon the experience, maturity, perceptiveness, and foresight of the organization's leaders. All too often, a symptomatic solution is adopted and accepted simply to expedite the recommencement of operations. The danger in premature recommencement of operations is the risk of immediate or eventual recurrence. Worse, the event that occurred may simply be a precursor of a more serious event soon to occur.

Documenting the Investigation

Since the purpose of investigation is to find out what happened so it won't happen again, documenting the investigation is important. Documentation should include records of evidence, reports of investigatory proceedings, description of the inquiry's conclusions, and summaries of the actions taken to preclude recurrence.

Elements of Event Reports

Event reports, in order to be complete, should address a few fundamental questions. They include:
- What happened?
- What were the conditions when it happened?
- What were the signs and symptoms that told us it was happening?
- What was the cause?
- What did we do about it?
- What are the lessons?

Preliminary Report

The first report of an event is typically a draft document known as a *preliminary report*. It is termed "preliminary" because investigation and cause determination may not yet be complete. It is issued to get advance information out to those for whom the event is important.

The preliminary report of findings should be issued as soon as possible after the critique. The report should summarize the event, describe the initial conditions of the event, detail the chronology of the event, recount the immediate corrective actions performed by equipment operators, diagnose the cause of the event (if possible without excessive conjecture), and delineate the lessons that should be learned from the event. This report will be modified and later finalized if further investigation is warranted.

Final Report

When investigation proceeds beyond the critique process, further reports may be necessary. A *final report* of an event should be issued as a revision of the preliminary report when an investigation is concluded. It should elaborate the facts of the event as necessary, detail the findings of the analysis, and clearly state the determined underlying causes. It should also recount additional corrective actions determined to be necessary based upon the analysis.

Disseminating the Lessons

Accidents are all the more tragic if they are not used to educate others in the mistakes that were made. Therefore, dissemination of *lessons learned* is a vital part of investigation.

Sometimes, incorporating the lessons into a required reading folder may be enough. Other times, a formal block of instruction should be created and presented to personnel for whom the event holds importance.

The lessons learned from an event should be distributed to everyone to whom they apply within a facility. Furthermore, if the events are also relevant to outsiders, wider dissemination is appropriate.

Unfortunately, abnormal event reports often receive inadequate attention within large organizations. Not only are they poorly distributed, leaders of other units are likely to read them without asking how similar problems could occur within their own jurisdictions.

Lessons from one's own industry are also not the only lessons that should be studied. The mechanisms of accidents are far more relevant than are the technologies in which the accidents occur. Industrial leaders should, consequently, be continuously alert to events holding lessons for their own personnel. Excellent sources of abnormal event discussions and lessons include the **NUREG**s from the Nuclear Regulatory Commission, **Operating Experience Weekly Summaries** from the Department of Energy, transportation accident reports from the National Transportation Safety Board, reports from the National Fire Protection Association, OSHA reports, and trade journal accident reports such as those contained in **Aviation Safety**.

Topic Summary and Questions to Consider

The dreadful experiences of those in the capsule compartment during the last seconds aboard Challenger were sealed with the deaths of seven astronauts. We know only that:

> *The Shuttle was going at nearly twice the speed of sound (Mach 1.92) and was passing through 46,000 feet of altitude. There were no alarms sounded in the cockpit. The crew apparently had no indication of a problem before the rapid break-up of the Space Shuttle system. The first evidence of an accident came from live video coverage. Radar then began to track multiple objects. The flight dynamics officer in Houston confirmed to the flight director that "RSO [range safety officer] reports vehicle exploded," and 30 seconds later he added that the range safety officer had sent the destruct signal to the Solid Rocket Boosters. During the period of flight when the Solid Rocket Boosters are thrusting, there are no survivable abort options. There was nothing that either the crew or the ground controllers could have done to avert the catastrophe.*

It is a tragedy that we hope must never again be experienced. But, it is a tragedy that may have been precluded through better abnormal event investigation when the O-ring problems began to raise warning flags.

Perhaps, though, the clearest lesson of Challenger is that *all* organizations, regardless of their stature and past history, are susceptible to errors of authority, complacency, poor communication, and the tyranny of production pressure overruling critical thinking. One of Admiral Rickover's oft pronounced principles was "Don't fool yourself; face the facts". He wrote:

> *If conditions require it, you must face the facts and brutally make needed changes despite significant costs and schedule delays. There have been a number of times during the course of my work that I have made decisions to stop work and redesign or rebuild equipment to provide the needed high degree of assurance or satisfactory performance. The person in charge must personally set the example in this area and require his subordinates to do likewise.*

Through objective investigation of abnormalities we are far more likely to stop such calamities before they cause disaster.

Ask Yourself

Do you formally investigate close calls as well as accidents?

Do you have people trained in event investigation on your team?

How (and to whom) do you disseminate the lessons learned from abnormal events?

Are the lessons learned from events within your own activity incorporated into your training program? What about events from other activities and other industries?

Chapter 21

Conducting Continuing Training

In Chapter 6, we learned that *continuing training*—advanced training in the knowledge and skills pertinent to improved industrial operations—is an important and never-ending part of developing alert, well-trained operators. In its absence, no operator (or operating team) ever reaches an elevated level of performance. Consequently, an effective continuing training program is an indispensable part of implementing systematic industrial operations.

What Is It?

Too often, the phrase *continuing training* evokes a picture of the monthly safety meeting or the yearly trek to a trade conference. But, continuing training, when understood and wisely employed, is far more. It is one of the most important tools an industrial team leader has available for individual and team improvement.

Definition of Continuing Training

Continuing training is an on-going training process for team members who are already qualified (or certified) at an initial level of equipment operation. Its objective is constant improvement of individual and team skills.

In one sense, continuing training is always occurring. For example, when a teenager successfully completes driver's education, he or she has achieved initial qualification as a motor vehicle operator. The driver's performance has been validated by an approved instructor who, through written and practical examination, has verified that the student possesses the minimum knowledge and skills to operate. The operator's license then issued (valid in all states) certifies that the driver has achieved minimum standards for safe vehicle operation. Yet, this young driver is only marginally prepared to take the road. In fact, his or her motor vehicle education has just begun. Much is left to learn about driving in adverse weather, driving vehicles equipped differently than the training vehicle, and driving in hazardous traffic conditions. The novice driver is embarking on a life-long experience of knowledge and skill advancement training.

For most drivers, continuing training is an informal process that depends primarily on the driver's motivation and acumen for improvement. In industrial operations, however, continuing training cannot be left to the individual alone. Though many excellent operators take it upon themselves to improve knowledge and skills, others will not. In addition, since operation is usually a team process, it requires a coordinated effort to improve not only the individual skills, but also the tactics and strategy that lead to team success. Therefore, for industrial operations, continuing training requires a formal process (to improve both knowledge and practical skills), administered on a weekly (or at least monthly) basis that addresses both individual and team improvement.

Purpose of Continuing Training

For motor vehicle operators, the purpose of continuing training is to improve driving skills beyond the minimum required by law. Similarly, the objective of continuing training in industry is to advance the knowledge and skills of qualified operators (and the teams of which they are a part) beyond minimum performance standards.

By sharpening the mind (through education) and skills (through practice), individuals and teams can learn to perform efficiently and safely. They become far better prepared to respond to abnormal operating conditions and to combat equipment emergencies and casualties.

Who Is It For? Perhaps a better question is, "Who needs to improve?" Clearly, continuing training is for everyone. "Everyone?" you ask. "Even clerks and secretaries?" Well, no—unless for some reason they are a part of your team!

Airline Industry Approach

Continuing training is a continuation—an advancement beyond initial training. Nowhere is this relationship better exemplified than in the commercial airline industry where, once having achieved a rating, training continues for the duration of a pilot's career.

Initial Training In commercial airline operations, anytime a pilot seeks a rating for a new type of aircraft, *initial training* is required. In the Air Florida Flight 90 accident report, NTSB writes:

> *Initial training is required and conducted for crewmembers who have not qualified in the type of aircraft and served in the same capacity on another aircraft of the same group.*

Though an aircraft pilot is familiar with the principles of flight (and the fundamental equipment and instrumentation), if the new equipment differs substantially from that for which the pilot was originally certified, initial training is again required. For example, it would be foolish to say that a pilot rated in a Cessna 172 is also qualified to fly a Boeing 737. The flight characteristics of the Boeing differ dramatically from those of the Cessna. Therefore, the Cessna pilot needs *basic* training in the Boeing 737.

Transition Training If a commercial pilot possesses a rating for a particular class of aircraft—twin turboprop aircraft, for example—and is preparing to fly another manufacturer's twin turbo, initial training is probably not necessary. *Transition training* is sufficient to prepare the pilot to operate the new aircraft. NTSB:

> *Transition training is required and conducted for crewmembers who have previously qualified and served in the same capacity on another aircraft of the same group.*

Between the two aircraft, the flight characteristics, controls, and instrumentation are similar enough that basic training is not necessary. Yet, though the two aircraft are alike, dissimilarities in controls, instrumentation, and flight characteristics dictate further education and skills training by a qualified instructor to ensure safe operation.

Upgrade Training In airline operations, though both captain and first officer are certified to operate their aircraft, the captain is assigned legal responsibility as the flight team leader. The captain is accountable for all that happens (or fails to happen) during ground and flight operations. The additional leadership responsibilities require another training component known as *upgrade training*. NTSB:

> *Upgrade training is required and conducted for crewmembers who have qualified on a type of aircraft and served as second-in-command before they are eligible to serve as pilot-in-command on that aircraft.*

Differences Training When equipment is modified, airlines require *differences training* for pilots designated to fly the modified aircraft. NTSB:

> *Differences training is required and conducted for qualified flight crewmembers on a new model of the same type of aircraft; for example, a 737-100 qualified crewmember would be required to take differences training for the 737-200 series.*

The intensity of differences training depends, of course, on the extent of modification to the original equipment. Minor modifications may simply require pilots to read about the changes. Comprehensive modifications, however, might well demand in-flight training under the supervision of another pilot familiar with (and certified on) the changed operating characteristics.

Recurrent Training   Even qualified and proficient pilots need periodic training in important topics and skills (known as *recurrent training*) as a fundamental part of their continuing training. NTSB:

> Once a flight crewmember is fully qualified and serves as either second-in-command or pilot-in-command on a specific type of aircraft, recurrent training is required. Such recurrent training consists of ground school for captains and first officers once a year. Recurrent training in the flight simulator is required every 6 months for qualified captains and once a year for qualified first officers. All training consists of a combination of video presentations, films, slides, and lectures. Training material is derived directly from the…Operations and Training Manuals. Video presentation used during each initial and recurrent…class include…winter operations, takeoff (rotation effects on initial climb performance) and landing performance, wet stopping, …windshear, upset, and landing illusions.

In the industrial setting, recurrent training should include training in infrequent or abnormal situations, seldom used systems, and the underlying theory of equipment design and performance.

Application to Industry

Leaders of every industry would be wise to pattern both initial and continuing training programs around the airline industry's model. Though *your* risks may be far different than those of aircraft operations, the fundamentals of training are the same. Rigorous initial training should be followed by needs-based continuing training (and testing), formally conducted and documented by skilled "pilots", all with the goal of constant improvement.

Goal-Directed   How must your team improve to accomplish the organizational mission with greater effectiveness and safety? The answer to that question is the basis for establishing continuing training program goals. Depending upon the maturity and sophistication of your team, you may need to focus on individual skills, team tactics, or overall strategy. Regardless of where you start, mission-directed improvement is the only legitimate reason for continuing training. If that isn't the objective, you're probably wasting time and money.

Needs-Based   For continuing training to be goal-directed, it must be *needs-based*. **Needs-based**, as the phrase implies, means that each element and subject of the continuing training process is relevant to the needs of the end user. Unless training addresses the needs of individuals or the team as a whole, it will have little value.

In the airline model, each kind of training—initial, transition, upgrade, differences, and recurrent—has a specific purpose. Some is directed at the individual pilot skills, while other training engages crewmembers as a team. Nevertheless, all is aimed at improving the safety and efficiency of aircraft control.

Taught by Skilled "Pilots"   *Credibility* is an indispensable part of the training equation. Instructors who "teach" subjects for which they have no proven capability discredit the entire training process. (They may also provide false or inadequate information.) Accordingly, instructors should be chosen from the ranks of the best performers who also have a penchant for teaching. To remain current and credibile, the instructors must also maintain proficiency in the skills and subjects that they teach.

Formally Conducted and Documented   Whenever an industrial team member learns something new, relevant, and helpful, continuing training has occurred. But, as Coach Bill Walsh admonished (Chapter 6), training must be approached in a structured, systematic fashion for teams to excel. Every class, every drill, every exercise must have an explicit purpose, always directed at team improvement.

In the Big Bayou Canot Bridge accident, we learned that, though towboat operators received initial training (and had to be certified before operating on their own), their training was loosely structured, incomplete, and poorly evaluated. Warrior and Gulf Navigation Company leaders were unable to say with surety what training towboat operators received. Further, little in the way of continuing training was documented. In short, W&GN had not established a formal, controlled training program for either initial or continuing training.

Conducting Continuing Training

Like it or not, documentation is an important part of developing and improving any training program. If you can't write what you envision, training will not be performed in accordance with your expectations. That means policies, schedules, and records of completion must be recorded. Do you suppose that Coach Walsh operates without succinctly defined (and recorded) objectives and schedules?

In hindsight, W&GN is easily singled out because of their involvement in the Big Bayou Canot Bridge accident. Yet, W&GN officials undoubtedly believed their program was at least adequate. In fact, their training program was probably better than many in industries with even higher risks. The issue, however, is not whether your training program is better than someone else's. Rather, it's whether your training program (both initial and continuing) has prepared your team for infrequent and abnormal events as well as normal, everyday occurrences. If you have not formally and systematically developed the training processes, your training is probably inadequate. Haphazard training can't produce distinguished team performance.

Choosing Continuing Training Topics

Far from a monthly staff meeting (often a catch-all for disseminating staff notes, completing administrative tasks, and eating someone's birthday cake), continuing training sessions should be focused educational and practice sessions, *always* aimed at improving individual or team performance. Every topic should have a stated purpose, and the method of delivery should achieve the best balance between effectiveness and use of time. The selected educational and practical topics should form an integrated approach to achieving the overall continuing training program goals.

Topics from Self-Assessment

When a continuing training program is truly needs-based, topic selection isn't difficult. Needs are easily determined by talking to team members, observing the team's performance during routine operations, and by testing the team through emergency preparedness drills. (Remember the Chapter 15 discussion of the need for frequent emergency and casualty drill training?) Of course, you won't know what your needs are unless you're looking and listening. As we shall see in the next chapter, that's where self-assessment fits in.

Self-assessment (Chapter 22) engages all team members in determining their own strengths and weaknesses. Team strengths must be systematically identified and augmented to create successful tactics and strategies. Conversely, individual and team weaknesses require recognition and rectification for the team to progress to its potential. Accordingly, needs determined through self-assessment are fundamental continuing training topics.

New Equipment Training

Incorporation of new or modified equipment into the operating environment is usually an occasion for continuing training. Sometimes, just reading about the change is sufficient. Other times, thorough hands-on training is necessary. As in the airline approach, the depth and degree of training required depends on the extent and complexity of the changes and the associated risk of incorrect operation. For major changes, it may be necessary to have vendor representatives instruct operating personnel in new equipment operation.

Emergency Preparedness

In Chapter 6, we learned that emergency and casualty drill practice was one of Rickover's six tenets of training. Industrial teams must be able to respond to emergencies and casualties originating both inside and outside their activities.

For emergency preparedness training to be effective, classroom studies must be integrated with practical exercises. This is not a simple process that can be delegated to unskilled personnel. Since emergency and casualty drills often produce abnormal operating conditions, they must be created, reviewed, approved, and coordinated by technically knowledgeable team members.

Infrequent and Abnormal Conditions

Emergencies and casualties are extreme instances of unusual operating circumstances. Far more prevalent, however, are the *infrequent* or *abnormal* conditions such as system startups and shutdowns—atypical situations that challenge the skills and knowledge of operating personnel.

Unusual equipment configuration often increases the probability of adverse events. When team members analyze, discuss, and practice for such circumstances, they are far better prepared to manage them when they arise. Accordingly, infrequent and abnormal condition studies (and practice) should be incorporated into the continuing training process. Further, when infrequent or abnormal events are scheduled, consider the benefits of conducting training for the event beforehand.

Lessons Learned One of the most important functions of continuing training is to cause team members to think about potential events in their own operating environments. Many events are postulated as a part of a plant's safety study. Yet, no safety study can anticipate everything that can go wrong.

Therefore, examination of accidents and close calls are a productive source of continuing training topics. In fact, the *lessons learned* from accident or close call reviews often suggest a need to reevaluate a team's preparedness to combat previously unidentified emergencies and casualties.

Don't limit the lessons learned to only those from a team's own activities. Events from other teams and other industries should also be considered. Also, the lessons learned shouldn't be only those from mistakes. When your team (or someone else's) turns in a remarkable performance, the lessons may be equally rewarding.

Emergent Problems Continuing training provides an excellent forum for solving problems that arise within an industrial team. Conflicts between operations and supporting groups such as maintenance or engineering can often be resolved through discussion with the "offending" parties. By inviting key officials from other groups to explain why they do things in particular ways, obstacles to efficiency can be identified and resolved.

Proficiency Another of Rickover's six tenets of training was to continually reinforce principles and procedures through recurrent and proficiency training. Operating skills deteriorate when they are not practiced. Thus, one objective of a continuing training program should be to maintain high *proficiency* levels.

Proficiency, in the context of industrial operations, refers to current ability to operate components and systems. Though a team member has been certified (rated) to perform specified operating tasks, lack of practice dulls the skills and related knowledge. Proficiency training (both classroom instruction and practice) keeps skill and knowledge at peak levels.

Proficiency training is particularly important for instructors and leaders. They are often isolated from the day to day operations by virtue of their additional job responsibilities, and may get "rusty". When they fail to remain proficient in their certified skills, they lose credibility with those they lead and teach.

Recertification **Recertification** is the process of periodically demonstrating to licensing authorities the requisite knowledge and skills required by law or policy to operate equipment.

When you first passed a driver's written and practical examination, you received official certification from your state (signified by a license) to drive a motor vehicle of a particular class. Yet, every four or five years, you must *recertify*. Though the recertification process may not be as rigorous as the original certification, it provides at least minimal assurance that you still possess the necessary knowledge and skills to drive. Further, if you allow your license to lapse, you may have to go through the entire process again.

Industrial operations are no different. Depending upon the risk and complexity of the operating tasks, recertification may be a necessary part of the continuing training program. The frequency of recertification should be tied to operating risk.

Integrating Continuing Training

Perhaps the most common mistake in developing a continuing training process is the failure to integrate education with skills. The U.S. Naval Nuclear Propulsion Program advocates a fundamental approach toward learning called *theory to practice*. The phrase is a simple acknowledgement that sound training is based upon excellent theoretical understanding followed by rigorous practice to reinforce theory.

Effective continuing training relies on the same two fundamental components: education and practice. Together, they create and reinforce mental pictures, stimulate thought, and guide the movements necessary to accomplish tasks.

Education

In strict terms, **education** is the process of transmitting and acquiring thoughts, models, pictures, and ideas. People learn from books, graphic images, music, motion pictures, conversation, watching others, personal experiences—there are dozens of ways to learn.

Continuing training education may include reading new or seldom-used procedures, studying abnormal event scenarios, classroom instruction in new technology, group seminar discussions of emergency preparedness dilemmas, or solving homework problems for engineering issues.

However addressed, educational topics must have the goal of advancing (for good reason) the knowledge of participants. Further, the media chosen to communicate topics must be effective. Reading, for example, has been shown through educational studies to impart approximately ten percent of the total of material read. Of course, effectiveness depends largely upon the reading skill of the person involved. Nevertheless, on average, very little is transmitted.

Practice

Though important and integral to effective continuing training, education alone is insufficient. Skill training—*practice*—must be incorporated to complete the training process.

A **skill** is the ability to perform a task. The task may be, primarily, a mental exercise or predominantly physical. Usually, it is a combination of the two. Regardless, the task must be conceived in the mind before it can be effectively performed.

It is through practice that coordination of neural and muscular activity is perfected. Whether the function is fire-fighting, soldering, backhoe operation, or reactor control rod manipulation, education must be followed by practice for improvement to occur.

Training as a Team

As we have already seen, industrial success depends on more than just individual skills. *Team* skills, tactics, and strategy must also be mastered. Therefore, continuing training must go beyond improving individual skills and knowledge—it must also create a team.

Team integration is best accomplished through a well-planned and conducted drill program. Through exercising and stretching the team's capabilities, strengths are enhanced and weaknesses are rapidly identified. But, to be effective, the drills must be challenging, realistic, and frequent.

Impediments to Continuing Training

Developing and delivering top-notch continuing training isn't easy. It takes time, commitment, money, and people to plan and conduct the training. Yet, advancement training for team members (and the team as a whole) is usually the most effective avenue for improved efficiency and safety. Industrial leaders are often too willing to accept more frequent injuries, equipment damage, and rework rather than invest the time and money to conduct proper training.

Costs of Training (and Not Training)

Good continuing training is expensive and time consuming. But, it's like using cash instead of a credit card. It costs up front, but you're eventually going to pay anyway. The advantage is in avoiding the interest and penalties associated with delayed payment.

For a team to practice, time must be allocated to the process. Probably the most common objection to continuing training is, "We don't have the time to train". Yet, those who have developed disciplined continuing training programs respond with a knowing smile, stating that they don't have time *not* to train. They find that continuing training is "axe-sharpening" time—an investment in future efficiency. It ultimately saves time, labor, and money, usually by avoiding *rework*—having to do something again because you messed it up the first time.

But how, you ask, can I start a continuing training program with people who are working a rotating shift? One possibility is to conduct training before or after the normal shift. Unfortunately, the lengthened days may cause the training to be less effective.

To accommodate better training, some organizations (as previously noted) have five or six operating shifts rather than the traditional four. With additional shift teams available, a crew can rotate through its normal work cycle—from first shift (day shift) to second shift (swing shift) to third shift (midnight shift)—and then, following long change (several days off before resuming work), rotate back onto day shift for a full week of training. In a six shift rotation, a crew would return the following week to perform duties in support of the operating and maintenance activities of the on-shift operating crew. Then, during the next week, the crew would resume the three shift operational rotation. Allowing operating organizations to train as teams rather than in a piece-meal fashion has proven of great value.

Is an extra crew expensive? Absolutely. If each crew has ten team members and the annual cost of each (including salary, pension, medical benefits, disability insurance, vacation, office space, equipment, etc.) is $100,000, the cost of the crew is about $1 million. On the other hand, how many high-dollar-cost accidents and errors has your team experienced in the last few years? How many could have been avoided by better prepared personnel? An extra crew may not be as expensive as you think.

Finding People to Lead Training

Another perceived impediment to continuing training is finding the people to lead it. Yet, the best continuing training leaders are the very people who are doing the work. They know the problems that they face and have a vested interest in solving those problems—there's no better way to get them involved and motivated. Get your team members involved!

But, you say, my workers aren't good stand-up instructors. That's OK. The best continuing training sessions are small informal group seminars, not formal classroom instruction. Present your team members with operating problems that they are likely to face and let your shift leaders facilitate the discussions. Remember, the objective of continuing training is to improve individuals and the team within the context of mission accomplishment.

Don't forget support organization leaders as candidates to lead training sessions for your operating team. They are experts in their own areas and probably have much to teach (and learn from) your operators. Remind them (frequently!) that they support your operating team.

Topic Summary and Questions to Consider

Clearly, improving the knowledge, skills, and capabilities of your team through continuing training is an important part of implementing the systematic approach to operations. But, even if you don't have a continuing training program in place, don't be overwhelmed. The important thing is simply to get one started. Start small. Work on one or two things at a time. Don't make the schedule cumbersome, but let people know what you're going to do and who's going to do it. And remember, continuing training is for team improvement.

Ask Yourself

Do you have a plan to systematically grow your team?

Is your continuing training program formally planned, conducted, and documented?

Do you include lessons learned from abnormal events? From inside *and* outside your organization?

Are both education and skill training integrated into your program?

Do you perform casualty and emergency drill training as part of your continuing training program?

Chapter 22

Evaluating Operating Performance

Initial and continuing training in the systematic approach to operations and maintenance, though indispensable, are only two parts of the continuing improvement process. Evaluation (especially in the form of *self*-assessment) is perhaps even more important, since it objectively measures all areas of performance to identify both strengths and weaknesses—showing where (and to what degree) improvements have been made as well as deficiencies requiring corrective action.

No team ever excels without constant evaluation—testing, coaching, and self-assessment. In fact, the specific needs for continuing training are determined primarily through the self-assessment and evaluation process. And don't forget the morale-building and motivational benefits of being able to say "look how much we've improved during the past year," and having objective data to back it up.

Of course, a comprehensive discussion of the evaluation process would require an entire book all by itself. [**Industrial Facility Self-Assessment: A Guide for Observation Program Planning and Implementation** by H. C. Howlett II is a recommended text on this subject. ED.] Thus, the objective of this chapter is only to provide an overview of the evaluation and self-assessment process: what it is, what it does, and how it fits into the overall strategy for operating success.

What Is Evaluation?

To answer this question, we must first differentiate between outside evaluation and self-assessment. Once that distinction is made, we will introduce some of the elements of evaluation. To some, the term evaluation means only *formal* testing or auditing. For industrial operations, though, evaluation is much more. It also includes the day-to-day coaching by team leaders, the formal and informal self-assessments at all levels within the facility, and outside evaluations (conducted by individuals or organizations from outside the facility).

Outside Evaluation   Outside evaluation is evaluation conducted by individuals or organizations from outside the facility. Such outside evaluators may be from corporate headquarters, government agencies, or outside consultants specifically hired to take a "fresh look" at facility operations.

If an effective self-assessment program is in place, outside evaluation should provide a verification or "cross-check" that operations are going well—just as a "routine" IRS audit should disclose no major problems with your tax return. But remember, an "outsider" can often provide a more objective look at your operations and disclose problems that you've overlooked. Sometimes you're too close to a problem to see it.

Self-Assessment   In its simplest form, **self-assessment** is everyone (at every organizational level) evaluating their own and their team's performance. But, for our purposes, we expand this definition to include all the evaluations—formal and informal, testing and auditing, coaching, etc.—initiated and conducted at each level in a facility's organizational structure. In other words, self-assessment is used primarily to distinguish between evaluations conducted by a facility itself and evaluations initiated and conducted by outside agencies.

In this context, self-assessment is the most important kind of evaluation—it is *pro*active, whereas outside evaluation tends to be *re*active. The best teams evaluate *themselves* critically—assess their own performances. They don't wait for outsiders to do it; they do it themselves. Every team member accepts responsibility for analyzing and improving both personal and team effectiveness.

Tests and Audits   Tests and audits are designed and intended as formal means to verify that (or determine whether) individuals (and teams) meet required performance standards. Those are certainly necessary processes. In previous chapters we saw that effective initial and continuing training must use evaluation in the forms of written and oral examinations (to test individual knowledge), on-the-job training evaluations (to test individual skills), and emergency and casualty drills (to test team skills).

Usually the term *audit* is understood to mean specifically the formal examination or verification of *financial* records. For industrial operations, however, we expand this definition to include the formal examination of performance and the records associated with all aspects of facility operations. To name just a few, the records evaluated during audits include training and qualification records, process data, operating logs and data record sheets, tagouts and tagout logs.

Tests and audits are elements of both outside evaluation and self-assessment. But, while outside evaluators rely almost exclusively on testing and auditing methods, effective self-assessment must go well beyond those limited, formal methods to ensure success.

Coaching   Although not normally considered, coaching is an essential part of effective evaluation programs. Successful development of *any* team, be it a sports team or an industrial team, requires not only the formal testing and auditing processes, but also constant, competent *coaching*—the daily observation, instruction, correction, and reinforcement of knowledge and skills by competent leaders—that provides immediate feedback for improvement. Football coaches constantly review game films with their teams not only to identify mistakes or areas needing improvement, but also as an objective "before and after" demonstration and reinforcement of improved performance results. Industrial coaching is no different. There, coaches have titles such as lead technician, shift supervisor, or plant manager—the leaders ultimately responsible for ensuring safe and efficient performance.

Coaching has, unfortunately, become a lost art in many industrial organizations—some are overstaffed with evaluators and understaffed with coaches. Yet, without competent coaching, teams are doomed to poor, uncoordinated performance.

By its very nature, coaching is, as we shall see, limited to internal or self-assessment performance. Outside evaluators generally do not provide—or are prohibited from—coaching.

MBWA   MBWA (management by wandering around) is yet another indispensable part of the self-assessment process. Whether conducted formally—e.g., as a pre-watch tour or as a scheduled supervisory tour using formal checklists—or informally, just getting out there and looking around is one of the best ways to see how your people and equipment are performing.

Note that "MBWA" in this context is used only as a term of convenience. There is no intent to imply that MBWA is solely a *management* function. To the contrary, *everyone* should be continuously looking at and evaluating all aspects of their facility.

Why Evaluate?

Identifying and establishing a strategy for operating success (Part II) and having a well-trained staff proficient in the twelve vital skills (Part III) are, of course, fundamental to the success of any industrial endeavor. But, if you stop there, how do you know it's all working? You check ("audit") your bank statement every month to make sure deposits have been properly recorded and no erroneous charges have been made, don't you? You periodically check the the oil and coolant in your car, and adjust or replace the fluids as needed, don't you? Shouldn't you take the same care, to check and adjust performance, in your industrial facility?

All too often in industry, training is conducted with the expectation that team members will return to their work stations and perform flawlessly. Yet, through subsequent audits, regulatory inspections, or (more likely) through abnormal events, leaders learn that nothing really ever changed.

General George S. Patton, Jr., understanding how easily communications are misinterpreted and how quickly people regress to original habits, admonished his field commanders:

> *Commanders must remember that the issuance of an order, or the devising of a plan, is only about five per cent of the responsibility of command. The other ninety-five per cent is to insure, by personal observation, or through the interposing of staff officers, that the order is carried out.*

Leading is more than just giving orders. All good coaches know that they must constantly observe, explain, demonstrate, and reinforce the concepts and skills that they teach. In that vein, five major purposes of evaluation are to:
- Verify that policies and procedures are properly interpreted and implemented,
- Evaluate the effectiveness of policies, procedures, and standards and their implementation,
- Provide objective, timely, and reliable performance data (for achievements as well as deficiencies),
- Provide recommendations for improvement where appropriate, and
- Provide opportunities to teach, correct, and reinforce proper work practices and principles.

Without proper evaluation, team improvement—and ultimately product or service improvement which is the real purpose of evaluation—cannot be achieved. Properly performed evaluation also provides additional benefits. It provides motivation for change. It can dramatically improve communications. And, it offers opportunities to correct and reinforce (to coach).

Improved Product or Service

The ultimate purpose of observation and evaluation is improved individual and team performance toward mission accomplishment. Dr. W. Edwards Deming reminds us in his fifth point:

> *Improve constantly and forever the system of production and service. Improvement is not a one-time effort. Management is obligated to continually look for ways to reduce waste and improve quality.*

No organization truly maintains the *status quo*. Technologies change. Environments change. People change. Teams are always either in a state of decline or a state of improvement. Coaching is a process of continuously monitoring and evaluating the team's *vital signs*; and providing constructive feedback on reasons, methods, and impetus for change. Based upon objective evaluation, the team can make the necessary corrections and adjustments to improve the product or service and meet changing customer needs and desires. The result of successful coaching is effective mission accomplishment *even when the mission is changing* and the assurance of employment for the members of the team through expanded market, reduced errors, reduced rework, and reduced overhead.

Long before Deming, Thomas A. Edison said:

> *Show me a thoroughly satisfied man—and I will show you a failure.*

Motivation for Change

People don't like change. Change is difficult. It's uncomfortable to change. So, without the impetus of daily evaluation and coaching, we all slip back into old habits, performing in the same comfortable ways that we knew before change was suggested.

When observant and insistent coaches (as well as seasoned players) are present to remind us of how (and why) tasks are to be performed through frequent, properly conducted evaluation, we have one of the most important tools for implementing change.

Improved Communications

Good communication is both an essential element and a valuable benefit of effective evaluation.

Nothing is more disconcerting than having an evaluator standing behind you, constantly looking over your shoulder as you work, and periodically writing furiously in a notebook—all without ever speaking a single word. All you think about each time the pen comes out is "What did I do now?" How can anyone perform well under these circumstances?

On the other hand, if active, non-adversarial dialogue is encouraged and implemented during evaluations, communication and understanding (especially between different organizational levels) are actually improved—to the benefit of everyone. Through such dialogue, the evaluator can gain a much better understanding of the day-to-day operations being observed by listening to the con-

Evaluating Operating Performance

cerns and suggestions of the operator than by merely watching what goes on. Similarly, the operator gains a better understanding of what is expected, of company goals, and of management's concerns about the overall operations.

Communications become distorted (shaped by perceptions and biases) as they travel through the levels of any bureaucracy—not by intent, but rather because we are all imperfect humans. Face-to-face interaction with members of the team at the lowest organizational levels is the only sure way to detect and correct such distortions and misperceptions. Encouraging and implementing open dialogue during evaluations will almost certainly improve communication as both evaluator and operator discuss more and more significant issues over time. And, as the openness of the dialogue increases, so does the effectiveness of the evaluation.

When the communication network is *closely coupled*, the result is efficient team performance. Through effective communication, team members are able to detect and correct problems in their infant stages rather than waiting until they become flagrant.

An Opportunity to Correct and Reinforce

Dr. Kenneth Blanchard and Dr. Spencer Johnson, co-authors of **The One Minute Manager**, remind us of an age-old prescription for successfully training winners:
1. Tell 'em.
2. Show 'em.
3. Let 'em try it.
4. Observe and evaluate performance.
5. Immediately and specifically provide feedback in the form of praise for improvement or redirection for incorrect behaviors.

It is obvious, then, that this coaching process relies on observation, evaluation, and feedback for success. Coaching at all levels involves a daily regimen of self-assessment—personal observation, praise, and correction of team members by organizational leaders as the team works toward mission accomplishment.

No true leader can escape the coaching responsibility. Joe Paterno, head coach and mentor of the Penn State Nittany Lions, writes:

> *A coach, above all other duties, is a teacher.... A coach's first duty is to coach minds. If he doesn't succeed at that, his team will not reach its potential.*

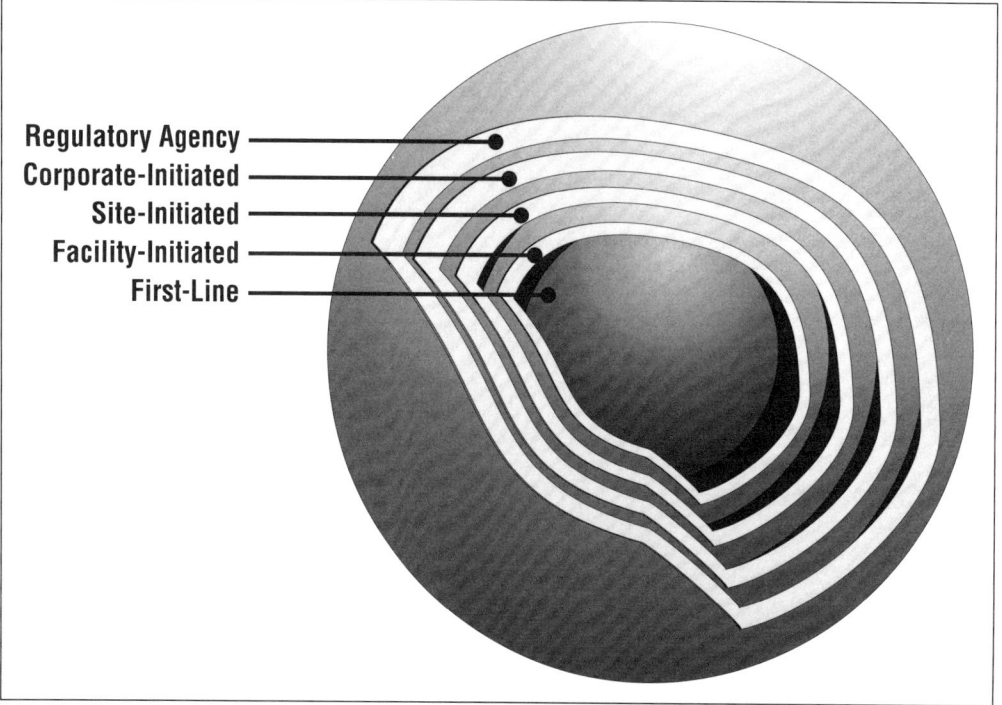

Figure 22-1: Levels of Evaluation.

A team cannot succeed and improve without leaders who are willing and capable to perform this most rudimentary and momentous of coaching duties. It requires daily evaluation of team member performance and on-the-spot feedback through which the team improves a little bit every day. Such coaching is the foundation of *total quality management*, whether in sports or industry.

Levels of Evaluation

As previously discussed, the observation and evaluation process is conducted at various levels in an industrial facility, ranging from individuals evaluating their own performance to outside evaluations by regulatory agencies such as OSHA or EPA. Each level of evaluation has its own advantages and limitations. Figure 22-1 illustrates the concept of evaluation as a multi-layered endeavor, with each successively higher level encompassing a larger scope of evaluation. For clarity in differentiating between levels, we have identified five major levels of evaluation:

- First-Line,
- Facility-Initiated,
- Site-Initiated,
- Corporate-Initiated, and
- Regulatory Agency.

Each of these five levels has a specific role in the assessment scheme, but there are significant differences in each level's ability to change the behavior of operating and supporting staffs. Assessment teams at all levels function to observe and evaluate, but coaching does not (and cannot) occur at all levels.

Moreover, receptivity to observation and evaluation is affected by factors such as credibility of the observers, legitimacy of the findings, ability of the evaluators to communicate their findings, and the timeliness of the resulting findings/recommendations report. The actual or perceived effect of each of these factors is influenced by the evaluation level.

The five evaluation levels—with the associated advantages and limitations of each—are discussed in the following sections.

First-Line Evaluation

At the very core of evaluation are those assessments conducted by first-line leaders and team members. Observation and evaluation are *everybody's* job. An industrial facility is affected by so many parameters in so many different ways that no one person or team can effectively observe the entire operation. So ultimately, the goal is to create a *neighborhood watch* program in which every operator, maintenance technician, engineer, health physicist, and janitor feels responsible and capable to see, report, suggest, and fix on a daily basis.

First-line evaluation is usually an informal process. Assessment performed by a crew shift supervisor, for example, on his or her own crew (perhaps with the assistance of key crew members) is, for our purposes, first-line evaluation. It has as its primary objective to assess (and correct) first-line unit and individual performance.

Though first-line assessment is often undocumented, important deficiencies (or strengths) that are observed should be recorded and forwarded to team leaders at the operations manager or plant manager level. In so doing, they can be factored into the overall evaluation process.

First-line evaluation is very effective, since the observations are made by those who can best formulate and oversee improvements. First-line evaluation has probably the greatest potential to change behavior. This is the level where the most coaching occurs, and where improvements or corrective actions can be most easily implemented on-the-spot. First-line evaluation is also the most difficult to establish. It takes a great deal of maturity and farsightedness for team members at the *doing* level to actively solicit and receive constructive criticism, and criticize themselves.

Despite the importance and benefits, first-line evaluation *does* have limitations. By its very nature, it is very focused, concentrating primarily (if not *only*) on the specific work group or function. Therefore, it doesn't always take the "big picture" into account. In other words, because of its focus (and the limited organizational responsibility of those doing the evaluation at this level), it doesn't assess how the team or operation is functioning and integrating with the rest of the facility. Which is why the next level of evaluation is necessary.

Facility-Initiated Evaluation

The second level of evaluation is organized at a facility or plant level. (For our purposes, we consider a facility to be the staff, grounds and buildings, and the plant machinery devoted to a specific process. A facility staff normally consists of a facility or plant manager and a supporting staff of subordinate leaders, operators, and support organization members, depending upon the size and function of the facility. There may be one or more facilities or plants on a site.)

In facility-initiated evaluation, key organizational leaders such as the operations manager, training manager, and key members of their staffs organize and conduct in-house assessments of normal and off-normal operations. It is desirable for members of operating crews or supporting disciplines (usually other than the one under evaluation) to participate in the assessment. In so doing, the discipline of observation and evaluation is promulgated through the ranks of the organization. Facility employees who participate in this process become acutely aware of errors and deficiencies within their own organizations by looking at the strengths and weaknesses of others.

Facility-initiated evaluation may be formally or informally conducted. For a formal assessment, the evaluation team normally compiles a written report of observed strengths and weaknesses as well as suggestions (or sometimes directives) for improvement. Most importantly, however, these experts, are able to provide on-the-spot guidance and suggestions while observing field operations. Care must be exercised by evaluators at this level to avoid unnecessarily usurping the position and the authority of first-line leaders while coaching unit team members.

To summarize, the benefits of evaluation at the facility level include:
- Involvement of higher-level management in the evaluation (facilitates communication between organizational levels; brings upper-level managers closer to the day-to-day operations)
- Broader scope of evaluation (multiple teams, processes, and their interactions)
- Retains the advantages of on-the-spot correction and coaching (evaluators at this level are still in the organizational chain of command)
- Involvement of "peer" evaluators from other work teams in the evaluation facilitates communication and transfer of ideas and techniques between work groups

But, some of these same benefits have associated limitations. The broader scope of evaluation also necessitates a more limited depth and detail of the evaluation. Evaluators from the higher-level ranks of management are also (necessarily) further removed from the details of day-to-day operation.

Site-Initiated Evaluation

Site-initiated evaluation, though still primarily focused on facility and crew performance, is organized at a site level. (A site, in this context, is a dedicated location within which one or more facilities reside. The site is the parent organization of the facilities and usually houses the major projects and administrative functions necessary to support sustained facility operations.)

Site-initiated evaluation is almost always a formal affair requiring scheduled activities and documentation of observed strengths and weaknesses. For small sites of one or two facilities, pivotal site leaders such as the operations manager, training manager, maintenance manager, and key members of their staffs organize and conduct assessments of facility operations for normal and off-normal conditions. Again, augmenting the evaluation team with members of other activities or facilities improves awareness and teaches the self-assessment discipline.

For sites with numerous facilities, it may be impractical to expect the site operations, maintenance, and training managers to function as full-time members of a site evaluation team. Such large sites are often better served by establishing a full-time evaluation cadre to move from facility to facility. Permanent evaluation cadre leaders should have proven credentials as former operations, training, or maintenance managers.

This interdisciplinary team of observers should not only provide a written report of findings, but should make on-the-spot suggestions for improvement. Obviously, interference in the chain of command should be avoided unless issues of immediate safety are involved.

The benefits and limitations of evaluations at this level are similar to those at the facility level. The scope of evaluation is broader (covering multiple facilities), and consequently more generalized. Higher-level managers are involved in the evaluation who are (again) further removed from specific day-to-day operational details. Opportunities for coaching and on-the-spot correction are more limited—primarily because of that "distance" from operational details and, since multiple facilities are involved, evaluators are not necessarily in the chain-of-command for the observed operation. But, at this level, there are valuable opportunities to comparitively evaluate performance at multiple facilities on the site.

Corporate-Initiated Evaluation

Corporate-initiated evaluation is assessment organized at the corporate level, usually in the form of formal audits and appraisals. (For our purposes, we are assuming an organizational structure with corporate headquarters at a location other than the site being evaluated.) Since primary evaluators at this level are normally not from the site being evaluated, corporate-initiated evaluation crosses the classification boundary from "self-assessment" into "outside" evaluation. The end product of corporate evaluation is usually a series of meetings and discussions with key organizational leaders and a final formal written report of findings and recommendations.

Corporate-initiated evaluation draws upon the expertise of facility, site, and corporate team members. Though corporate evaluators are usually well-qualified experts—and are usually accompanied by local facility experts during the evaluation—the success of this type of assessment in improving performance depends essentially on the willingness of those observed to listen and learn. If the evaluated parties enter into such evaluation with a learning attitude, corporate-level evaluation may be very effective. Conversely, if it is conducted in an "adversarial" atmosphere, the success of corporate evaluation depends on the evaluation team's ability to coerce compliance—which, of course, can't really be called success. Thus, the outcome of higher levels of evaluation is often grudging (usually temporary) change and adversarial, destructive relationships.

Advantages and limitations of evaluations at this level combine some of those at the site-initiated level with more influence from the "outside evaluation" aspects.

Some of the specific advantages include:
- Provides a "second-check" of performance and internal evaluation (at the first three levels)
- Opportunity to evaluate performance comparitively against other sites within the company
- Corporate evaluators ("outsiders") can provide a fresh, sometimes more objective look at operations
- Opportunity to keep up-to-date on corporate goals and new technologies—if communication is open

And some of the limitations of corporate-initiated evaluations are:
- Corporate evaluators are of unknown credibility
- Opportunities for real coaching are absent at this level

It should be noted that the corporate evaluator credibility issue can be overcome if a close working relationship is maintained between the corporate and site levels—if communication is open, and the corporate evaluators aren't "strangers" to the site.

Finally, corporate evaluations (as indicated above) should be viewed as valuable second-checks of self-assessments at the first three levels. If the first-line, facility, and site evaluation programs are effective in identifying and correcting problems and in implementing improvements, there will be few (if any) deficiencies for corporate evaluators to find.

Regulatory Evaluation

Regulatory evaluation refers primarily to federal, state, or local regulatory compliance audits and appraisals. Examples include assessments performed by the EPA, OSHA, or Department of Health.

These evaluations are the broadest in scope (in that they evaluate multiple sites and companies) but, at the same time, are relatively focused (concentrating only on those performance areas relevant to the evaluating agency).

Unfortunately, evaluations at this level are nearly always conducted by evaluators of unknown (and sometimes questionable) credibility, and are nearly always conducted in an adversarial atmosphere. But, however dreaded these regulatory agency assessments are, they are necessary and do offer some benefits:
- They provide a measure of performance relative to other sites and companies in the industry
- They provide a second-check to ensure compliance with industry standards and regulations
- By verifying compliance with standards and regulations, liabilities can be limited

But, again, the limitations:
- Regulatory agency evaluators are outsiders with unknown credibility
- No coaching occurs during these evaluations
- The evaluations tend to be adversarial in nature

Note, again, that if self-assessment (and corporate evaluation) programs are effectively conducted, regulatory agency evaluators shouldn't find any significant deficiencies. Even the adversarial nature of these assessments can (at least partially) be overcome. If you have developed a tradition of self-assessment, you become both used to evaluation and confident in your performance; and such evaluations are not feared. Being confident in your performance and being used to evaluation, you can be open and congenial with even these evaluators—and it's hard for even the toughest evaluator to remain adversarial when those being evaluated are confident, congenial, and open.

Which Level Is Most Valuable?

Each of these five levels of evaluation has a specific role in the assessment scheme; all are necessary in any overall evaluation program; no one type of evaluation can be used to the exclusion of others. If any level of evaluation is excluded, frustration and failure will surely be the end result.

As we discussed, each evaluation level looks at different aspects of performance and also provides a second-check on assessments at lower levels. But again, there are significant differences in each level's ability to change the behavior of operating and supporting staffs.

Assessment teams at all levels function to observe and evaluate, but not all teams can (or even are permitted to) coach. Top tier evaluators are further hampered by an inability to provide *immediate* feedback. Regulatory agency evaluators, for example, cannot normally interact with operators on the floor to provide on-the-spot correction and guidance. Rather, these evaluators are obligated to observe and then present findings to leaders at the facility and site management level. Subsequently, first-line operators and technicians receive guidance only after it has been filtered through several levels of management—usually days or weeks after the time of observation. The result is often complacency or hostility toward the observation and the observer.

Corporate and regulatory evaluators (besides being separated from front-line workers by position) are usually of unknown credibility and character to the employees being observed and evaluated. Consequently, whether competent or not, top tier evaluators are separated from those observed by a huge positional barrier. The further removed in position and credibility an assessment team is from those who are observed, the less likely are the observed personnel to understand, accept, and implement corrections based on the assessment team's findings and recommendations. On-the-spot coaching (probably the most important element of the observation and evaluation process) is nearly non-existent at the top two tiers of assessment.

In most cases, the coaching advantage accrues to leaders at the first, second, and third levels. Coaches in the bottom three levels—unit, facility, and site levels—over which industrial leaders have the most control, are particularly well-suited to on-the-spot coaching. When assessment is done well at these levels, an industrial team is far more likely to succeed.

Yet, many industries rely primarily on top tier evaluation, and some have entirely given over the process of observation, evaluation, and feedback to a host of detached agencies whose interest in success is abstract rather than personal. In terse terms, some have abrogated the most important responsibility to the team—to coach them into winners.

What Should Be Evaluated?

The short answer is everyone and everything at every organizational level should be evaluated by one means or another. Since *every* team function is important to mission accomplishment, no part of the organization should be exempt from the evaluative process. At a minimum, critical self-examination should be an integral part of everyone's job.

Yet, most people (and teams) seem also to need some degree of formal, structured evaluation—a point on which the esteemed Dr. Deming might take issue. We tend to fool ourselves into believing that we have no faults. It often takes a cold, outside look to wake us up. Leaders, in particular, must be careful to avoid this pitfall. Theoretically, the higher in the organization a leader is, the more mature he or she should be. The skills of ego submersion, solicitation of criticism, critical self-examination, and listening must be well-developed. Self-examination should be sufficient. Unfortunately, experience often proves otherwise.

Regardless, both self-examination and outside examination are desirable for all functions at all levels. Clearly, however, those parts of the organization which are exposed (or expose others) to highest risk must must also have the highest priority in the evaluation scheme.

Choosing *what* to evaluate is crucial if leaders wish to successfully determine the health of an organization. If operating evaluations focus on the wrong elements, incorrect conclusions are likely. Hence, it is important to determine the *vital signs* or pulse points of an organization.

Organizational Vital Signs

Every organization is characterized by basic functions which must be correctly performed every day for the organization to succeed. In collegiate and professional sports, coaches have learned to isolate, monitor, and measure a host of parameters that affect individual and team play. Individual player speed, strength, agility, and basic skills are targeted as initial selection criteria prior to choosing team players. Once the team is formed (and as the team develops), the ability of individual players to learn and execute assignments is closely monitored. Successful execution of each aspect of the game is dependent upon each player performing a multitude of seemingly mundane tasks correctly. These "mundane tasks" are actually the critical parameters upon which success is based. They represent the vital signs of the organization.

Industrial teams are little different than sports teams. Successful team performance usually depends upon each team member performing a multitude of mundane tasks well. Each player must effectively execute basic shift routines, formally communicate plant information with other crew members in verbal and written form, understand and follow operating procedures, institute safeguards such as tagging and locking equipment when unusual or dangerous equipment and operating configurations occur, investigate and report abnormalities, respond intelligently to equipment casualties and emergencies, and train new team members in proper equipment operation. As we learned in Part III of this book, such skills are the fundamentals of excellent industrial operating performance. They are also the core upon which operating evaluations should rest.

Beyond these vital signs, experienced observers have learned that serious underlying problems are sometimes masked by strengths in other areas. A facility's operators may have a deserved reputation for meeting production schedules while, at the same time, their concern for safety wanes. Thus, observers must also be sensitive to indicators which belie underlying issues.

Operator Knowledge

Operators who are well-trained and work in a motivating environment strive to attain high levels of knowledge. They pride themselves in knowing more about their equipment and procedures than is required. Knowledge is a source of satisfaction, confidence, and security.

Whenever serious knowledge deficiencies are detected in facility operators, observers should begin probing for the depth, seriousness, and causes of the deficiency. Causes may include unidentified needs for training, inadequately performed training, insensitivity toward the need for training by managers, or an environment of low motivation and dissatisfaction among the operators.

A seasoned observer has attained enough knowledge through experience and preparation to ask probing questions that will identify the depth and seriousness of the deficiencies; and, if not knowledgeable in a particular area of facility operations that appears deficient, will recommend followup surveillance preceded by a period of upgrading study.

Operator Attitude

The attitude of personnel involved in operations, maintenance, training, or supporting tasks is the most reliable indicator of the overall organizational health. An experienced observer is constantly aware of the comments, demeanor, and behavior of personnel within a facility. Inattentiveness is a common indicator of insufficient supervisory interaction with the facility operators. Operators who don't feel like part of the team often adopt a "just-doing-my-time" attitude. Indifference or belligerence are reliable indicators of unjust treatment, ever-changing objectives, or management inconsistency. An observer who fails to establish a trustworthy and reliable reputation is unlikely to learn the root of this type of problem because the underlying causes are usually gleaned from informal conversations with operators.

Operator attitude problems must be viewed with the deepest concern. The alert, well-trained operator is the *primary* safeguard for preventing problems. When individuals begin to fail during routine tasks, more serious problems are inevitable.

Record-keeping Practices

Facility records document the design, construction, operator training, daily operational history, and maintenance history of an industrial complex. Without well-kept records, there is no proof of compliance with established standards for design, construction, operations, maintenance, and training; the standards upon which sound engineering and management decisions are based.

Poor record-keeping practices deny future managers the opportunity to make informed decisions concerning equipment maintenance, training effectiveness, and operational trends. Deficient records also present a legal liability when they are insufficient to prove good work practices. By following a task from beginning to end, an observer can usually determine if the record-keeping process effectively attests to the work that has occurred.

Patterns of Personnel Errors or Injuries

When a pattern of similar injuries or errors begins to develop in a facility, it is usually not by coincidence. A common cause or poor practice is frequently at the root. Minor errors or injuries are often warnings (or precursors) of potentially more serious incidents. For such "small" events, observers need to ask the question, "What *could* have happened as a result of this error?" rather than "What *did* happen as a result of this error?" A dead operator is doubly unjustifiable if the warning signs were apparent but not heeded.

Housekeeping and Safety Practices

Poor housekeeping practices are often closely related to safety problems. Unkempt workspaces often indicate inadequate planning, lack of self-discipline, and deficient thought processes. When these indicators appear, there is a high potential for injury in areas such as electrical work, rigging tasks, and careless operations around machinery. A seasoned observer will recognize these symptoms as warning signs and focus attention on potential hazard areas.

Don't Forget to Look for Excellence

All too often when discussing evaluation, the emphasis is only placed on looking for deficiencies. Sure, one important objective of evaluation is to identify and correct deficiencies, and we all need to learn from our mistakes. But can't you learn at least as much from your successes? Besides, how will you be able to determine whether previously implemented corrective actions are effective if you fail to look at what *is* working. Don't forget that the ultimate objective of evaluation is improvement of product and service; so you must (through self-assessment) be able to see what is effective toward accomplishing that objective in order to apply similar strategies to other areas of operation.

And don't forget the impact on worker attitude. How can you expect to develop a habit of constant improvement and self-assessment if the perceived objective of evaluation is only to find fault? When the IRS field agent knocks on your door, you *know* he isn't there to tell you what a good job you did on your last tax return; that raw fear of a tax audit comes from a long-standing, well-developed perception that the IRS is *always* going to find something wrong. A similar attitude is fostered toward police officers in many communities where there is a perception of implied quotas for the number of traffic citations issued.

It's no different in industry. If success and excellence are documented and recognized during audits, evaluations will not be feared, and workers will strive for improvement—and the attendant recognition of excellence. But, workers will actively resist evaluation when there is a perception that management measures the success of an audit only by the number of deficiencies found; and evaluators are pressured to find deficiencies (even if there are none), thus getting reputations as "nit-pickers".

The bottom line? Evaluations and audits must be fair and honest if they are to be respected. You can't allow the pendulum to swing too far either direction. Record deficiencies (and suggestions for improvement) when they are found, but don't "nit-pick" just for the sake of finding something wrong. Similarly, recognize and document excellent and improved performance, but don't invent or reward trivial success. Remember, evaluation is the yardstick for measuring improvement. If the successes aren't recorded as well as the failures, it's difficult to demonstrate a trend of improvement.

Cultivating the Self-Assessment Habit

To be really effective, evaluation must be a continuous, never-ending process; self-assessment must become a habit. Coach Tom Landry, former head coach of the Dallas Cowboys, an organization that posted winning season every year from 1965 to 1985, writes:

Perhaps the biggest factor in the Cowboy's reputation as strong finishers was our quality control program. When I first hired a coach for the sole purpose of studying our films and analyzing our own performance, we waited till the end of the year to evaluate which plays we ran well, which required fine-tuning, and what things needed to be thrown out or forgotten altogether. But I quickly concluded, "Why wait until after the season when it's too late to improve this year?" That's when we began evaluating the effectiveness of all areas of the team every four or five games.

Since coaching seems to have the greatest capacity for causing change, effective evaluation at the first three levels—the heart of self-assessment—is paramount to industrial team improvement.

Many organizations have institutionalized self-assessment. Hewlett-Packard calls it *management by wandering around* or MBWA for short. The Department of Energy community uses the phrase *walking your spaces* for a similar practice. Whatever you call it, it means just getting out and looking around.

But, getting people to look critically and carefully at themselves, their colleagues, and their work doesn't happen by itself. It takes a lot of effort in planning, training, and leadership example. If the boss says, "Do it," but doesn't do it himself, self-assessment will never take root. Rather, it's a habit that must be cultivated.

One way to establish self-assessment at the first-line level is to first institute it formally at the facility and site tiers. When team members learn that formal and informal assessments are both conducted and expected—and aren't just fault-finding expeditions—they, too, will soon take up the practice.

Facility- and site-initiated evaluations that focus on operations, maintenance, and training are really mini-readiness reviews that help create and reinforce the values and standards for the systematic operation strategy. (Nearly all facility functions are governed ultimately by operations, maintenance, and training. Therefore, by systematically reviewing those areas, most significant facility problems will probably be detected.)

Once again, evaluation is the yardstick for measuring the health and growth of the organization. Cultivating a habit of self-assessment is the key to establishing a tradition of continuing growth and improvement.

Evaluation Strategy and Logistics

As stated in the chapter introduction, the nitty-gritty details of establishing and implementing a comprehensive evaluation and self-assessment program are beyond the scope of this text. But, in emphasizing the importance of effective self-assessment in the overall operating strategy, we would be remiss if we failed to provide at least an overview of the process. Therefore, this section discusses:
- Establishing an Evaluation Strategy
- Written Evaluation Policies
- Evaluation Teams and Leaders
- Training Evaluators
- Scheduling Evaluation
- Evaluation Reports

An Evaluation Strategy Establishing a comprehensive, effective evaluation and self-assessment program is not an easy task. As for any other complex activity, the first step is to identify and develop a strategy. For evaluation and self-assessment, the process of strategy development includes:
- Understanding and defining the purpose and objectives of evaluation
- Delineating the scope and methods of evaluation
- Identifying and allocating the resources necessary to implement the program (time, personnel, etc.)
- Determining the means of reporting and communicating evaluation results
- Developing a means for tracking and evaluating corrective actions and improvements
- Making provisions for analyzing program effectiveness and for making necessary adjustments

A strategy must be established for *each* of the planned evaluation levels (particularly site, facility, and first-line) since the scope, methods, needed resources, reporting and tracking requirements will vary. Once again, since the ultimate objective is improved industrial performance, the strategy for evaluation at all levels should focus on systematic operations. As a minimum, evaluations should assess whether the eight elements of a sound operating strategy are present and if individuals and teams are adept in the twelve vital skills presented in Part III of this text.

The evaluation strategy must take into account that change will not come easily or quickly. As Dr. Deming was fond of saying, "There is no instant pudding."

An Evaluation Policy

The evaluation strategy must be communicated through *written* policy if it is to be effective. The policy should clearly define the purpose and scope of evaluation (at each level), the mission and composition of evaluation teams, evaluation methods, and what evaluation is expected to yield. It should define the general format, duration, and periodicity for each type of assessment. And, it should delineate the requirements and expectations for deficiency correction and implementation of recommendations for improvements.

A well-written policy allows evaluation team members to envision the assessment process clearly. It also helps facility leaders and their employees to obtain a clear picture of the mission and methods of site, facility, and first-line evaluation. And, it helps ensure fairness and consistency in evaluations, so they don't degrade into simple fault-finding exercises.

Though the evaluation policy may require change as the process develops and matures, an initial policy will help get the evaluation program organized and started.

Evaluation Teams and Leaders

Establishing a permanent, interdisciplinary evaluation cadre (with operations, maintenance, and training expertise) is beneficial, especially for large sites or facilities where evaluation can become a full-time job. The primary purpose of such a team is to organize and conduct formal assessments at the facility and site levels. But, full-time evaluators also develop unique observation and assessment skills, and can therefore serve to train others in the techniques of effective self-assessment.

Remember, coaching is an essential element of effective evaluation; so, selection of appropriate evaluator-coaches is critical to success of the process. An evaluator observing (but not competent to perform) a process or operation has limited credibility when offering advice or recommendations.

Evaluation cadres (like all teams) need leaders to direct and coordinate the team effort. Although formal site or facility evaluations can be led by cadre members, consideration should be given to having top level managers take turns chairing these evaluations. Their presence lends extraordinary emphasis to the meaning and importance of the evaluations. Moreover, these senior managers gain additional insight into the details of routine operations and are able to advise with authority on corrective action and new direction.

Training Evaluators

Whether evaluation teams are permanent or temporary, individuals selected to be evaluators should ideally have some experience in observation and evaluation, and all should receive training in the specialized skills necessary for effective evaluation. With reasonably established evaluator selection criteria, training should focus on evaluation objectives, strategies, policies, scheduling, techniques, and reporting. Special attention should be given to *diplomatic* methods for on-the-spot coaching and correction.

Every evaluator must understand that the primary objective of evaluation is to facilitate continued improvement through identifying and correcting deficiencies, recognizing and measuring success, and evaluating the effectiveness of implemented changes. "Find me five deficiencies and someone to hang" must never be even perceived as an objective of any evaluation.

Evaluation Schedules

Unless evaluations are scheduled (especially at the facility-initiated level and above), self-assessment programs have little chance to successfully become habitual practice. As with any other activity, without a firm, religiously followed schedule, performance will be intermittent, haphazard, and inefficient. Scheduling should establish both periodicity and specific performance dates. Assessments at the facility and site levels are usually divided into annual, semi-annual, and quarterly evaluations, depending on the scope of assessment—with the more encompassing audits performed at lower frequencies. Focused (shorter) first-line or facility assessments are usually done monthly or weekly.

Of course, evaluation schedules are extremely facility-dependent, and must be tailored to individual industry needs. Major assessments will probably need to be more frequent:

- When an evaluation program is first being established
- If there are many significant deficiencies
- During the startup phase for a new facility
- After major plant or process modifications

But, once the habit of continuing self-assessment and deficiency correction becomes firmly established at the first-line level, the frequency of major assessments at the higher levels can (and should) usually be reduced significantly; those major evaluations become merely a second check to verify continuing excellent performance. Remember, the facility mission is to produce a product or service, and the evaluation objective is to facilitate improved production—not to substitute for it.

Evaluation Reports

What good is an evaluation without a means to communicate the results? That's the purpose of the evaluation report. Of course, the formality and detail of the report will vary, depending on the kind of assessment—ranging from possibly a single handwritten page for a supervisory tour to a weighty formal document for an annual site readiness review. The written report provides a valuable record for discussion and formulation of improvement plans, as well as a vehicle for tracking and documenting corrective actions.

The most important part of the report is documentation of observed strengths and weaknesses. If relevant and responsibly written, they provide a basis for evaluating performance trends.

Emphasis in reporting should *not* be on writing a rigorous formal report. Probably the greatest value in the evaluation process is the on-scene coaching provided to individual employees and their leaders. During every assessment, each evaluator should provide verbal feedback to the personnel being observed regarding observed strengths, weaknesses, and suggestions for improvement. For major evaluations that last several days, the evaluation team should meet at least once per shift to discuss observations and then present those observations to the leaders and key representatives of the work unit observed.

Responding to Evaluation

Evaluation and self-assessment are not ends unto themselves. Rather, they are only the starting points for additional action—whether deficiencies are found or not. Remember, the ultimate objective is *improvement*. If deficiencies are found, they obviously need to be fixed. But, if observation of a process or operation shows only excellent, superior performance, shouldn't you take a closer look to see if what's working right there can be used to improve another aspect of operation?

Just as on-line instrumentation is used to monitor and adjust critical plant parameters, evaluations are the instruments to gauge team, facility, and site performance parameters. But self-assessment is only the measurement. You still need to respond to the indications and make the necessary adjustments to bring performance "back in the band" or to reach new levels of excellence.

Given that response to evaluation is essential, *how* you respond determines the degree of success.

The Right Attitude

Evaluation is worthless if those evaluated are unwilling to listen and act on the results, or (worse) if evaluators use the process just to exercise authority. The *right attitude* toward evaluation is essential.

Robert Waterman states in his book **The Renewal Factor** that the best organizations have learned to both actively solicit evaluation and then (more importantly) listen intently to the advice:

> *The pleasant surprise of this research was how the renewing organizations have made curiosity an institutional attribute. They listen to their customers, of course. They also listen to competitors, first-line employees, suppliers, consultants, outside directors, politicians, and just about anyone else who can reflect a different view of who they are than the one held inside....They seek a different mirror, something that tells them that the world has changed and that, in the harsh light of the new reality, they aren't as beautiful as they once were. The mirror also tells them that unless they change, they're in for a crisis.*

A proper attitude is essential for evaluators as well as those they evaluate. The key to *constructive* criticism is providing cogent comments with an attitude of respect and concern. Similarly, the key to receiving constructive criticism is to shed the natural attitude of defensiveness, determining how the comments of others can help in the continual quest for constant improvement.

Unfortunately, the ability to shed that attitude often is unattainable without first acquiring self-confidence. A poor, verbally battered organization isn't likely to respond well to criticism. Rather, such an organization will usually respond with defensiveness and rationalization. Therefore, evaluators must be coaches as well as critics.

The Right Leaders

As we said, openness to criticism is based primarily upon self-confidence. And an organization's self-confidence is very leader-dependent. Without confident, competent leaders, an organization has no chance for improvement through listening to criticism.

Dr. Deming's seventh point states:

Institute leadership. The job of a supervisor is not to tell people what to do or to punish them but to lead. Leading consists of helping people do a better job and of learning by objective methods who is in need of individual help.

Furthermore, one's ability to render constructive criticism depends on his or her leadership skills. Thus, as stated earlier, proper selection of evaluators is fundamental to successful evaluation.

A Focused Response

Fixing one thing at a time will require some patience on the part of the evaluators. The self-assessment cadre will be able to find far more than the facility team members will be able to quickly fix.

The self-assessment cadre should remember the advice of the Roman philosopher, Publius Syrus:

To do two things at once is to do neither.

Similarly, the facility team members should avoid simply working on the areas in which their performance is already satisfactory. Coach Tom Landry states:

Too many teams contentedly spend their practices working on the areas in which they're already good. Everyone likes to do that. It's easier and more immediately satisfying. But if you want a football team to improve, you need to identify the areas in which you aren't as good and practice those. Then you'll have a shot at the championship.

Factoring Evaluation Results Into Training

The last chapter, discussed how the continuing training curriculum is based upon many inputs and how continuing training is goal-directed and needs-based. Evaluation results provide some of the most important training inputs by helping to establish those goals and needs—goals for improved performance and needs for correcting deficient performance.

Just as abnormal event investigations and critiques (Chapter 20) provide valuable lessons to be learned and disseminated through continuing training, evaluations highlight deficient performance areas that often can be corrected by simple additions or adjustments to established training. But, the benefits of self-assessment go well beyond the lessons learned from abnormal events. Evaluations also provide the much more valuable opportunities to learn from what *does* work. And isn't it much more satisfying to learn from your *successes* than it is from your mistakes?

Topic Review and Questions to Consider

If you don't know where you are, how can you possibly know which way to go?

In this chapter, we have been able to only very briefly introduce the importance of evaluation and, especially, self-assessment to the success of any industrial endeavor.

Evaluation and self-assessment are the instruments for measuring the status of industrial performance. More than just indicating areas of deficiency and excellence, properly used self-assessment is a primary tool for indicating trends and growth of the organization—and for facilitating the continuing improvement process. But, like any tool, it must be properly used to be effective.

Improper use of evaluation and self-assessment can be as dangerous as improper use of a power saw. Properly used, a power saw is an essential tool for building a house; properly used evaluation is an essential tool for building an industry. Improperly used, a power saw will sever limbs; improperly conducted assessments will be equally crippling to the industrial organization, particularly through divisive effects on worker attitude.

And remember, evaluation and evaluation reports are not end products. Just as using ruler and saw to trim framing lumber is only a starting point in building a house, using self-assessment to measure and adjust performance is only a starting point in building the industrial team. If you develop and conduct extensive customer surveys (another form of self-assessment), but then fail to use the results to improve your product or service, haven't you just wasted lots of time, money and effort?

If you use self-assessment only to identify and correct deficiencies, you're only using half the potential of this essential industrial tool.

Ask Yourself

Have you established a tradition of self-assessment?

Have you organized formal site-initiated and facility-initiated evaluations?

Do your managers from operations, maintenance, and training lead and participate in evaluations? Does the CEO lead one every now and then?

Have your first-line leaders learned to look at themselves critically?

Does coaching occur during evaluations or are they "bludgeoning" sessions?

Do you look for (and document) excellent performance during evaluations? Or, are evaluations only fault-finding expeditions?

Is self-assessment properly used as a starting point for further action to continually improve and grow? Or, are evaluations only used as a tool for correcting deficiencies?

Chapter 23

A Case Study in Implementation

Even with the tools of event investigation, continuing training, and self-assessment, implementing the systematic approach is no easy task, especially for a team that has never experienced disciplined industrial operations. It requires vision, patience, persistence, and devotion.

This chapter provides a unique opportunity to test your ideas and knowledge of the systematic approach in a difficult (but not unusual) industrial situation.

As you assess this hypothetical case study, compare your approach to the author's as you develop a plan for correcting the course of a wayward industrial team.

The Situation

You have recently been appointed plant manager of a chemical production plant in the state of Wyoming. The plant is owned and operated by a well-known major corporation and employs 450 personnel on the average, 375 of whom are members of the operations, maintenance, engineering, and technical support staffs. Other supporting organizations include health and safety, training, human resources, procurement, quality assurance, and plant security. The mission of the plant is to produce a chemical compound used in the production of semiconductors. The compound is a moderately toxic, volatile liquid which is stored in 5,000 gallon double-shell tanks. When ready for transport, the compound is pumped into tankers for shipment to over twenty different customer corporations. Several of the byproducts of production are classified as hazardous waste. Some are toxic, some are corrosive, and some are explosive. Most require special care in packaging and storage.

The plant has a generally good safety record over its thirteen years of operational history. Recently, however, an explosion occurred in the plant which killed two operators and a maintenance technician. An investigation of the event identified several serious deficiencies in the handling and storage of hazardous materials, one of which was the direct cause of the explosion. In addition, the investigation of the event revealed deteriorated training of hazardous material handlers, inadequate procedural controls for the handling of hazardous materials, and a generally complacent approach to operations on the part of the operating staff. The investigation further disclosed a series of safety-related events over the last five years of operation in which eight people received minor injuries on five different occasions. The previous plant manager of the facility has been reassigned and the plant operations manager is currently serving as the interim plant manager.

Background

You are a 42 year-old executive with a bachelor of science degree in chemical engineering and twenty years of varied experience with three different military and business organizations. After graduating from your university, you served for four years in the Air Force in flight operations having received your commission through the reserve officer training program in which you participated throughout your undergraduate studies. You left the Air Force as a captain having last served as a pilot ferrying equipment overseas on Air Force transports. You left the Air Force to join your current corporation in order to pursue your interests in chemical engineering. Since coming to this corporation, you have served as a process operator, junior and senior production engineer, shift engineer, shift supervisor, maintenance supervisor, and operations manager in two similar plants in the state of New Jersey. Your performance has been excellent. You have enjoyed the respect and admiration of your subordinates, your colleagues, and your managers and you have been on the *fast track* toward plant management. When the accident occurred in Wyoming, your plant manager and the corporate vice president for operations met with you and expressed their desire that you take the job as plant manager in the Wyoming plant. After talking it over with your family and friends, you have accepted the position, packed your household goods, and made the long trip west.

The Problems

You have had three weeks of turnover time with the interim plant manager and have spent long hours observing in the plant, reviewing the processes, talking with operators and technicians, and analyzing the strengths and weaknesses of the organization. You have observed that the equipment and the operation are fairly sound, the crew and technical support leaders are, for the most part, competent and treat their people well; but, you are disturbed at what you perceive to be a loose or casual attitude on the part of most of the leaders and their employees toward their responsibilities.

Your tours in the plant, especially on backshifts, have identified imprecise communications, poorly kept station logs, and incomplete recordings in the data record sheets required of the station operators. The shift turnovers between on-coming and off-going crews are sometimes ineffective in communicating the shift plans, equipment status, and current plant hazards and abnormalities. Most of the crew leaders don't go to the shift change meeting but, rather, go directly to the control room and have coffee and conversation with the offgoing crew leaders. Many of the crew members don't arrive in time for the meeting at all. Few operators seem to walk through their stations and observe the status of station equipment prior to assuming responsibility. Also the verbal turnovers between operators are unusually brief or non-existent. Procedures seem seldom to be used to perform operational and maintenance functions. In fact, many of the procedures that you reviewed were out of date and had handwritten notes in the margins, renumbered component identifiers, and penciled-in additions to system drawings.

As you proceeded through the plant on your nightly tours, you noticed an inordinate number of unidentified and uncorrected material deficiencies. Operators seemed in some cases to be unaware of or oblivious to minor deficiencies and even some obvious safety deficiencies. You had to point out some of them personally to the interim plant manager because they were of such danger.

You also have begun to have growing concerns about the coordination of the operations staff and the supporting elements. Both plants at which you worked in New Jersey seemed to be *closely coupled*. They were more like teams. The operations organizations were better supported by the support organizations. People frequently asked one another about the quality of service they were providing. Support organization leaders often approached the operations leaders and queried them as to how they might better support the operating team. Here in the Wyoming plant, however, members of the various staff elements appear to be interested primarily in the survival of their own organizations rather than supporting the facility mission. Also, the plant leaders and supporting staff do not coordinate well with the crew operations staff on backshifts. In fact, some do not coordinate well among themselves, either. For example, the maintenance staff and the operations staff don't interact well when preparing for performance of tasks such as preventive maintenance. Post-work tests are sometimes incomplete or not performed. In fact, there often seems to be confusion regarding who is responsible for performing the tests.

And it wasn't just operations and maintenance in which you observed weakness. The plant training regimen also appears to be weak and inconsistent. Training is poorly planned, undirected, and ineffective. The training sessions that you observed were characterized by boring instructors, inattentive participants, and unimaginative presentations. Training sessions often started late and ended early. The instructors that you observed appeared to be poorly prepared.

Finally, you have noticed that something that went on in the New Jersey plants regularly doesn't seem to be happening here. In New Jersey, the plant leaders walked around their plants often, talking with operators, spotting deficiencies, patting people on the back for particularly well-done jobs, and asking employees how better to address problems that were cropping up. That process seems to be conspicuously absent here in Wyoming.

Your Task

Before your departure, the vice president of operations called you to corporate headquarters and discussed her growing concern about the generally undisciplined approach to operations identified by the last two years of safety audits at the corporation's eleven plants around the nation. She spoke of a presentation that she attended at an industry conference in which a member of another petro-chemical manufacturer discussed a systematic approach to operations for industrial facilities. She stated that several members of the corporate operations and training staff had also attended and that they were very impressed by what they heard. After the conference, they had agreed among themselves to consider instituting a similar operational approach for the corporation at its eleven facilities. The vice president gave you a copy of the presentation brief and asked you to review it thoroughly.

The brief outlined a series of guidelines for safe operation of industrial facilities covering a range of topics including *procedural compliance and control, clear communications, shift routines, equipment status control, continuity between operating groups, protection of operators and equipment, on-the-job* training, investigating abnormalities, and learning from mistakes. You have been studying the presentation brief over the last few days and it brought back a lot of memories of your Air Force service and flight operations. So much of what you did on those Air Force base flight lines seemed at the time to be too conservative. In fact, you were somewhat relieved to be away from those rules and the rigorous enforcement of the practices and conventions. But over the last several years, you've seen (and even been involved in) a number of *near misses* in your job assignments that resulted from simple, relatively minor mistakes—little mistakes that piled up. Most of the mistakes, you have realized, could have been avoided by observing some of the same rules and practices that you had applied on those flight lines. You've had a nagging feeling that the *Air Force way* actually needs to be applied to your industry as well.

Last night you received a call from the corporate vice president. She asked if you had studied the presentation brief. When you said that you had, she asked what you thought of it. You replied that, though some of the guidelines didn't seem to be applicable to the plants with which you are familiar and some seemed to be too restrictive, you had come to the conclusion that the approach was not only workable, but necessary. She asked if you would consider initiating such a program at your plant as a pilot program for the corporation. You told her that you had already been thinking along those lines and would like to try it. But you also stated that you had reservations about how long it would take. You said that you didn't believe that it was going to be either quick or easy. The vice president said that she understood, but she stated that a lot of concern about performance at the Wyoming plant had been expressed at the corporate level, especially considering the bad publicity that the explosion had caused. She said that, not only was it important that plant performance improve rapidly, but also that the plant be able to demonstrate within approximately six months whether the approach should be considered for implementation at the other ten facilities. You said that you would do your best.

What Should You Do?

All over our nation, executive leaders are being confronted by situations disturbingly similar to this one. Many of these leaders are recognizing the need to adopt a systematic approach to operations and maintenance. The cost of injuries to personnel and plant shutdown time are dictating a *slow down and do it right the first time* approach.

The task with which you are faced is daunting; but, the situation is recoverable if you approach this problem correctly. Assuredly, there is no *quick fix*. It's going to take some time to get this plant back on track. You can, however, show progress toward your goal quickly and overcome in particular the safety hazards which threaten you and your team members.

The Obstacles

Acceptance by Subordinate Leaders

The obstacles which you will encounter in this task are formidable. As you have recognized, it is unlikely that you, as an untried newcomer, are going to be received with open arms by the immediate members of your subordinate staff. You will be perceived as *that guy from back East*. You are infringing upon occupied territory and your presence will be seen as a threat to the power and careers of other leaders in the facility.

Acceptance by Employees The next major obstacle that you will face is the acceptance of the rank and file employees. Many will have the same feelings as the plant leaders and will be fearful of the future under your leadership. All will recognize that major changes are in the offing and a general feeling of uncertainty and insecurity will prevail. Some will respond initially with distrust. Others will be purposefully uncooperative. A third contingent will be militantly disrespectful and abusive. *You* must be unperturbable.

Denial of the Problems The third major obstacle that you will face is convincing your subordinate leaders that a significant problem exists. Many will be attuned to the need for some change because the deaths of the three employees will have made an impact. But some will perceive the event as a disconnected occurrence and will resist any change (beyond superficial rearrangements) as *meddling*. You may, however, have the full support of one or two perceptive individuals who recognize the depth of the problem.

Resistance to Change Even when faced with the reality of this problem, many plant leaders and employees will have great difficulty visualizing and implementing the necessary changes. *"We've always done it like this before"* will be a common refrain that must be overcome. You will also probably be advised by some of your subordinate leaders that the fatal incident was *just a run of bad luck*. Yet, the data indicates otherwise.

Time You face a fifth major obstacle—time. Great momentum is building at the corporate level regarding the problems in this plant. You *must* demonstrate your abilities and you *must* turn the direction of this facility. You enjoy a large degree of confidence from the corporate vice president and others or you wouldn't have been asked to take this assignment. They believe in you but they need to see signs that their confidence is well-placed.

In the early going, you will be forgiven much. You need only show *some* progress. Remember, it is likely that no one else covets your current job under these circumstances, so the people at corporate headquarters are looking for a champion. Even minor successes, if properly advertised, will instill confidence and earn the gratification of your leaders. That means that you can afford to lay a foundation here *so long as you can demonstrate positive changes in the performance of the facility*. Later, however, your leaders won't be so forgiving. After you become entrenched, they will expect more progress. But for now, you are the pilot project. You are charting new territory and your leaders know that the going is likely to be slow and bumpy at first. If you do this well, you're going to be seen as somewhat of a folk hero. You will be the example that people talk about and the person most likely to be chosen to spearhead the implementation of these methods throughout the corporation. So, though there is significant risk now, the potential gains are great.

Most importantly, you have the opportunity to make a difference for the better in the lives of the 450 employees over whom you have been given authority. Three have already died. If your actions prevent the deaths of any more, you will have done well. Adhere to the principles that you know to be true and effective, treat people with respect, educate and coach your subordinates toward high standards, and you will succeed.

The Approach

There are many methods of attacking this problem. But in every case, a few key steps must be accomplished in order to successfully implement a systematic approach to operations at the Wyoming facility. Here are eight steps that you should consider. Note that some of these actions must be initiated co-incidentally. You also may see the need to add other steps to this process for your own situation.

Prepare to Take Over First, you must prepare to assume the leadership role in this facility. Your mental preparation and vision for the future are requisite to your success. If you cannot visualize the potential of this team and a method for achieving it, you cannot communicate that vision to your subordinates.

Control Immediate Safety Risks Second, you must consider shutting the facility down or curtailing production to limit the danger of an accident recurrence until the team has been retrained and retried in fundamental operations and safety issues. If any more are injured or killed, the legal liabilities will be enormous. They threaten the future existence of the facility and its employees.

Establish Leadership Confidence Next, you must initiate the process of gaining credibility, trust, and the respect of your employees and subordinate leaders. The problems that you face cannot be solved by you alone. This will have to be a team effort, or it won't succeed. You will need every leadership tool that you have ever learned to get this facility turned around.

Identify the Problems The fourth step is to identify to the leaders of the facility the nature and symptoms of the problem. Your assistant coaches must understand the problem before you (and they) can formulate a solution. If your subordinate leaders do not attain the same understanding of the nature and magnitude of the problem as you, they will not wholeheartedly support the solutions. Implementation will fail.

Once your subordinate leaders understand the problem, you and they must convince the rank and file team members of the facility of the danger in continuing in the current method of business and the need for change. Only with the support of the majority in this facility can the necessary changes be made that will solve this problem.

Evaluate the Problems Early in the problem-solving process, you and your team should perform an initial self-assessment of current performance in areas of concern. This baseline self-assessment will help you to determine your plant's strengths and weaknesses. It will provide to you valuable insight in planning the restoration of a disciplined approach to operations and maintenance.

Following completion and evaluation of the self-assessment results, you and your team must determine those areas which require immediate improvement. Your decisions will have to be risk- and resource-based. In other words, you will have to put your resources to work immediately in those areas of greatest risk exposure with the best opportunity for improvement.

Establish Written Guidance Having determined the importance and sequence of issues to be addressed, you and your team must establish written policy guidelines to which you intend to train your team members and hold them accountable. The rules of performance and conduct must be rational, achievable, and clearly stated. If you are unable to justify them, they won't be followed.

Plan and Conduct Corrective Training It will be ineffective and destructive simply to issue an updated set of guidelines and policies for operations (and maintenance). You and your team must develop and execute training on the systematic approach to operations that will explain, demonstrate, practice, and test team members on the new guidelines.

Implement and Assess Effectiveness of the Solutions Once the team members have learned the new guidelines and their bases, you and your team must evaluate team performance and coach team members continuously in proper execution through a process of ongoing self-assessment. Without coaching, you will eventually end up again in the mire of complacency and sloppy performance that resulted in the deaths of three of your employees.

Preparing to Take Over

Until you take control of this facility, support the interim plant manager, if at all possible, and counsel with him or her. He or she will likely be your operations manager. You need this person on your team. You have a wonderful opportunity to coach and educate this person if you win his or her confidence early. Coordinate with the interim plant manager to pass the baton of command in an orderly fashion. Let everyone know when the baton pass will occur. Don't make a production of it, but don't allow any confusion as to who is in charge.

Meet with your subordinate staff. The interim plant manager probably introduced you early in the process and your subordinates are taking bearings on you. Everybody knows why you're there. They just don't know if you're going to use them as ladder rungs or whether you are truly interested in

A Case Study in Implementation

them and the plant. Act like a team player. Don't be arrogant. Show respect for their opinions and expertise. Analyze the staff to determine who is honest, who is talented, who is a team player, and who is a back-stabber. Be careful. Some first impressions will be wrong and some of the people that you don't initially like will become your best supporters. The ones who say what they think in a straightforward, honest manner that is not ego-driven are worth their weight in gold.

Controlling Immediate Safety Risks

You have an immediate safety problem to consider in this facility. If you believe that another fatal or injurious situation is likely to occur, you must consider stopping or curtailing production to make immediate corrections. Such an action will be difficult and unpopular, perhaps even career-threatening, especially if the dollar costs will be large. The corporate staff has probably already foreseen this as a possibility or may have already ordered the action taken. But, in any case, your moral responsibility to the employees of this company may require a courageous action in this instance.

If you must shut down for safety considerations, there will be great pressure to return promptly to operation. You must make the best use of your time. You, your staff, and a team of your employees need to determine those areas of greatest danger and vulnerability within the facility. You must then move rapidly to put extraordinary safeguards in place to protect against further incidents until you can shore up the damage with long-term solutions.

An external safety review performed by a team of experts may provide you with valuable insight and guidance. It may also confirm your position to members of the corporate staff. You may, as a result, be in a better position to obtain more time and additional resources if necessary. You'll have to implement a crash course in safety in those areas of greatest vulnerability. You and your management staff are probably going to have to literally live in the facility until the danger period is past.

Establishing Leadership Confidence

You must quickly begin the arduous process of establishing credibility. Without the trust and respect of your employees and subordinate leaders, you cannot coach this team back into winning ways. Hopefully, you have gained some credibility as a result of your prompt actions in regard to the immediate facility safety considerations. But you may also have done some damage since you likely have had to use a coercive style and make some unpopular decisions to get this situation under control. Your character, for the most part, is unknown to your subordinates. At best you may have some referent power as a result of your good reputation established in New Jersey.

Perhaps one or two people at the Wyoming facility are transfers from the New Jersey plants and are acquainted with you. If so, they will know that you are a fair and honest leader. The word will get around. But, in the final analysis, most people will reserve judgment until you have proven yourself through action. Your actions will bear you out over the long haul; but, for now, there are a few things you can do to improve your standing:

Walk the Plant Floor Walk the plant floor on all shifts as you have been doing. Make this your habit. Don't let it die after the initial rough going is over. Your face and your personality must become as comfortable as an old shoe to your subordinates. That doesn't mean that you're easy on people regarding standards, but it does mean that they can trust you and be honest with you. They know you won't "shoot the messenger" if he or she happens to be the bearer of bad tidings. Meet as many people as you can. Learn their first and last names, their career experiences, and a bit about their histories. *Listen!!* Even though it will take time and you may be detained, a leader who listens will be greatly appreciated. Remember, you win an organization over to your cause one person at a time. And you do it by persuading each person that *your* cause is actually *their* cause.

State Your Concerns As you talk with your subordinate leaders and employees, state your honest concerns and ask for opinions and advice on the topic. Physically point out examples of deficiencies that you observe and describe why they are of concern to you. Always relate them to your desire to improve the safe working environment and the efficiency of the facility. Point out that good business guarantees good jobs and good pay.

Involve the Team As you tour the plant spaces and engage in conversation, ask the leaders and employees what problems *they* see. Ask for their ideas on how to fix the problems and ask for their help. Get them involved. If someone has a legitimate pet project that will help solve your problems, encourage them to pursue it. Provide resource help when you can. Express your appreciation for their willingness to assist you. Show genuine interest in *their* expertise and knowledge. Treat them with respect and professional admiration. You will be well-rewarded for that investment. Remember that team building is nothing more than a process in which a group of people solve hard problems together. They may not be a team in the beginning, but with a challenge and a little leadership, they will become a team.

Meet Everyone If you have done your work well in this phase, you will not be meeting your subordinates for the first time when you hold your initial all-employee meetings. You will already have advocates of your cause and supporters of your personality and leadership style in the audience. They will offset the inevitable few who will be your detractors regardless of what you do or say.

Identifying the Problem

Despite the fact that three deaths have occurred in this facility, some leaders and employees will be unable or unwilling to see that the problems are more than minor issues. Problems, as we all know, cannot be solved until they are first identified. Since you cannot solve these problems by yourself, you need a team effort to succeed in this process. Your subordinate leaders and employees must all be committed to making this a better and more efficient work place. Consequently, one of your most important tasks is to convince your subordinates of the nature and symptoms of this problem. Here are some steps that will assist you in this process:

Conduct Leader Indoctrination Conduct management and key employee indoctrination on the problem. Introduce the concept of the systematic approach to operations, and the function of a viable self-assessment program for finding and fixing problems. Include shop stewards, foremen, and lead technicians. Even though they may not officially be members of management, these employees are key members of the management team. You cannot be successful without a solid team effort. If you leave out the *enlisted* men and women, you have lost the rank and file. They are the ones who must eventually embrace this philosophy. *You* should be the leader of these sessions. In so doing, you will establish your expertise in the technology, your attitude of concern and professionalism in the minds of your subordinates, and your position as a coach and a mentor. This step is actually the beginning of a broad training process that must reach to every member of the facility.

Engage Leaders in Plant Tours Perform *mini*-assessments in the form of plant tours that *you* lead. The purpose of these tours is to show subordinate leaders and key facility employees examples of the problems that you have observed over the last several weeks. Avoid the perception of arrogance. Engage your subordinates in identifying problems themselves. Ask questions such as: "What does that thing do?" "This is set up differently than I'm used to. Why is it done that way?" "That doesn't look right to me. Is that the way it's supposed to be?" Your subordinates will get an opportunity to demonstrate their expertise and educate you. They will also begin to see problems that they hadn't seen before. Many of your subordinates will be shocked that they didn't see what you are identifying. It's easy to become accustomed to deficient and even dangerous conditions. Others will be ecstatic that, finally, there is a leader who is seeing the problems that they have vainly been trying to expose for years. And some of your subordinates will continue to resist through challenging the validity or seriousness of your findings. If you are well-armed with reports of incidents and accidents from other facilities and other industries which illustrate how these or other seemingly innocuous deficiencies contributed significantly to injury or property damage, they will either begin to listen or at least they will shut up. Your expertise will be an imposing and impressive tool.

Show Examples of Deficiencies As you indoctrinate your facility leaders and subordinates, make extensive use of visual examples of problems with related case studies from your own and other industries in which these types of problems have led to injury and property damage. These will lend credence to your cause. The goal is to persuade all of the facility team members that these seemingly small problems and innocuous habits lead to hazardous situations which could ultimately affect the health and safety of the employees and the profitability of the plant. The approach is to solicit the help of the employees in creating a safer and more efficient workplace.

Consider dividing the staff into interdisciplinary teams of two or three persons after you've conducted initial plant tours and indoctrination. Provide to each team a polaroid camera and a roll of film. Assign each team to tour the plant (or, perhaps, one another's areas of operation) and take pictures of situations which are deficient. On the reverse side of each photograph, require the team members to record the location of the photographed condition and the date of the photograph. Require also that they state the nature of the deficiency and why they believe it to be significant. If possible, the team members should reference the deficiency to documentation which identifies it as deficient and state why they believe the condition to be significant. A similar result can be obtained through the use of video cameras and short team presentations which convey the same information.

Lead All-Employee Meetings

It is at least as important to convince the rank and file employees of the existence and nature of the problem as it is to gain the support of your subordinate leaders. One of your primary objectives must be to create an environment in which the employees feel that it is their duty and responsibility to identify and assist in correcting deficiencies within the facility. Without an understanding of the reasons for implementing a conservative operating philosophy, it is unlikely that the facility employees will support this endeavor.

Therefore, *you* lead all-employee meetings to identify the same concerns expressed in the managers' meetings. You can employ assistance from among the ranks of the believers on your staff, but *you* must set the tone. These are critical meetings. You will either gain great credibility or lose the confidence of your people based upon how you present yourself. If you have done a good job of interacting with the employees during your plant tours, the audiences will not be nearly so resistive and hostile. Keep the size of the meetings as small as you can afford. You need them to be personal so that you can have eye contact with individuals and answer questions. You will, in so doing, have a better chance of avoiding the *mob mentality*. People are less responsive in large groups and are more likely to resist when surrounded by friends who might encourage them.

Visit Other Facilities

If at all possible, conduct one or more field trips with as many of the subordinate leaders and key employees as possible to other industrial locations to show how the best organizations in your business or similar businesses are operating. This type of comparison is the most powerful of all tools to change the vision and mindset of people. If field trips cannot be conducted, attempt to accomplish a similar function using videotapes and interviews of leaders from other organizations. Consider inviting leaders of related industries who are performing at high standard levels to come and address your leaders on the need for implementation of a systematic approach to operations and maintenance, the advantages of implementation, the approach to implementation, and the problems associated with implementation. You may find it valuable to continue such an exchange program long after your problems are solved.

Evaluating the Problem

One of the most important tools that you have available for evaluating the full extent of the problem is comprehensive self-assessment. You have already begun the process of self-assessment within this facility when you initiated your own tours. You should now perform a comprehensive baseline self-assessment to establish the strengths and weaknesses of your facility.

Self-assessment, in its purest form, is the process of coaching the team members of a facility toward continuous improvement through daily observation and feedback on the strengths and weaknesses of operations and supporting functions. Self-assessment is one of the most important functions of facility leaders. When properly done, the process should eventually become a part of every employee's job.

Within the Wyoming facility, it appears that the self-assessment habit has never been established among the facility leaders. It certainly has not become a part of each employee's work regimen. The facility is in need of a baseline self-assessment to establish current strengths and weaknesses in the areas of systematic operations and maintenance. A comprehensive self-assessment will also serve to show that you are serious about the process and will allow you an opportunity to see how well your subordinate leaders understand their facility. Once started, the self-assessment process should become a never-ending habit of daily evaluation and coaching.

Here are some steps to guide the development of a successful self-assessment process:

Select a Cadre Select an initial self-assessment cadre. This cadre should consist of an interdisciplinary team of managers and employees who have shown a high level of interest in this process. Attempt to use volunteers but ensure that you have chosen a group of knowledgeable, well-informed people who will provide a critical self-assessment. This cadre will not only be performing the initial self-assessment, but they will also likely become the core of teachers that you will use to train others in the self-assessment process. Many may become members of a permanent self-assessment group if you determine that it is necessary to maintain such an organization.

Choose Assessment Topics Choose and assign self-assessment topics based upon your personal evaluation of the facility's areas of greatest vulnerability. The industry council brief that your corporate vice president provided you identified several critical areas of industrial operations for your review and consideration. They included *procedural compliance and control, clear communications, shift routines, equipment status control, continuity between operating groups, protection of operators and equipment, training new team members, investigating abnormalities, and learning from mistakes*. Nearly every industrial operation must address these issues, so they are probably a good place for you to start as well.

Avoid assigning unit leaders to assess their own areas of responsibility. Use interdisciplinary teams, each of which is led by a manager who has no axe to grind against the unit being assessed. Compose the team of managers, union stewards, lead technicians, and others who have the greatest interest in ensuring the future success of the facility.

Conduct the Assessment Conduct a baseline self-assessment to determine initial strengths and weaknesses. This assessment will also provide an opportunity to begin the practical training for a cadre of observers that will perpetuate the program. For a facility of your size, you should allot approximately one week for completion of the assessment.

Evaluate the Results Evaluate the results of the baseline self-assessment to determine areas in need of special emphasis. The baseline self-assessment is likely to identify several important areas in which you will receive the most return on investment as you begin to implement disciplined operations. As an example, operations turnover is often the source of a large percentage of shift-related problems. If the baseline self-assessment confirms that operations turnovers are being poorly conducted, you may be wise to isolate operations turnover for the first training objective in the coaching process.

Establishing Written Guidance

Based upon the results of the baseline self-assessment, you will have a much better perspective of the entire problem and its related causes. The causes may, in fact, range far beyond the scope of operating and maintenance discipline; but, there is little question that you must address that area early in your tenure.

Having identified the areas of operations upon which you must focus, you must now establish written guidance for future operations and maintenance performance. The industry council guidance that you received from your corporate vice president may have much that you will find useful; but, the rank and file members of the facility operations and supporting teams are accustomed to operating by facility procedures, not a presentation brief from an industry council. It is, therefore, imperative that policies or procedures which spell out the standards and expectations of performance for the systematic approach to operations be established for *your* facility.

You have several options in this regard. First, you can gather a team to write the policies or procedures and simply present them in draft or final form to the employees. Second, you can use policies or procedures established within other organizations and customize them to fit your facility. Finally, you can conduct workshops as a part of the training process in which the personnel undergoing training take a part in writing the policies and procedures.

Using previously established procedures is faster in the short run but has the disadvantage of not involving the user in the creation process. Without initial ownership, implementation may actually take longer. Involvement of the employees, on the other hand, in the writing process is slower in the short run but has the advantage of creating the feeling of ownership. In the long run, implementation is likely to be easier.

A good compromise is to provide baseline documentation and have the employees in a workshop environment do the customization for their own facility. Little time is spent on the initial development and the feeling of ownership and involvement still accrues.

Regardless of which option you choose, you must determine which areas of performance require written guidance and then you must develop the guidelines. If possible, gather example procedures from other industries that are already successfully employing similar conventions and practices. They will provide a good starting point and keep you from "reinventing the wheel".

Teaching the Systematic Approach

In actuality, you have already begun the training process for implementing the systematic approach through your indoctrinations and walkthroughs. You cannot have an effective training program until you have achieved plant-wide recognition of the problem and have curried an attitude of acceptance or at least open-mindedness among the facility team members.

You must now enter, however, into the formal process of educating the facility leaders and employees in the philosophy and mechanics of the systematic approach. You must develop and execute training on the systematic approach to operations and the guidelines that you are developing for its implementation.

A common mistake often made in such a process is in believing that a compact classroom training program on the subject will fix the problem. Seldom is this the case. Although it is both valuable and necessary to conduct classroom training on the concept and mechanics, the training will fail if presented only in a classroom environment.

There are five major steps that must occur if the training process is to be successfully implanted in an industrial operation:

Teach the Philosophy of Systematic Operations

The rules, codes, and practices of the systematic approach to operations are designed to protect against worst-case conditions. As such, they are often seen as too conservative for most situations an operator encounters during daily operations. There will naturally be resistance to the implementation of such rules. But operators who receive proper education in the reasons behind the rules will be far more likely to embrace and execute such rules on a daily basis.

The process of teaching the philosophy of the systematic approach is usually best accomplished in a classroom environment in which the modes of presentation include lecture, class discussion, and seminar case studies. Depending upon the background and level of industrial experience and maturity, this process is likely to take from two to five days. Excellent resource materials include accident case studies from sources such as the National Transportation Safety Board, incident and accident reports from your industry and similar industries, and positive examples of how the conservative practices and principles of the systematic approach are used to improve both safety and efficiency.

Educate Team Members in Individual and Team Skills

Once operators and managers understand why the rules are so important and how they relate to maintaining the integrity of the safe operating envelope, the next step is to teach the mechanics of the systematic approach. Usually, the best approach to this step is to conduct workshop-style seminars on the specific topics of concern. The purpose of these workshops is to determine in detail the practices and conventions of, for example, properly performed turnovers.

Demonstrate the Skills

Skill demonstration is a step that is often deleted from the training process. Most people have a difficult time visualizing how to actually implement rules that have only been discussed. Operators must be *shown* how the conventions of each area of guidance are applied. The training process, therefore, must demonstrate what the desired behavior looks like.

For example, your instruction on logkeeping must be reinforced by showing operators what a good set of logs looks like. The training can be creatively implemented by presenting a written situation to the students and tasking them with creating a set of log entries representative of the situation. Similarly, written and oral communications, recording of instrumented parameters on data record sheets, shift turnovers, control area activities, operator tours, procedures, operator aids, and equipment and piping labeling can all be modeled and reinforced with creative exercises to show the students the proper methods of implementation.

It is both possible and desirable to combine the steps of education and demonstration in the same workshop. For example, the essentials of log writing can be both taught and demonstrated in a four-hour seminar.

Practice the Skills

The transition from classroom to on-the-job application of the skills is a crucial one. Once employees have been taught and shown the expected performance standards, the next step is to practice. As team members practice the skills, they must be coached in their performances. Individual logkeeping must be reviewed and corrected; shift turnovers must be stopped and modified when they are incorrectly performed; and verbal communication must be rectified if improperly executed. Without knowledgeable coaches drilling the team on the job, this process will fail.

Reinforce the Skills Through Daily Coaching

Classroom training and initial practice in philosophy and mechanics will prove to be only marginally effective unless it is followed rigorously with inspection and review by facility and crew leaders on a daily basis. Coaching is a daily job. Performance monitoring and coaching is the method by which the rules, conventions, and practices of the systematic approach are reinforced. It is a step that never ends. Facility leaders must interact with the employees of the facility and provide on-the-spot feedback regarding the strengths and weaknesses of their performances. Excellent performance or notable improvement must be praised while undesirable performance must be corrected.

As new concepts and guidelines are added to the team's systematic approach to operations and maintenance, new segments to the self-assessment process should be added as well to check performance. Checklists or reminder cards may be usefully employed during this coaching process to assist employees in practicing the rules and conventions learned during the classroom training.

Guidelines for Conducting the Training

Perhaps the most common mistake in planning and conducting training in the systematic approach to operations is trying to develop and teach all of the guidelines in a short period of time. Team members who have been accustomed to performing in one manner will find it difficult and sometimes undesirable to change to a more controlled way of operating.

Fix One Thing at a Time

A better approach is to employ a step-wise training process in which one aspect of performance is isolated at a time. Focus on one topic, get it right (or at least on the road to being right), then shift focus. Joe Paterno, legendary coach of the Penn State Nittany Lions football squad, recounts his own early failure to learn this coaching lesson:

> I'd look at a kid and see that when he was going to throw he had a habit of holding the ball lower than when he was going to hand it off or run, or his footwork was bad, or this or that was bad. He might make five or six kinds of mistakes and I wanted him to fix them all in one day. But I began to learn…to get one thing right at a time. Spend a couple of days with a kid on his footwork and get it right. Then go to the next part and get that right. [That] taught me to analyze a problem, put down a specific plan for how to get from here to there, step by step, in the time we had available.

The training modules required to teach the systematic approach to operations cannot effectively be taught in a single course of instruction. Most of the training discussions and workshops for topics such as shift turnover will require at least half a day to conduct depending upon the content. The seminars should be conducted over a period of time to allow participants time to digest and apply the contents. As a result, the initial course of instruction will span a period of several months.

The sequence in which the training modules are presented is also important. Many of the topics such as logkeeping and data record sheet entries, or radio/telephone communications and face-to-face communications are best grouped together since they address closely related issues.

Reinforce Classroom Training with Field Evaluation One possible scenario for executing the training process is to conduct orientation training in one area of instruction (shift turnover, for example) for selected team members and then use the same team members to perform a plant evaluation on that topic. The purpose of the evaluation is to determine the current practice in the area of concern. Based upon the evaluation, the team members can determine what changes, if any, are appropriate for their facility's policy.

Use this process not only as a basis for establishing operating policy, but also as an opportunity to learn and practice the self-assessment process.

Finally, engage the participants in evaluating and rewriting proposed guidelines. Don't reinvent the wheel. Consider using the policies from other facilities as baseline documents from which you may draft your own customized documents. Place the guidelines on a company standard word processing software program so that they can be revised as necessary to fit the plant's needs.

Implementing and Evaluating

Once the philosophy has been taught, the skills instructed, and the mechanics demonstrated, implementation and evaluation must begin. First line leaders must initiate operation using the new guidelines. You must, however, expect confusion, resistance, and setbacks. Team members will still be suspicious of the systematic approach and, without excellent daily coaching, nothing will change.

It is imperative that the self-assessment process become a cornerstone of daily operations, maintenance, training, and technical support. Only through rigorous, daily self-assessment by members of your self-assessment cadre and, eventually, by the other key team members will you be able to determine whether the team understands the systematic approach and whether first line leaders are requiring their teams to adhere to the guidelines.

Here are some steps that you should consider to ensure implementation of a vital self-assessment program:

Develop Integrated Self-Assessment Self-assessment without planned objectives is not as effective as integrated assessment. Develop a long-range self-assessment plan that envelopes operations, maintenance, and training. Remember, the self-assessment process applies to far more than conduct of operations. Expand the process into a *neighborhood watch* program through training and indoctrination of all managers, supervisors, and employees. Ensure that employees at all levels understand and embrace its purpose and value.

Implement the Plan Develop an implementation schedule. The self-assessment plan should normally span a period of two or three years. The implementation schedule extracted from it will probably cover a period of one year. It can then be subdivided to span a month or a quarter just like a preventive maintenance schedule.

Initiate the self-assessment process in accordance with the schedule. Don't allow the process to die. Once it is established as a habit, subordinate leaders and employees will recognize the value of the process as a coaching and training tool.

Look for Trends Conduct frequent benchmark meetings with the self-assessment cadre, facility and project managers, and facility employees to debug the system, continue familiarization, and market the program. Analyze the data content and quality produced from the initial self-assessments and assist in formulating corrective actions. Certainly, individual observations will require correction; but, look for trends. The sum of the observations is your greatest concern.

How Long Will It Take?

The process upon which you have embarked is a never-ending process. Your goal is continuous improvement of facility operations and maintenance as well as all of their supporting teams. But you must be able to demonstrate substantial improvement within a period of six months to a year or your efforts may be declared a failure.

Within One Month Since you will be unable to fix each of the facility's deficiencies simultaneously, you will clearly have to establish priorities and short-range goals that support the overall improvement plan. Your first concern obviously must be to correct the serious safety hazards in a very short period of time. You cannot allow additional injuries or deaths due to slipshod operating or maintenance practices. You must reinstitute safe material handling practices within this facility within a month or less.

The second objective for this first month should be to re-establish in the minds of each leader and team member the mission of this facility and how each unit and each employee supports that mission. By the end of the first month, you should have clearly constructed in the minds of the team members a vision of this facility's future and the major milestones necessary to achieve it.

The next major goal should be to stop errors that are occurring as a result of poor shift turnovers. Inadequate turnovers account for an inordinate percentage of crew errors in industrial operations. A forceful effort to establish clear shift turnover standards, checklists, and guidelines will pay immediate dividends for relatively small cost. Since the training process appears to be inferior, you will probably need to take this opportunity to develop and teach this module yourself. If you do it well, not only will you be making headway in the area of shift turnover, you will be setting the standards for your training department by showing them what you expect. Shift turnovers should be placed on track within the first month.

Above all, within the first month, you must begin to change the leadership environment within this plant. You want leaders who are knowledgeable coaches—coaches who respect their subordinates, demand high standards, and teach their team members how to achieve those standards while performing as a team. The process of leadership improvement must never end.

Within Two Months You have already begun the process of self-assessment, but you want it to become a matter of daily business. Upon completion of the baseline self-assessment, you may find it prudent to establish an *ad hoc* assessment cadre that will keep up the self-assessment momentum for the first six months or so. Over time, your goal will be to constrict the size of the self-assessment cadre while expanding the process through training to include many more key facility team members. You will always need a cadre to keep the process vital, but your desire is to get literally everyone involved in looking at themselves and one another with an eye toward improvement.

In addition to developing a vital self-assessment process, you must revitalize the training program. Without a training process that meets the needs of your facility, you cannot improve. The example that you set in developing the shift turnover module will go far toward setting a training standard, but if your training department has been filled with castoffs, you may have to develop a training department from scratch. Seek out the best operators and technicians that have a desire and the raw skills to teach others. Then get them some help from outside the plant, perhaps from your New Jersey colleagues, to help them bootstrap the department. If you can develop a strong training department that has an intense desire to support the needs of operations and maintenance, you can solve most of the remainder of your problems. Never forget, however, that *you* are the chief trainer. Your training manager is your primary staff specialist for training, but you will always be ultimately responsible.

Within Six Months With major safety hazards under control, shift turnovers in place, and a self-assessment cadre established, you can now continue to add new operating skill disciplines such as formal communications, operator tours, investigation of abnormalities, equipment status control, and training new team members to the capabilities of your facility employees. Remember to focus on one or two things at a time. This is a long-term project. Following the accident at Three-Mile Island, it took the commercial nuclear industry the better part of ten years to raise operating and maintenance standards to high performance levels.

As each new team skill is trained and evaluated, you must advertise your successes to your corporate headquarters and invite their participation in the assessment process. With each acquired skill, your team members will gain new confidence and pride in their capability as an operating team.

By the end of the six month time-frame, you should have begun in earnest the process of training other key facility personnel in self-assessment skills. You should be able to reduce the size of the original self-assessment cadre and begin utilizing other interdisciplinary team members. Keep the self-assessment standards high and never let the process subside. *You* must be a frequent participant on the self-assessment teams if you want this skill to continue and grow.

Within a Year

By the end of the first year, a rudimentary, yet solid training program should be in place. The training department should not only be capable of providing good, fundamental skills training with the assistance of subject matter experts throughout the facility, they should also have embarked upon the process of developing written qualification standards for plant operators, technicians, and engineers. They should have mapped out a master training plan for the next year (or preferably two) designed to advance the knowledge and skills of plant personnel on a monthly basis. Inherent in the plan should be the elements necessary to develop, practice, and test the facility's emergency response capability for plausible accidents.

Upon conclusion of the first year, you should have in place a reliable, repeatable method of investigating abnormal events and extracting lessons from your facility's mistakes and the mistakes of other industries. Such a process will necessarily require special training for some event investigators in root cause analysis.

Further Long-Range Goals

Beyond the one year mark, you should have a clear idea of the level of performance and integration that you desire of your team. The level that you can achieve within the first year is very rudimentary. Even if you experience no turnover of personnel, a year of success will merely get you out of the danger zone. Significant advancement of skills is still necessary to reach excellent performance. Nevertheless, the achievements of this year of success will probably appear nearly miraculous in the eyes of your corporate headquarters. But don't rest on your laurels. Continue your path of improvement.

Measuring Improvement

Measuring your improvement will be important not only to you as the leader of this facility, but also to your corporate headquarters. They will need proof that your efforts are effective and that this program is worthy of consideration for each of the other ten plants.

The first and most important measurement of your success will be how many additional accidents and injuries occur within your plant. Any further accidents resembling the one that led to your appointment as plant manager will seriously impair your capability to lead this team. Also, the confidence and support that you currently enjoy from your corporate headquarters will be damaged. As previously stated, it is of utmost importance that safety in this facility be rapidly brought under control.

A second important measure of your success will be your ability to bring this facility back to normal production levels while conducting business safely. If production has been stopped or curtailed, there will be tremendous pressure to restore operations within a few weeks or months. Though commercial contract obligations may be initially met by your other sister plants, your plant probably won't be able to stay off line for a long period of time. Therefore, you are faced with the challenge of upgrading knowledge, skills, and teamwork while, at the same time, improving production. In the long run, your actions to develop a systematic approach to operations will improve teamwork, eliminate many serious mistakes, and enhance production output. But, in the short run, production will suffer as you attempt to remold this team. Again, focus on one or two major improvement topics at a time. Don't overwhelm your team members with new ideas. As you demonstrate the viability of one or two new ideas such as improved shift turnover, team members will gain confidence in your leadership and teamwork will improve.

Another important means of measuring success will come from the process that you established within the first year to investigate abnormal events and disseminate the lessons of those events. Prior to your assignment as plant manager, you learned of a series of dangerous events that predated the fatal accident. Eight injuries occurred as a result of those plant incidents. As you reviewed them, you found within them the warning signals of a more serious upcoming event. If plant leaders had only studied those events and taken their lessons to heart, the fatalities might have been averted.

As you read the text of those other events, you wondered how many other incidents *almost* occurred; but you could find no record of *near misses*. You determined to improve the plant's ability to report and investigate not only actual events that result in injury or equipment damage, but also the events that almost occur and are narrowly avoided. Through improved sensitivity to the development of dangerous conditions and through the training of your team members in the lessons of actual or near events, you will be able to stop the big problems more often when they are small.

Finally, you have already put in place the best of all tools to measure, evaluate, and coach your team's performance—self-assessment. By conducting frequent self-assessment and evaluating the team against criteria that you establish for excellent operational performance, you will be able to demonstrate the team's improvement from year to year. Coupled with outside evaluation by corporate headquarters, OSHA, and EPA, a record of your team's operating performance will be established.

Topic Summary and Questions to Consider

The complex problem with which you have been faced has no single, simple solution. It can be resolved only through the development of disciplined habits of operation, maintenance, training, and technical and administrative support.

Ask Yourself Does this case study feel just a little "too close to home"? Do you see similar signs and activities in your own operation?

If so, it's probably time for some interdictory action. Start now. Don't wait. A dead employee is a devastating experience.

Chapter 24

Your Challenge

As stated at the outset, this book is dedicated to the study of improved industrial performance through a team approach to operation—an approach dependent upon alert, well-trained operators controlling equipment and processes within specified operating bands in accordance with approved procedures.

Yet, as we learned in Chapter 3, no team is successful for long without superb leaders—leaders who understand both their technology and their people. Accordingly, we wish to close this book with some final thoughts, ideas, and admonitions for those who are conscientiously engaged in improving their teams.

It Won't Happen By Itself

Systematic industrial operations isn't going to happen in your organization by itself. Team strategy and team*work* don't develop without a visionary leader who understands the goal, shows the way, and steers the team. That's *your* job.

If the text that you just read made sense, you need to evaluate your team's strengths and weaknesses in comparison to the standards presented. Where you find deficiencies, talk about them with the team and, together, develop a plan for improvement. Remember, it won't work unless it's *their* plan too.

It Won't Happen Unless You Show Them

People need pictures—mental models—of how things look and work in order to reproduce them. Don't expect your team members to read about shift turnover and then flawlessly implement it. They need to see it performed correctly, talk about it, try it, change it, and try it again until they understand it. If they don't understand it, they won't sustain it.

Accordingly, *you* must understand it. If you don't know what you're talking about, your team members will recognize your weakness in an instant. Your credibility on this issue will be lost, and no one will offer the effort necessary to make the change.

It Won't Happen Overnight

Implementing the systematic approach is not an overnight task. (Coach Dale didn't correct his team's habits or instill his philosophy overnight!) Rather, it is a process of improvement that takes months and years, just like building a top-flight sports team. In fact, it's a process that should never end.

That means a two-day class isn't going to fix this problem. You need both long-range and short-range training objectives, a good idea of how to meet them, and a plan that will still allow you to accomplish your daily work load.

Plan it out. Focus on fixing one thing at a time. Get shift change right and then work on logs. Then go to formal communications. Pretty soon, your team will begin to see how they all fit together. You'll reach a point where your team takes pride in showing others how a good industrial team performs.

It Won't Happen Without Persistence

As you begin each modification or change in operating habits, remember that old ways die slowly (and not at all if what you're advocating doesn't make sense). You must anticipate set-backs. Team members are going to gravitate back to old ruts.

Therefore, when new operating habits start to falter or succumb to short cuts (and they will), shore them up. Re-establish them as habits. Don't let them die. Correct the team, not once or twice, but as many times as it takes to reassert the habits. Then, keep the habits sharp through self-assessment.

Coach Joe Paterno writes in his autobiography:

> *Sloppiness is a disease. Nobody ever built a great organization just worrying about the big things. It's the little things that give you the edge. If the equipment man in the locker room doesn't check his equipment properly, the player senses it and the sloppiness gets into his bloodstream and the disease spreads. And the most difficult thing is to find people on every level who understand that and who are completely committed to detail and to standards of excellence.*

How would Coach Paterno rate *your* team?

More to the point, how do *you* rate your team?

It Won't Happen Unless You Get Started!

Finally, none of this is going to happen unless you get started. Do something *today* to improve your team's operating habits. Don't allow yourself to become overwhelmed with the enormity of the task. Think about it, talk about it, and work on it. Every day. One step at a time. That's how good teams get to be great teams.

It *can't* happen by itself; it *won't* happen if you wait for someone else to do it; it will only happen if *you* get it started. Remember, good habits are contagious just like the bad. Lead by example!

You have the tools. You have the skills. Are you up to the challenge?

Part V

Process Safety Supplement

○ **A Study in Process Failure**
○ **Process Safety Management Overview**
○ **Dissecting the Law**

This book has offered a philosophy and guidance for developing a systematic operating strategy in more or less general terms in order to demonstrate the applicability of the underlying principles to—and facilitate their implementation by—*any* industrial operation.

But, what about applying these principles to *specific* industries—especially those which come under additional, more strict, government regulation and control?

Responding to that question, we developed a special edition of the text that includes this "Process Safety Supplement" to address the particular needs and concerns of the petroleum and chemical industries. Specifically, this supplement correlates the principles of the systematic approach with the requirements of OSHA Regulation 1910.119, "Process Safety Management of Highly Hazardous Chemicals."

This supplement is organized into three chapters. Chapter 25 describes a disastrous failure in a chemical process complex, identifies the causes in context with the nineteen common factors (from Chapter2), and missing elements of the operating strategy (from Part II).

Chapter 26 defines process safety management for the petroleum and chemical industries, correlates that definition with our "systematic approach", and introduces OSHA's guidelines and recommendations for compliance.

And finally, Chapter 27 presents a point-by-point discussion of the OSHA Standard, correlating it with the principles and practices of the systematic approach presented in the main text.

Again, as you read and study this text and supplement, the basic question you should be asking yourself is: "How do the elements, principles, and conventions in this book apply to my specific operation?" Hopefully, you will find that by understanding *and implementing* the systematic approach you will be providing a solid foundation of operational safety upon which to build. A foundation that will take you—beyond mere compliance with minimum regulatory requirements—to process and product improvement, to the kind of operational excellence that is the ultimate goal of the systematic approach.

Chapter 25

A Study in Process Failure

Chemical processes have been around in one form or another for several centuries. With the processes came inherent hazards, some of which led to injury, death, and environmental damage; but, in most cases, the consequences of processing were accepted as a cost of doing business.

As the scale of chemical processing increased, however, so did the potential for widespread damage. In 1974, a gaseous release from a process plant at Flixborough, England, exploded, killing 28 people. Because of an improperly devised equipment modification, flammable cyclohexane escaped to atmosphere and ignited. The process plant was destroyed.

Though the toll of the Flixborough accident paled in comparison to dozens of other more costly tragedies, it became a milestone on the road to a new awareness of chemical process hazards, primarily as a result of instant, visual communication. And with added awareness has come reduced tolerance. Though some injury, death, and environmental damage are still accepted as a real (albeit highly undesirable) part of the cost of doing business, acceptable risk within industrial operations has greatly diminished.

With the advent of nuclear power (as well as rapid sophistication in most other modern processes),

Figure 25-1: Aftermath of the Phillips 66 Explosion and Fire. [From OSHA Report]

A Study in Process Failure 275

the need for refined hazard assessment, risk assessment, and risk management became apparent. The need was punctuated in March of 1979, when the accident at Three-Mile Island catalyzed concern within the nuclear community for improved systematic safety analysis. Though this wasn't the U.S.'s first major reactor accident (a prompt criticality at the Army's Stationary Low Power Reactor [SL-1] killed three at the National Reactor Testing Station in Idaho in January, 1961), it *was* the first large scale event in a U.S. commercial nuclear power station. Unit 2's misfortune shined a bright light on the interrelation of humans, machines, and the environment. In particular, the accident called attention to the human factor and the need for rigorous training in situations where risk was high.

Though Flixborough (and a chemical release of dioxin in Seveso, Italy, in 1976) had raised concerns over process safety in the petroleum and chemical process industries, many of the lessons of Three-Mile Island remained within the nuclear community. Then, in late 1984, two events struck a nerve in all three industries. In November, a massive LPG explosion outside of Mexico City killed 542 people, injuring thousands more. Just a few weeks later, the Mexico City tragedy was overshadowed by the disastrous release of methyl isocyanate at the Bhopal pesticide production facility in India. Clearly, rapid improvement was needed for all operations having the potential for widespread damage.

Corporate consortiums responded to the call. It was apparent that, without tough self-regulation, government *over*regulation would prevail. In the United States, organizations such as the Institute of Nuclear Power Operations (INPO) and the Center for Chemical Process Safety (CCPS) were born; organizations which, since their inception, have become tremendously instrumental in improving the operating performance of their constituent members.

Yet, the lessons of systematic operations sometimes spread slowly, for 1989 was a banner year, not only for maritime disaster, but also for chemical process failure.

Phillips 66 Chemical Complex Explosion and Fire, October 1989

Early in the afternoon of October 23, 1989, a disastrous explosion and fire destroyed the Phillips 66 Houston Chemical Complex, a polyethylene production plant, on the Houston Ship Channel near Pasadena, TX. The accident, described by the Department of Labor as "one of the worst industrial workplace accidents in the United States in the past 20 years", resulted in the deaths of 23 employees, 130 injuries, and $750 million property damage. The accident was caused by an improperly performed maintenance action, resulting in a rapid, uncontrolled release of over 85,000 pounds of highly flammable hydrocarbon gases. The gaseous mixture exploded upon contacting an undetermined ignition source.

Historical and Geographical Context

Prior to October 23, the Phillips 66 Houston Chemical Complex was a vital link in the national production chain for polyethylene, manufacturing nearly 1.5 billion pounds yearly (15% to 20% of the nation's yearly output). But, it was also just one of a number of large chemical process operations located along the Houston Ship Channel, north of Pasadena, TX.

Houston lies at the center of one of the largest oil production and refining regions in the entire world. In fact, the area between Houston, TX, and Lake Charles, LA, is home to hundreds of petroleum and chemical process operations. One can little travel the corridor along U.S. Interstate 10 between those two cities without observing the facilities and the machinery of chemical process operations. A world seaport, Houston is also a major shipping terminus and embarkation point, served by Galveston Bay and the Houston Ship Channel, a canal opened in 1914 for shipping operations, developed by dredging out Buffalo Bayou (leading into the city) and San Jacinto River, wending its way from the Northeast through what is now Lake Houston.

Galveston Bay and the Ship Channel are no strangers to chemical fires and explosions. With such a concentration of process operations, this region has seen its share of industrial accidents. One of the most famous occurred on April 16, 1947, when a cargo of ammonium nitrate fertilizer caught fire in the hold of the transport *Grandcamp*, a converted Liberty Ship, docked at the waterfront near the Monsanto chemical plant in Texas City, TX. Improper control of the fire resulted in a massive explosion of the 2,300 ton cargo, destroying the Monsanto plant and setting the city ablaze. *Highflyer*, another cargo ship berthed near *Grandcamp* (also loaded with sulfur and ammonium nitrate), caught fire and detonated the following day, April 17. The worst harbor explosion in American history, the cost of the Texas City disaster was great. Five hundred fifty-two people were known to have perished,

3,000 were injured, and 200 were never found. Every window in the city was broken (as well as half of the windows in Galveston, ten miles away). The explosion was heard at a distance of 160 miles.

Plant and Process Description

Within the context of this region's rich industrial history, the Phillips 66 Complex was not unusual, nor was it a newcomer. Engaged in the production of high density polyethylene (used primarily for manufacturing plastic containers), it had a 33 year operating record, having been in production since 1956.

On a daily basis, the facility employed 905 regular company workers and another 600 contract employees (engaged, for the most part, in plant maintenance and construction activities). Plant workers labored in two active polyethylene plants (IV and V) to react a mixture of isobutane, ethylene, hexene, and hydrogen in process equipment to form polyethylene, the end product. A report to the President published in April, 1990, by the Department of Labor described the process operation in succinct terms:

High-density polyethylene is manufactured in Plants IV and V of the Phillips Complex [see Figure 25-2] from ethylene gas dissolved in isobutane, which is reacted in long pipes under elevated pressure and temperature. Various chemicals are added to the process to modify the polyethylene to meet the desired product characteristics. This combination of process gases at elevated pressure and temperature is extremely flammable. The dissolved ethylene reacts with itself to form polyethylene particles that gradually come to rest in settling legs [see Figure 25-3], where they are eventually removed through valves at the bottom. At the top of each of these legs, there is a single ball valve (DEMCO® brand) where the legs join with the reactor pipes. The DEMCO® valve [see Figure 25-4] is kept open during production so that the polyethylene particles can settle into the leg.

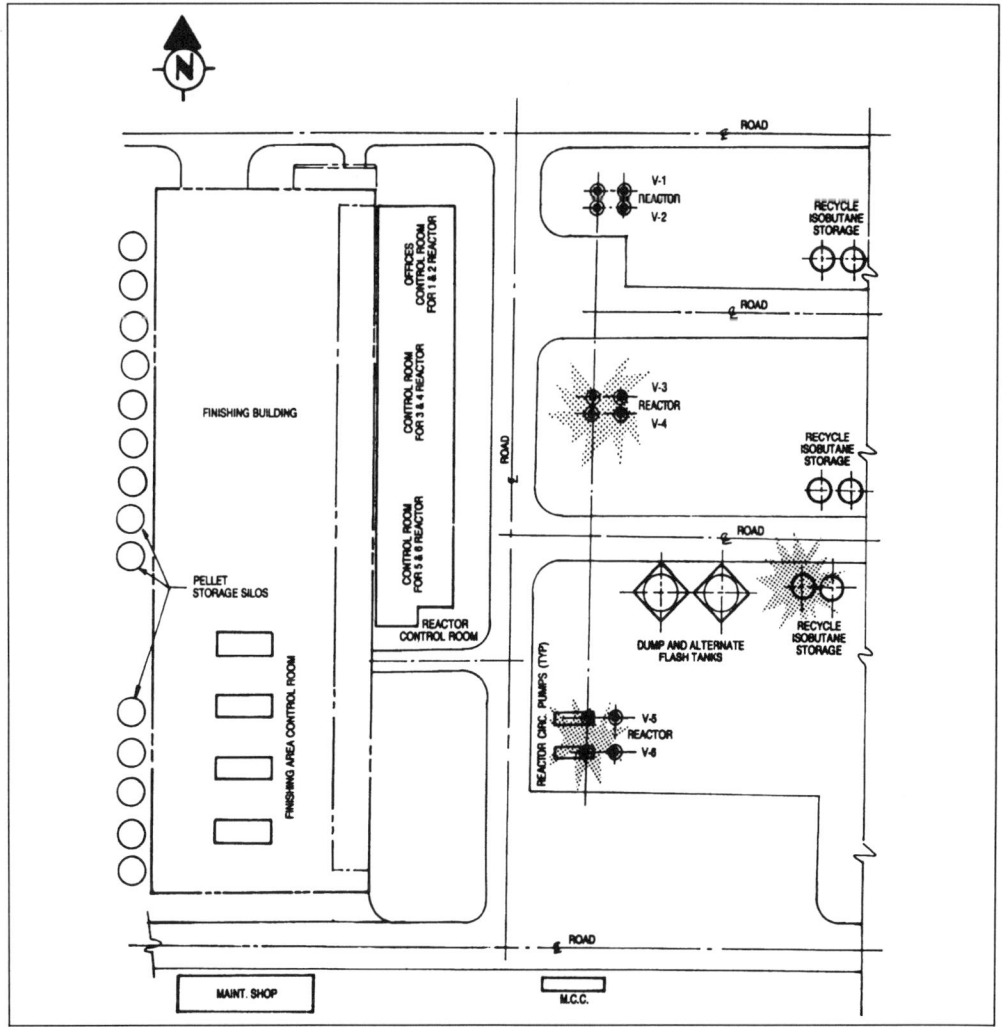

Figure 25-2: Partial Equipment Location Plan for Plant V Reactor Area. [From OSHA Report]

The reactor settling legs serve as conduits for removing plastic polyethylene particles. If a settling leg becomes clogged with particles (a common event), the leg must be isolated, the settling leg connection broken, and the clog cleared. To isolate the leg, an air-actuated DEMCO® globe valve must first be closed and secured in the closed position. If this maintenance task is improperly performed and reactor containment is breached, a dangerous release of flammable gases is possible.

Certainly, anyone familiar with chemistry will recognize the volatility, flammability, and (under the right conditions) the explosiveness of the gases used in this process. Plant design, construction, operation, and maintenance all have to be intelligently controlled to avoid the inherent hazards. In particular, cleaning out a clogged settling leg attached to an operating chemical reactor requires strict controls to maintain the leg isolated from the reaction.

Accident Sequence As we learned in Chapter 17 (Isolating Energy Hazards), piping isolation includes aligning valves and installing blank flanges to prevent fluid flow, locking and tagging at isolation points, disconnecting motive sources, and independently verifying that isolation is properly accomplished.

Yet, at approximately 1:00 P.M. on October 23, the protective barriers in Plant V failed. During settling leg maintenance, the containment for chemical reactor V-6 was breached, releasing 85,000 pounds of flammable gases, culminating in a fuel-air mixture explosion less than two minutes later. The force of the explosion approached 2.4 tons of TNT equivalent.

Here is the text of the Labor Department's report to the President describing the accident sequence:

At the time of the event, a settling leg was undergoing a regular maintenance procedure: the removal of a solidified polyethylene blockage. Under Phillips' written procedures for this maintenance function, which was usually performed by a contractor, Phillips' operations personnel were required to prepare

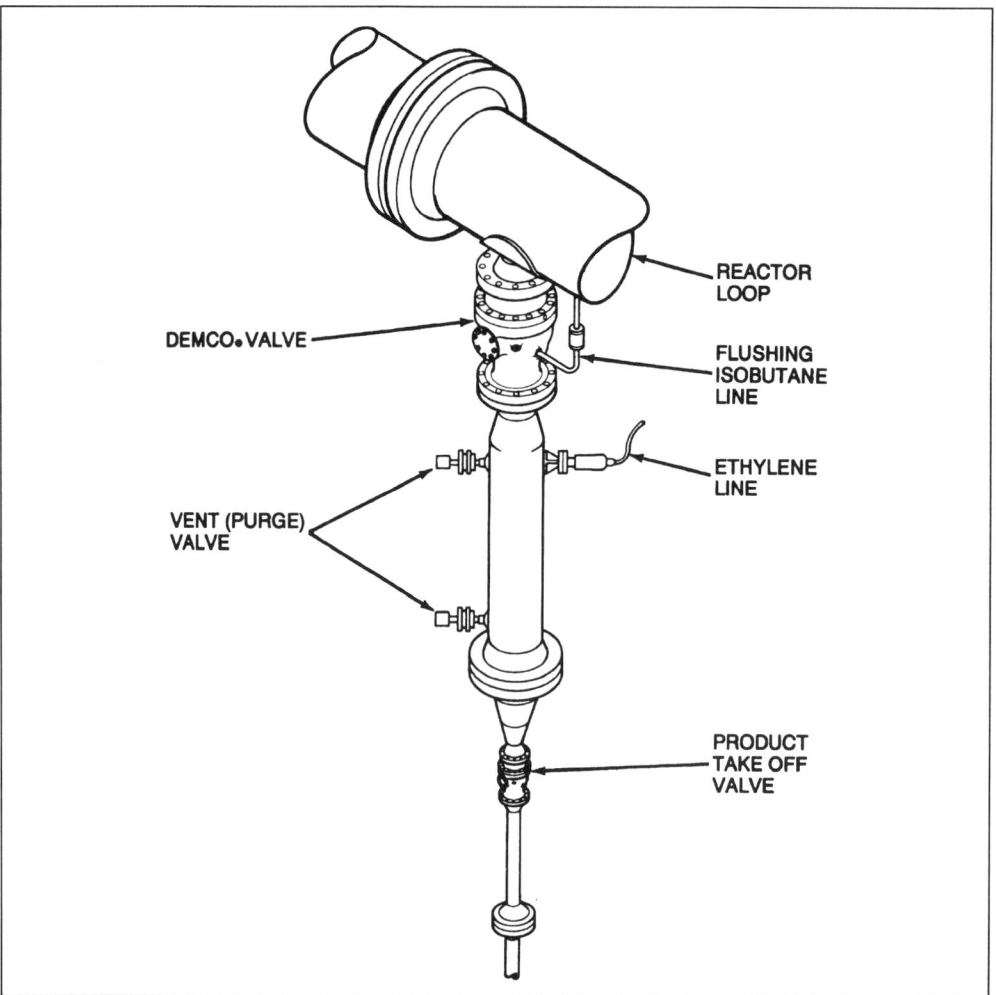

Figure 25-3: Typical Settling Leg Arrangement. [From OSHA Report]

the product-settling leg for the maintenance procedure by isolating it from the main reactor loop before turning it over to the maintenance contractor to clear the blockage.

On Sunday, October 22, a Fish Engineering crew began work to unplug three of the six settling legs on Reactor 6. According to witnesses, all three legs were prepared by a Phillips operator and were ready for maintenance, with the DEMCO® valve in the closed position and the air hoses, which are used to rotate the valve, disconnected. Number 1 leg was disassembled and unplugged without incident. At approximately 8:00 on Monday morning, work began on Number 4 leg, the second of the three plugged legs.

The Fish Engineering (contractor) crew partially disassembled the leg and managed to extract a polyethylene "log" from one section of the leg. Part of the plug, however, remained lodged in the pipe 12 to 18 inches below the DEMCO® valve. At noon, the Fish employees went to lunch. Upon their return, they resumed work on Number 4 leg. Witnesses then report that a Fish employee was sent to the reactor control room to ask a Phillips operator for assistance. A short time later, the initial release occurred. Five individuals reported actually observing the vapor release from the disassembled settling leg.

Because of the high operating pressure, the reactor dumped approximately 99 percent of its contents (85,200 pounds of flammable gases) in a matter of seconds. A huge unconfined vapor cloud formed almost instantly and moved rapidly downwind through the plant.

There were several potential ignition sources: a small diesel crane used by a Fish maintenance crew, but not in operation at the time of the blast; an operating forklift; a gas-fired catalyst activator with an open flame; welding and cutting-torch operations; 11 vehicles parked near the polyethylene plant office building; and ordinary electrical gear in the control building and the finishing building. The actual ignition source has not been identified.

Within 2 minutes, and possibly as soon as 90 seconds, the vapor cloud came into contact with an ignition source and was ignited. Two other major explosions occurred subsequently, one about 10 to 15 minutes after the initial explosion when two 20,000-gallon isobutane storage tanks exploded, and another when another polyethylene plant reactor catastrophically failed about 25 to 45 minutes into the event.

Figure 25-4: DEMCO® Valve. [From OSHA Report]

Fuel-air mixture explosions frequently generate widespread devastation. The Phillips 66 accident was no exception. Production plants IV and V, occupying an area of approximately 16 acres of the Phillips 66 complex, were both destroyed. The heat and concussive energy contorted steel support beams into eerily twisted shapes, and the explosion cast debris as far away as six miles.

Impaired Emergency Response

Following the 1:00 P.M. explosion, an immense fire engulfed the complex, resulting in a series of as many as nine other explosions. Within 15 minutes of the initial blast, two Plant V isobutane storage tanks exploded as heat from the fire intensified. Later, the Reactor V-3 and V-4 processes would also detonate, spreading the fire throughout the complex.

Fire suppression was an immediate priority. Firefighting response, however, was impaired by explosion damage. Investigators concluded that:

> The Phillips Complex did not have a dedicated water system for fighting fires. Water for that purpose came from the same water system that was used for the chemical process. Consequently, when the process water system was extensively compromised by the explosion, the plant's water supply for fighting fires was also disrupted. Fire hydrants were sheared off in the blast, and because of ruptures in the system, water pressure was inadequate for firefighting needs.

Fire hoses had to be laid to remote sources including a cooling tower basin, a settling pond, a water treatment plant, and a water main from a nearby plant. Electrical power, of course, was disrupted, necessitating reliance upon emergency diesel pumps for hose pressure. Unfortunately, of three diesel fire pumps, one was out of service at the time, another ran out of fuel prematurely (because it hadn't been completely fueled), and the third failed during firefighting operations—a serious deficiency in the complex's emergency preparedness. Nevertheless, with the help of the surrounding community, the fire was brought under control within 10 hours. Because of the intense residual heat and structural damage, search and rescue efforts were delayed until the following morning.

Improper Energy Isolation

As Phillips personnel and the community began the horrific task of identifying victims, others turned to the responsibility of investigation. The most pressing concern for investigators was to determine how and why the gaseous release which led to the explosion had occurred. Were there sufficient protective barriers to prevent this accident? If so, how had they been breached?

Since the maintenance task in progress involved clearing the Number 4 reactor settling leg, the investigation focused on how the leg was isolated prior to disconnection. Suspecting an open DEMCO® valve, the valve was recovered and examined. Inspection revealed that it had, indeed, been open at the time of release.

The DEMCO® valves employed in the settling legs are air-actuated globe valves (Figure 25-4) with control and indicating devices located in the reactor control room. Operation of a valve's remote positioning switch electrically realigns compressed air line valves which are hooked to the DEMCO® valve's air-actuation mechanism. Compressed air subsequently causes the valve either to open or close. Prior to maintenance, proper isolation would normally require electrical disablement of the remote actuating switch and physical disconnection of the actuator air lines. If possible, the DEMCO® itself should also be locked in the closed position. Further, for high pressure systems involving great risk in the event of a breached containment, double valve isolation (or the installation of blank flanges) is desirable.

In this instance, the isolation barriers were either not in place or were negligently breached. It is unclear what action was accomplished by the control room operator who was summoned by the Fish Engineering contractor. But the investigative report states that:

> According to witnesses, all three legs were prepared by a Phillips operator and were ready for maintenance, with the DEMCO® valve in the closed position and the air hoses…disconnected [as of October 22].

If the witness accounts are correct, the Number 4 settling leg's isolation valve air supply was, at some point, reconnected but reversed. Applying closing air opened the valve. High pressure, flammable gases rushed out.

The Labor Department's investigative report cited four specific deficiencies in the isolation process that contributed substantially to the accident:

(1) the DEMCO® valve actuator mechanism did not have its "lockout" device in place,

(2) the hoses that supplied air to the valve actuator mechanism could be connected at any time even though Phillips' operating procedure stipulated that the hoses should never be connected during maintenance,

(3) the air hose connectors for the "open" and "close" sides of the valve were identical, thus allowing the hoses to be cross-connected and permitting the valve to be opened when the operator might have intended to close it, and

(4) the air supply valves for the actuator mechanism air hoses were in the open position so that air would flow and cause the actuator to rotate the DEMCO® valve when the hoses were connected.

The accident investigation showed that the valve was, in fact, capable of being physically locked in the closed position, an action that would by itself probably have prevented the accident.

Procedural Violations

The determination that Phillips' corporate operating policies did not allow connection of the actuating mechanism air hoses during maintenance is particularly disturbing. As we learned in Chapter 2 (Common Components of Accidents), failure to comply with established procedures is a frequent accident contributor. In almost every event that we studied, failure to follow procedures played a major role.

Unfortunately, in the Phillips 66 event, it seems that a local maintenance procedure which diverged significantly from corporate instructions had been substituted. Investigators reported that:

Established Phillips corporate safety procedures and standard industry practice require backup protection in the form of a double valve or blind flange insert whenever a process or chemical line in hydrocarbon service is opened. Phillips, however, at the local plant level, had implemented a special procedure for this maintenance operation which did not incorporate the required backup. Consequently, none was used on October 23.

A part of the basis for OSHA's citation of Phillips (ultimately settled for $4 million) states that:

Phillips' existing safe operating procedures for opening lines in hydrocarbon service, which could have prevented the flammable gas release, were not required for maintenance of the polyethylene plant settling legs. The alternate procedure devised for opening settling legs was inadequate; there was no provision for redundancy on DEMCO® valves, no adequate lockout/tagout procedure, and improper design of the valve actuator mechanism and its air hose connections.

Whatever the cause, procedural compliance or the procedures themselves were inadequate; moreover, those charged with implementing the procedures ignored standard energy isolation practices and the barriers failed.

Absence of Safe Work Permits

When conducting work on hazardous systems, a standard industrial practice is to employ safe work permits. For example, an electrical safety checklist—a special kind of safe work permit—is commonly employed before working on or around electrical systems. The checklist, when executed by knowledgeable team members, assists them in consciously considering hazards and protections prior to work performance. Also, since it is a permit system, a cognizant team leader must review the checklist and authorize the proposed work before the activity can commence. In essence, the permit is a safety addendum to the work control document.

Hot work permits apply the same concept to activities which could lead to fire or explosion. Welding, cutting, and grinding jobs are the most common *hot work* operations; but, any activity that raises the specter of fire or explosion is also a candidate for hot work analysis. The permit, of course, provides an opportunity for the work team and its leaders to step back, look at what they are about to do, and consider the potential hazards. The critical question raised by the permit is simply this: "Is my team reasonably protected from the potential hazards of this activity?" In this manner, safe work permits raise an effective barrier in the construction of the safety envelope for maintenance activities.

Sadly, the permit system at the Phillips complex was deficient, probably contributing to the accident. OSHA cited both Phillips and Fish Engineering for neglect:

An effective safety permit system was not enforced with respect to Phillips or contractor employees to ensure that proper safety precautions were observed during maintenance operations, such as unblocking reactor settling legs.

Fish Engineering, in particular, was cited for "willful violations for failing to obtain the necessary vehicle and hot work permits when working in the polyethylene plant". Proposed penalties exceeded $700,000.

Inadequate Supervision of Maintenance

The actions which led to the release (and subsequent explosion) clearly call into question the supervision of both the operations and maintenance work groups. As we learned in Chapter 16 (Overseeing Maintenance, Modification, and Testing), maintenance activities rely extensively on a competent operating team. Neither maintenance nor operations can conduct their activities excluding the other. Both bear responsibility for the success of maintenance.

Nevertheless, we also learned that the operating staff must retain responsibility for plant activities, even when maintenance teams are at work. Obviously, in this event, the work control process failed. And, by implication, the operating staff failed. Remember, we're not suggesting that the maintenance team is free of responsibility. They failed as well. We're simply saying that ultimate responsibility must reside with someone—and that "someone" is the operating team.

Deficient Team Training

Within the Phillips complex, the maintenance function seems to have been accomplished primarily by contract workers. It is no surprise that businesses have begun to contract work to outside suppliers. Capital costs have long been overshadowed by the cost of employees as medical insurance, retirement benefits, and other associated costs of labor have risen. By out-sourcing, another employer must carry the overhead burden of employees.

But, as we learned in Chapter 3, teamwork is critical to industrial operations. Even if a team chooses to use personnel who are not regular team members, team leaders cannot neglect to ensure that *all* team members are trained and following team procedures. In military terms, when a specialty group is attached to a host unit (to tailor the unit for a particular task), the attached unit is still the immediate responsibility of the host unit commander. Yes, the attached team is still administratively bound to its parent. That's where they learned their trade and that's where they get their paycheck. But, team success is the responsibility of the host. Therefore, appropriate checks and supervisory leadership are required. A host employer that implores after an accident, "But they weren't my people!" is not absolved of responsibility.

In the Phillips event, teamwork was deficient. Therefore, we must at least consider whether failed training contributed to this operations/maintenance error. Considering the actions that led to the accident, it seems likely.

Unheeded Warnings

As we learned in Chapters 3 and 7, most accidents are preceded by warning signs—*accident precursors* as we called them. As unsafe conditions develop over time, people become comfortable with them, often ignoring them to their detriment. Safety conditions at the Phillips complex were suspect. Department of Labor's report states that:

> OSHA's investigation revealed that a number of company audits, which were done by Phillips' own safety personnel as well as by outside consultants, had identified unsafe conditions, but had been largely ignored.

If safety was viewed by plant personnel as an impediment to production, "cutting corners" may have become an accepted practice. As a minimum, the approach to energy isolation was cavalier.

Ineffective Emergency Signal

The Department of Labor report identifies another possible indication of a less-than-vigilant safety attitude at the Phillips complex—multiple fires and alarms. OSHA investigators learned that:

> In the months preceding the explosion, according to the sworn testimony of an employee, there had been several small fires, and the alarm had sounded as many as four or five times in one day. A siren was used to warn company and maintenance contract workers to vacate the plant. Some of the workers in the finishing building may not have heard the siren because of the ambient noise level inside the building. Consequently, those employees may not have been aware of the impending disaster.

Was the warning inaudible to some? Or, possibly, had team members become desensitized by many "false" alarms?

In the Phillips 66 event, one of the critical questions is "Who died and why?" Curiously, only two of the six maintenance technicians working at the settling leg job site—the source of the explosion—were fatally injured. Yet, twenty-one other workers in the vicinity lost their lives, apparently unaware of the impending hazard. (The production control rooms to the west and adjacent to the chemical reactors were particularly hard hit.)

In most explosions, loss of life is confined to a relatively small radius around the epicenter of the blast. Of this event's dead, all were within 250 feet of the seat of the initial explosion. Fifteen were within 150 feet. Running away from the leak was the safest and most expedient course.

Witnesses reported that workers who were near the gaseous release ran from the area immediately upon detecting the leak. They obviously understood the seriousness of the situation. Why hadn't more people escaped? Was there time?

Fuel-air mixture explosions are peculiar in that the explosion cannot occur until the flammable gases have had time to mix with oxygen and find an ignition source. For example, a volatile, flammable gas (such as hydrogen) explodes only when oxygen-balanced—and then only within a limited mixture range. In a normal atmosphere, hydrogen explodes (in contact with ignition source) within a range of 4% and 75% concentration. The more volatile the gas, the more quickly explosive conditions are reached. For heavier and less volatile gases (ether, for example), the longer the time to reach oxygen balance. (Of course factors such as temperature and pressure directly affect the mixing time.)

A boiling liquid evaporative vapor explosion (BLEVE) is a classic example of a fuel-air mixture explosion. Should a railroad tanker transporting liquid natural gas derail and rupture, fire is probable. As the fire engulfs the tank car, pressure is raised, leading to rapid and widespread dissemination of the gas through the breach. When properly oxygen-balanced, a devastating explosion is apt to occur. Yet, some time is usually available to clear the area.

Usually, then, the conditions necessary to support a fuel-air explosion take time to develop. At the Phillips 66 complex, the time for the mixture to balance and reach an ignition source was between 90 and 120 seconds—a substantial amount of time for warned personnel to exit the danger area if they are well-trained and well-drilled.

Considering the distribution of victims, it is probable that some didn't hear the warning. But it is also very common for people to ignore warning signals, especially if they are frequent and false. You will recall that, at Bhopal, the warning siren had more than one meaning and sounded frequently. Consequently, some citizens did not heed the siren as a danger signal. The same thing may have happened at the Phillips complex. If so, emergency unpreparedness played a devastating role in this accident.

Incomplete Safety Analysis

Perhaps the most important lesson to be drawn from the Phillips 66 accident involves thorough safety analysis—examining (and compensating for) the potential dangers of operation in advance (and throughout the life) of a facility. When safety studies are not diligently performed (and rigorously executed), the likelihood of industrial calamity is magnified. Successful operations—both for routine and emergency situations—demand that team members understand the potential dangers and the steps necessary to combat them.

In Chapter 3 (Systematic Industrial Operations), we cited well-defined boundaries as one of eight essential elements in the strategy for operating success, emphasizing the need to define a safety envelope beginning with equipment or process design. Then, in Chapter 4 (Boundaries of Safe Operation), we described how safety envelopes are developed, creating protective barriers constructed of physical, procedural, and administrative constraints.

We also learned that hazard analysis—the first phase of a serious safety study—is a process of determining the dangers of equipment operation and the consequences of failure. Our study demonstrated that, without thorough identification of hazards, effective protection can never be developed—nor can a responsive emergency plan.

At the Phillips complex, hazard analysis was incomplete. The investigative report states that:

> A process hazard analysis or other equivalent method had not been utilized in the Phillips polyethylene plants to identify the process hazards and the potential for malfunction or human error and to reduce or eliminate such hazards.

Granted, the plant was built long before the recent advances in safety study sophistication. But, we must remember that safety studies are not static. Frequent reevaluation must accompany changes in technology and improvement in hazards knowledge.

One glaring design deficiency that may have been identified by better safety analysis involved site layout. Accident investigators wrote:

> The site layout and the proximity of normally high occupancy structures, such as the control room and the finishing building, to large capacity reactors and hydrocarbon storage vessels also contributed to the severity of the event. The large number of fatally injured personnel was due in part to the inadequate separation between buildings in the complex. The distances between process equipment were in violation of accepted engineering practices and did not allow personnel to leave the polyethylene plants safely during the initial vapor release; nor was there sufficient separation between the reactors and the control room to carry out emergency shutdown procedures. The control room, in fact, was destroyed by the initial explosion.

Other deficiencies included "the lack of permanent combustible gas detection and alarm systems...to provide early warning of leaks or releases", "ignition sources...located in proximity to...large hydrocarbon inventories", "ventilation system intakes for buildings in close proximity to...hydrocarbon processes...not designed or configured to prevent the intake of gases in the event of a release", and, as we learned earlier, "the lack of a water system dedicated to firefighting".

Certainly, in old complexes with long operating histories, there is great pressure to avoid reevaluating and compensating for hazards that haven't caused problems in the past. And we are well aware that gross regulatory over-reaction sometimes result from high visibility accidents in our overly-litigious industrial environment. Nevertheless, the concept of reasonable risk minimization requires vigilance toward probable dangers and cost effective protection. If risk of disaster is high, it doesn't make good business sense (at least in the long term) to avoid protective action.

A Preventable Accident?

Was it preventable? There is probably no more important question that can be asked in the wake of any accident; for if the answer is "Yes", the next question is "How?"

One effective approach in determining preventability is to compare the conditions and actions that led to the accident with the nineteen common components studied in Chapter 2. Another is to determine which of the eight essential elements of the strategy for operating success (Chapter 3) were missing. We already know from our recently completed study that missing elements can be remedied and common components can be eliminated. (The twelve vital skills assist us immeasurably in dissolving accident conditions.) See, then, if you agree with the following analysis:

Contributing Common Components

Consider how many of the common components of accidents directly contributed to the Phillips complex explosion and fire. *Non-compliance with procedures* resulted in inadequate isolation of the Number 4 reactor settling leg (a *deficient equipment maintenance* practice and a severe *degradation of operating/maintenance limits*). Isolation of the settling leg was not adequately substantiated prior to commencement of maintenance (a neglectful *failure to independently verify critical tasks*). The maintenance team (which, by necessity, includes operations) suffered from *insufficient management oversight*. Had oversight been adequate, it is doubtful that the team would have been so careless in isolated the settling leg. Maintenance personnel, operators, technicians, and their leaders clearly failed to control this important task, a demonstration of *faulty teamwork*. Once the release had occurred, an ineffective emergency warning signal (*incomplete communication*) left many plant personnel without sufficient warning of the impending explosion (an indication of *inadequate emergency preparedness*). Finally, unsafe conditions (as evidenced by numerous and frequent small fires and alarm warnings and unheeded safety audit findings) exemplify routine *acceptance of abnormalities*. That's nine of nineteen—not a very good percentage, but not at all uncommon.

What about components that *probably* were involved? Were members of the maintenance and operating teams *unfit for duty*? (Remember, unfitness is not just sleep-related. It's also training-related.) How about *ignorance of equipment operating characteristics*? Did one or more of the team members not understand the air hose connections on the DEMCO® valve? And what about keeping track of process status? Did they *fail to monitor equipment status*? (We already know that atmospheric monitoring equipment wasn't installed.) What about complacency? Was there a *sense of invulnerability*? Had the team members made these same errors before and gotten by with them? Were they *ignorant of warning signs* of an impending problem? Should the frequent fires and alarms have raised their awareness?

Were they hurrying to get the job done—*capitulating to production pressure* and not balancing schedular needs with good critical thinking? Were the *organizational structure and staffing deficient*? (In other words, were maintenance and operations put together in a way that they communicated and coordinated well?) And how about team guidance? Was poor leadership a contributor?

These are all difficult questions to answer without more information. But, we know from our study of the systematic approach that a dozen or more accident contributors is not unusual.

Missing Strategic Elements

Also, consider the requisite eight elements in the strategy for operating success. Was this team's performance ruled by an *underlying philosophy* of operations and maintenance? (Would Coach Dale praise them for their knowledge of the "defense and fundamentals" of industrial operations?) Were they working with *reliable equipment and facilities*—good facility design, quality tools, and functional equipment that performed consistently in accordance with design specifications? (The investigative team certainly doesn't think so, at least where site layout and critical separation distances are concerned.) Were their *operating and maintenance boundaries* well-defined? (Apparently not, considering the *policy and procedural discrepancies* between local and corporate guidance for isolating a settling leg.) Were team members *alert, and well-trained*? (It hardly seems so in retrospect.) Did they have *superb leaders*, an *efficient operating structure*, and a *team approach*? (We've already concluded that this "team" didn't function successfully in the performance of a critical maintenance task.)

The Most Important Lesson

In retrospect, the Phillips accident looks very much other accidents that we have studied. Sloppy operating and maintenance habits were allowed to develop and persist. In the end, those deficiencies exacted a heavy toll.

Moreover, as the case studies of this text indicate, through lapses in judgment and concentration, the best team members (as well as the worst) are susceptible to making dangerous errors. If we take any general lesson from this tragedy, it must be this: **If risk-laden operations are not approached with** discipline and constant vigilance by skilled leaders and team members, errors that injure people **and damage equipment or property are easily committed!**

Chapter 26

Process Safety Management Overview

It seems clear that, at the Phillips 66 complex, many of the elements in the strategy for operating success were either missing or conspicuously deficient. Further, numerous accident components had taken root and grown. Yet, for all its deficiencies, the Phillips complex was little different from other industrial processes. The accidents at Flixborough, Seveso, Mexico City, and Bhopal all harbored similar flaws. Those teams had neglected to erect (or consistently implement) integrated schemes of protective barriers based upon systematic hazard analysis.

The Phillips event spurred OSHA to hasten work (already begun in early 1989) on a standard for systematic chemical process operations known as *process safety management*. Its purpose was to establish minimum operating, maintenance, staffing, and training standards for high hazard chemical process facilities.

Today, that standard is codified in Title 29 of the Code of Federal Regulations, Part 1910, Standard 119 under the title "Process Safety Management of Highly Hazardous Chemicals". The remainder of this supplement is devoted to a review and comparison of that standard with the systematic approach to operations that you have just studied.

Definition and Purpose

In the first four parts of this text, we examined in detail the elements and skills necessary for superior industrial operations. We termed our strategy a *systematic approach*. The task before us now is to evaluate the petroleum and chemical industry's methodology for establishing safe process operations (known as *process safety management*) and see if it stands scrutiny.

What Is It? Simply stated, **process safety management** is the management of chemical process safety. We certainly don't mean to insult your intelligence; but, for a change, a term has been chosen that (fortuitously) states its meaning well. OSHA Regulation 1910.119 Appendix C (Non-mandatory Compliance Guidelines and Recommendations for Process Safety Management) defines the phrase in this way:

> Process safety management is the proactive identification, evaluation and mitigation or prevention of chemical releases that could occur as a result of failures in process, procedures or equipment.

The American Institute of Chemical Engineers' **Guidelines for Hazard Evaluation Procedures** (Second Edition with Worked Examples, published by the Center for Chemical Process Safety, New York, NY, 1992) defines it a little more generally:

> A program or activity involving the application of management principles and analytical techniques to ensure the safety of process facilities.

Do these definitions sound familiar? They should. In Chapter 4, we went to great lengths to develop the purpose, the importance, and the pieces of the boundaries of safe operation. Remember the three major steps? Identify the hazards. Determine the risk. Establish the protection. We called it *safety analysis*—a critical part of the systematic approach.

But, we didn't stop there. Safety analysis isn't enough. As we learned in our recently completed study, comprehensive protection consists of excellent design, superior construction, and operation within design constraints (always supported by precise maintenance and modification). As the CCPS definition states, process safety management involves "the application of management principles" as well as "analytical techniques" to ensure safety.

We find, then, that there is little (if any) difference in the systematic approach to operations as described in this text and the spirit of process safety management portrayed in OSHA 1910.119. It doesn't really matter what you call it. It only matters that you implement it correctly.

What's It For? Once we agree on definition, we need to compare substance. The topic question becomes: "What is process safety management for?" In other words, what is the objective?

Appendix C of OSHA 1910.119 describes the objective of process safety management in this way:

> *The major objective of process safety management of highly hazardous chemicals is to prevent unwanted releases of hazardous chemicals especially into locations which could expose employees and others to serious hazards.*

Remember the goal of the safe operating envelope from Chapter 4? To protect the operators, the environment, and the general public from hazards characteristic of the technology. When equipment and process operations involve hazards (and we can think of few that don't), the second part of mission is always the same—to protect people, environs, and other equipment from unacceptable harm.

How Does It Work? It's pretty clear that the objectives of the systematic approach and that of process safety management are the same. Though one addresses specifically the control of hazardous chemical production processes, both are designed to systematically identify and control industrial hazards.

We must look a little deeper, however, to ensure that they are truly similar in substance. Appendix C summarizes the process safety management approach in straightforward terms:

> *An effective process safety management program requires a systematic approach to evaluating the whole process. Using this approach the process design, process technology, operational and maintenance activities and procedures, nonroutine activities and procedures, emergency preparedness plans and procedures, training programs, and other elements which impact the process are all considered in the evaluation.*

We rest our case. The same combination of design, construction, operation, maintenance—the same thorough training of team members for routine and non-routine activities, including emergencies—and the same development and use of procedures for process and equipment operations described in the systematic approach also underlie process safety management. In retrospect, how could it be otherwise?

The Law and Process Safety

There can be little doubt of the similarity between process safety management and the systematic approach. They have the same purpose and, for the most part, they are accomplished in the same way. Process safety management is the chemical process industry's way of discussing the systematic approach to operations.

But, process safety management and process safety management *law*, as we are about to see, are separated by some important differences. We need to understand them before we can proceed.

Minimum Standards Though OSHA Regulation 1910.119 is titled "process safety management", a truly systematic approach to hazardous chemical process control (or any other high risk industrial operation) is likely to exceed the legal requirements contained in regulation. As we would expect (and demand), the law simply prescribes *minimum* standards and controls which must be attained for each process covered under the regulation.

True process safety, however, is likely to go beyond the law in a way similar to the difference between law and ethics. Law sets maximum boundaries beyond which citizens, if they proceed and are caught, will be punished. Ethics are more restrictive, self-imposed boundaries that are, perhaps, best known and accepted as "good form".

Laws are devised primarily as a last line of defense against wrongdoing. They are valuable for dealing with the "miscreants" of society. Ultimately, however, they cannot substitute for good citizens doing the right thing. If a society must rest upon laws rather than good citizens, it will fall of its own weight.

An "industrial society" is little different. The limits of industrial "behavior" are established in law, but self-imposed boundaries of ethical behavior are more restrictive. Moreover, they are understood by good citizens to be necessary for long-term success.

Please don't misunderstand. We are of the opinion that there has been too much unintelligent law passed in the regulation of industry and business—law supported by "junk science" and not rooted in logical cost-benefit analysis. Nevertheless, there is a right way and a wrong way to approach industrial operations. We support the right way. We believe it is nothing less than the systematic approach described in the text.

Spirit of the Law

To one who understands the systematic approach to industrial operations, the elements of process safety management law are almost trivial. They are met naturally as a part of responsible design, construction, operation, and maintenance. In fact, Appendix C to OSHA 1910.119 helps explain the *spirit of the law* upon which the regulations themselves are based. As the appendix states in its introduction:

> *This appendix serves as a nonmandatory guideline to assist employers and employees in complying with the requirements of this section, as well as provides other helpful recommendations and information.*

Though not mandatory, the guidelines help us to fully envision how an excellent process operation should be designed and operated.

The difference between process safety *law* and process *safety* can also be quickly determined through a review of the Center for Chemical Process Safety's **Plant Guidelines for Technical Management of Chemical Process Safety** (Revised Edition). This excellent discourse shows in great detail what is required to achieve process safety, not just legal compliance.

Our immediate goal, however, is to review process safety management law, determining whether it supports the concepts which underlie the systematic approach to industrial operations that we just studied. Let's evaluate the fourteen points of protection prescribed under OSHA law for process safety management of highly hazardous chemicals, understanding as we should that these are the minimum elements of process safety required by law.

Chapter 27

Dissecting the Law

OSHA Regulation 1910.119 identifies fourteen points of protection—fourteen areas of law—that hard-won experience has proven essential to safe (and efficient) process safety management. They include:

- Employee participation
- Process safety information
- Process hazard analysis
- Operating procedures
- Training
- Contractors
- Pre-startup safety review
- Mechanical integrity
- Hot work permits
- Management of change
- Incident investigation
- Emergency planning and response
- Compliance audits
- Trade secrets

In the pages that follow, we will analyze each area of process safety management law, relating it to what we learned in the first four parts of this text, recommending actions for implementation, and citing other helpful resources. First, though, we need to determine to whom this law applies.

Application

The first paragraph of Standard 119 delineates which chemical process operations are covered under this regulation. If yours falls within these guidelines, it is called a *covered process* and these regulations apply to you.

1910.119(a)

(a) Application.
(1) This section applies to the following:
  (i) A process which involves a chemical at or above the specified threshold quantities listed in Appendix A [not reproduced in this text] to this section;
  (ii) A process which involves a flammable liquid or gas (as defined in 1910.1200(c) of this part) on site in one location, in a quantity of 10,000 pounds (4535.9 kg) or more except for:
    (A) Hydrocarbon fuels used solely for workplace consumption as a fuel (e.g., propane used for comfort heating, gasoline for vehicle refueling), if such fuels are not a part of a process containing another highly hazardous chemical covered by this standard;
    (B) Flammable liquids stored in atmospheric tanks or transferred which are kept below their normal boiling point without benefit of chilling or refrigeration.
(2) This section does not apply to:
  (i) Retail facilities;
  (ii) Oil or gas well drilling or servicing operations; or,
  (iii) Normally unoccupied remote facilities.

You can see that these regulations were designed specifically for chemical *processes*—operations which combine hazardous chemicals for production of other end products. They do not apply to hydrocarbon fuel storage depots (such as an airport fuel tank farm), retail outlets for chemicals which are in their end-product form, most oil and gas field operations, or unoccupied facilities which are not near population centers. Further, they do not apply if your process chemical inventories fall below the limits specified in Appendix A of the process safety management standard.

Employee Participation

Employee participation is the first of fourteen elements incorporated into law under OSHA Regulation 1910.119. At first glance, this may seem to be an odd, nondescript, and, perhaps, not easily achieved section of law. Yet, without employee participation in the analysis and communication of hazards (as well as the protection afforded against them), safe and efficient performance is probably not possible. Here's what the law says:

1910.119(c)

(c) **Employee participation.**

(1) Employers shall develop a written plan of action regarding the implementation of the employee participation required by this paragraph.

(2) Employers shall consult with employees and their representatives on the conduct and development of process hazards analyses and on the development of the other elements of process safety management in this standard.

(3) Employers shall provide to employees and their representatives access to process hazard analyses and to all other information required to be developed under this standard.

Is It Reasonable?

If you just finished reading Parts I through IV of the text, you probably recognize that, without employee involvement, your team can never achieve high levels of performance. In Chapter 2, we noted failed teamwork as a common component of accidents. In Chapter 3, we emphasized the importance of teamwork as one of eight essential elements for successful industrial operations. In Chapter 6, we examined the value of the alert, well-trained operator conducting operations within well-defined boundaries (Chapter 4) in accordance with clearly established operating principles and practices (Chapter 5).

Since rank and file employees are usually the team members who must actually operate equipment to produce a product or service, it doesn't make sense to neglect their involvement in hazard analysis, hazard communication, and hazard protection. Remember the Bhopal pesticide production facility (Chapter 7)? One of the chief complaints of crew members was inadequate training in the hazards of methyl isocyanate and the protective systems, processes, and practices. This deficiency extended even to many of the supervisors!

Clearly, if team members are well-trained and superbly led, they will have not only good theoretical knowledge of hazards and hazard protection, but also an invaluable practical knowledge of whether the established design, construction, operating, and maintenance constraints are adequate to protect against those hazards. That practical knowledge allows an operating team to constantly analyze for hazards and periodically update protective systems, processes, and practices.

The requirement for employee participation, then, is both reasonable and necessary. So, rather than being intimidated, let's examine the mandates of paragraph (c), determine how they can be met, and, more importantly, used to our advantage in building high quality industrial teams.

Involving Your Employees

Paragraph (c) invokes three requirements for employee participation. First, it requires that you write a policy stating how you involve your employees in conducting process hazard analysis (in particular) and process safety management (in general). The second requirement states that you must follow your policy, actually engaging your employees in process hazard analysis and the other thirteen elements of process safety management described in the standard. The third requirement states that you must make accessible to your employees your facility's process hazard analysis and the supporting information known as *process safety information.*

Let's address the last (and easiest) requirement first. Making process safety management information (including process hazard analysis) available to all team members is trivial. Once the process hazard analysis is done (and it can't be done very well without process safety information), you need only have the analysis and related information on file and accessible to your team members. (If you're reluctant to share *trade secret* information with your employees, take a look at paragraph (p)).

The first requirement of paragraph (c), writing a policy of how employee involvement is to be achieved, is a little harder; but, in reality, it amounts only to putting your ideas and decisions on paper once finalized. So let's get to the hard part—actually involving your employees.

Perhaps the first consideration is this: If you haven't been operating as a team before, this is going to be difficult. On the other hand, if you already have a team—leaders and team members working together to accomplish a mission—this may seem ridiculously simple. Here are some implementing suggestions for you to ponder—six things that you can do to expedite team member participation.

1. Use process safety management! You can't involve your employees in process safety management unless you use it. And, if your activity falls within the applicability guidelines of this regulation, you don't have a choice. So, whether you're working on your initial process hazard analysis, validating one already in place, or improving your entire process safety management program, *determine to involve your team members*!

2. Engage employees in process hazard analysis. When performing the process hazard analysis, include (as analysis team members) operators, maintenance technicians, engineers, and others who are knowledgeable and experienced in the process. Such *subject matter experts* provide indispensable practical viewpoints.

3. Incorporate process hazard analysis and process safety management into initial training. Process hazard analyses and process safety information that are "on file and accessible" don't do much to help your employees avoid a Bhopal. They have to be *trained* in the hazards, the risks, and the protection. Paragraph (g) of this standard states:

> *Each employee presently involved in operating a process…shall be trained in an overview of the process and in the operating procedures…. The training shall include emphasis on the specific safety and health hazards, emergency operations including shutdown, and safe work practices applicable to the employee's job tasks.*

How can it be otherwise? Without a thorough understanding of hazards and hazard protection, few operations can succeed for long.

Is your initial training program nailed down? Do you already provide thorough hazard and hazard protection training to your team? If you're training your team members well in these subjects, you should be taking credit for employee participation. Remember, process safety management relies ultimately (as do all industrial operations) on alert, well trained operators and their supporting teams.

4. Upgrade hazard and hazard protection knowledge through continuing training. We learned in Chapter 21 that, for an industrial team to be successful, initial training must be supplemented with continuing training. Continuing training, you will recall, aims at constantly improving the knowledge and skills of team members.

Certainly, one of the highest priority continuing training topics is recognition and response to hazards. Chapter 14 of the text was dedicated to the study of abnormality recognition. Chapter 15 focused on combatting emergencies and casualties. Both skills are closely related to hazard recognition and protection—and both are fundamental to process safety management. If not challenged through continuing training (including drills), the skills and knowledge of hazards and hazard protection soon degrade.

5. Investigate abnormal events. It is undesirable to have accidents; but, it is unacceptable not to learn from them. In Chapter 20 of the text, we studied the investigation of the Challenger accident in which a solid rocket motor joint failed to seal, leading to the explosion of the external tank and the loss of the shuttle. Through a thorough review of the event, equipment flaws and human errors—both of which are critical factors in process safety management—were detected, assessed, and corrected. All of NASA was involved—not just the leaders.

Investigation of abnormal events (and dissemination of lessons learned) are critical parts of raising hazard awareness. By evaluating hazards (including ones previously not considered) and improving knowledge of the protection employed against them, team members sharpen their skills. If employees are not involved in this process, how can it succeed?

This standard's non-mandatory guidance (Appendix C) reminds us that:

> Section 304 [of the Clean Air Act Amendments]…requires employers to train and educate their employees and to inform affected employees of the findings from incident investigations required by the process safety management program.

And, as we will see in paragraph (m), investigation of abnormal events—utilizing the skills and knowledge of all team members—plays a key role in process safety management.

6. Perform self-assessment. There are few better ways to involve your employees in process safety management than through self-assessment. In Chapter 22 of the text, we defined self-assessment as the process by which individuals and teams evaluate their own performances through formal and informal means (including the processes of testing, auditing, and coaching). We suggested that both formal and informal assessments should occur frequently and that they utilize team members from many disciplines.

Not surprisingly, paragraph (o) requires that compliance audits be performed periodically "using knowledgeable team members" to ensure that the requisites of the standard are being met. In fact, paragraph (o)(2) requires that "The compliance audit shall be conducted by at least one person knowledgeable in the process."

Why not involve *many* of your employees in compliance audits? There are few better ways to teach team members to be more critical of their *own* performances. Also, if you develop a self-assessment program along the lines of the one described in Chapter 22, not only will your employees be involved, they will be invaluable in improving your process safety management program.

Process Safety Information

The second element of process safety management addressed in OSHA Regulation 1910.119 is called *process safety information.*

The term **process safety information** (or *process knowledge and documentation* as the Center for Chemical Process Safety prefers) refers to the body of recorded design, construction, and operating information associated with hazardous chemical processing components and systems (as well as the chemicals themselves). For example, the design specifications for chemical reactors, pumps, piping, valves, and electrical systems of chemical processing equipment clearly constitute process safety information. So do the as-built drawings which provide an engineering record of process assembly. Similarly, material safety data describing the dangers of process chemicals are also considered under this regulation as process safety information.

Scrupulously recorded process safety information sets the stage for proper hazard analysis, risk analysis, and risk management. Without it, we can only guess at what potential problems our processes and chemicals present. Here's what the law says:

1910.119(d) (d) **Process safety information.** In accordance with the schedule set forth in paragraph (e) (1) of this section, the employer shall complete a compilation of written process safety information before conducting any process hazard analysis required by the standard. The compilation of written process safety information is to enable the employer and the employees involved in operating the process to identify and understand the hazards posed by those processes involving highly hazardous chemicals. This process safety information shall include information pertaining to the hazards of the highly hazardous chemicals used or produced by the process, information pertaining to the technology of the process, and information pertaining to the equipment in the process.
(1) Information pertaining to the hazards of the highly hazardous chemicals in the process. This information shall consist of at least the following:
  (i) Toxicity information;
  (ii) Permissible exposure limits;
  (iii) Physical data;
  (iv) Reactivity data;
  (v) Corrosivity data;
  (vi) Thermal and chemical stability data; and
  (vii) Hazardous effects of inadvertent mixing of different materials that could foreseeably occur.

Note: Material Safety Data Sheets meeting the requirements of 29 CFR 1910.1200(g) may be used to comply with this requirement to the extent they contain the information required by this subparagraph.

(2) Information pertaining to the technology of the process.

(i) Information concerning the technology of the process shall include at least the following;

(A) A block flow diagram or simplified process flow diagram (see Appendix B to this section); [not reproduced in this text]

(B) Process chemistry;

(C) Maximum intended inventory;

(D) Safe upper and lower limits for such items as temperatures, pressures, flows or compositions; and,

(E) An evaluation of the consequences of deviations, including those affecting the safety and health of employees.

(ii) Where the original technical information no longer exists, such information may be developed in conjunction with the process hazard analysis in sufficient detail to support the analysis.

(3) Information pertaining to the equipment in the process.

(i) Information pertaining to the equipment in the process shall include:

(A) Materials of construction;

(B) Piping and instrument diagrams (P&ID's);

(C) Electrical classification;

(D) Relief system design and design basis;

(E) Ventilation system design;

(F) Design codes and standards employed;

(G) Material and energy balances for processes built after May 26, 1992; and,

(H) Safety systems (e.g. interlocks, detection or suppression systems).

(ii) The employer shall document that equipment complies with recognized and generally accepted good engineering practices.

(iii) For existing equipment designed and constructed in accordance with codes, standards, or practices that are no longer in general use, the employer shall determine and document that the equipment is designed, maintained, inspected, tested, and operating in a safe manner.

Is This Requirement Necessary?

The primary requirement of this section of law states simply that, prior to performing a hazard analysis, the minimum process safety information listed in paragraphs (d)(1), (d)(2), and (d)(3)(i) shall be assembled. Paragraph (d)(1) addresses the characteristics of the chemicals used in the subject process; paragraph (d)(2) treats the process itself; paragraph (d)(3)(i) relates to process systems and components.

This is clearly not a trivial requirement. As you can imagine, developing process safety information could be a difficult and costly task if you don't already have it available. In fact, for old processes that have not been well-documented, you are faced with a situation not unlike trying to restore a civil war era home to modern day electrical codes. (Fortunately, for most modern hazardous chemical processes, this data is already on hand. You just have to know what you're looking for. But, we'll talk more about this later.)

If you have a questioning mind (and if you don't, you shouldn't be in industrial operations), you may legitimately ask, "What is the purpose of this requirement?" In fact, unless there is a reasonable basis for this section of law, we should raise objection. So, let's test it against reason.

A Prerequisite to Hazard Analysis

As we have already concluded, the objective of process safety management is to protect the operating personnel, the environment, and the general public from the potentially harmful effects of hazardous chemicals. That objective cannot be achieved without first identifying and then evaluating the hazards characteristic of the technology.

In Chapter 4 of the text, we learned how the boundaries of safe operation for industrial processes are developed. Hazard analysis laid the foundation for risk analysis, the predecessor of risk management. And risk management, as you will recall, is the process of establishing protection. (For a detailed discussion from the chemical industry perspective, review Chapter 6—Process Risk Management—of **Plant Guidelines for Technical Management of Chemical Process Safety** [Revised Edition]. Also CCPS's **Guidelines for Hazard Evaluation Procedures** [Second Edition with Worked Examples] gives in-depth instruction on hazard analysis.)

Yet, hazard analysis is difficult unless the characteristics and capabilities of the process equipment and process chemicals are clearly documented and understood. Accordingly, process safety information must be assembled before comprehensive hazard analysis can commence. In other words, detailed process safety information is a prerequisite to thorough hazard analysis.

Necessary for Operations and Maintenance

We doubt whether anyone is stunned by the revelation that process safety information necessarily precedes hazard analysis. Anyone in the industrial operations business has learned long ago that most of engineering rests upon the ability to predict the performance of materials, components, and systems under specified conditions. That's where design specifications come from. So, let's consider in greater detail the importance of having process safety information available and up-to-date to support routine operations, maintenance, and engineering.

Chemical Characteristics

As we just learned, paragraph (d)(1) requires process chemical characteristics to be available. The characteristics include:
- Toxicity information;
- Permissible exposure limits;
- Physical data;
- Reactivity data;
- Corrosivity data;
- Thermal and chemical stability data; and
- Hazardous effects of inadvertent mixing of different materials that could foreseeably occur.

None of this information is difficult to acquire. The material safety data sheets (provided by chemical manufacturers) which accompany the compounds provide it. You can also find much of the information in the **Emergency Response Guidebook** provided by the Department of Transportation.

How much sense, then, does it make to use chemicals (or equipment) that we don't understand? Again, we need only look to Bhopal to see the consequences of unintelligent process control. That operating staff was oblivious to the dangers presented by the compound they were processing and, more importantly, the state of the process equipment.

To put it bluntly, if you're using hazardous chemical compounds and you don't understand their potentials, you're making a mistake. That's the point of the Hazards Communication Law (OSHA Reg 1910.1200). It makes no sense *not* to have this information on hand (and understood!) at all times. Maintenance and operations depend on it.

Process Technology Information

Understanding the characteristics of chemicals is only part of the requirement, though. Paragraph (d)(2) pertains to information regarding "the technology of the process". Specifically, it identifies the need to have available:
- A block flow diagram or simplified process flow diagram;
- Process chemistry (i.e., what are the intended and expected interactions of the chemicals within the process);
- Maximum intended chemical inventory;
- Safe upper and lower limits for critical parameters (e.g., temperature, pressure, etc); and,
- An evaluation of the consequences of deviations from critical parameter or process controls.

Do you recall our discussion in Chapter 10 regarding critical parameter observation? We learned that equipment is controlled by monitoring important indicators—physical characteristics such as temperature, pressure, speed, distance, fluid level, concentration, and purity—and making the necessary adjustments to keep these factors within specified operating bands.

We called such indications *critical parameters*—operating properties that must be controlled to avoid equipment failure. (There's an exceptional discussion of this concept in Appendix 12B—Example of Critical Operating Parameters: Interpretation Guidelines—of **Plant Guidelines for Technical Management of Chemical Process Safety** [Revised Edition].)

This section of law simply states that you must have records and flow diagram illustrations of how the process is supposed to operate, what maximum amounts of the hazardous chemicals are allowed to be on hand, what the critical parameter limits are, and what could go wrong. That doesn't seem an irrational expectation.

On a more practical level, consider this question: "How can you run a training program to produce competent team members without this fundamental information?" In the absence of thorough chemical and process characteristic knowledge, an operator or maintenance technician cannot hope to intelligently control processes under the spectrum of normal and abnormal conditions he or she is likely to face. Can you imagine an airline captain who does not understand how his or her equipment sustains flight? What the equipment limitations are? What is likely to happen if the limits are exceeded? Remember the premise of the text! **There is no substitute for the alert, well-trained operator controlling equipment within specified operating bands in accordance with approved procedures!**

Process Equipment Information

Hmmm. Hard to argue that process technology information shouldn't have to be accessible. So what about the next paragraph? The law requires "information pertaining to the equipment in the process" to be available. It includes:
- Materials of construction;
- Piping and instrument diagrams;
- Electrical classification;
- Relief system design and design basis;
- Ventilation system design;
- Design codes and standards employed;
- Material and energy balances for processes built after May 26, 1992; and,
- Safety systems (e.g., interlocks, detection or suppression systems).

What we're looking at here are the design specifications and as-built drawings for the process—the records of the standards to which the process was originally built (and is currently maintained). (*As-built* drawings differ from the original design drawings in that they reflect the actual configuration of components and systems as they were installed.)

In Chapter 3, we determined that the second element in the strategy for successful industrial operations is reliable equipment and facilities. We learned that *reliability* is based upon conservative design, superior construction, and precise modification and maintenance. Without the process equipment information identified in paragraph (d)(3), reliability cannot be sustained. The pedigree of parts, components, and systems will be lost and the ability to maintain the process in a reliable state will be, at best, difficult. Further, plant engineering and maintenance departments operating without this information can be little more than "shade tree mechanics".

Material and Energy Balances

Before we proceed further, one item from this paragraph—material and energy balances—deserves a special note of explanation. If you are reading this supplement, you are probably well aware that hazardous chemical processing is often subject to great quantities of energy transfer, usually in the form of heat. *Exothermic reactions*—chemical compound interactions resulting in the release of heat—are particularly common. The law requires that, for modern processes (those which have been built since May 26, 1992), calculations regarding how the energy is expected to be transferred (and whether the process equipment and systems have sufficient capacity to accommodate such transfers) must be completed. Such calculations should account not only for routine operations, but also for foreseeable abnormal conditions. That's one of the reasons that the law also requires an accounting of maximum allowable chemical inventory.

What If I Don't Have This Information?

In retrospect, it seems clear that chemical characteristics, process technology information, and process equipment data must be available if a hazardous chemical processing plant (or any other high risk facility) is to be competently operated and maintained, regardless of what process safety management law says.

Fortunately, as we noted before, most modern hazardous chemical processing facilities already have process safety information on hand. It is an inherent part of design and construction, and it is also necessary to obtain operating permits. But what about older processing activities that have been in operation for many years? For the most part, they were legitimately designed, constructed, operated, and maintained under the standards of their day; yet, under current standards they may face compliance difficulties.

Usually, compliance problems arise where as-built drawings have not been kept up to date or where temporary modifications were inappropriately analyzed and documented prior to (and after) installation. As a result, what's in the plant and what's on the drawings are very different. (By the way, this is a public sector problem as much as private sector.)

Apart from posing legal dilemmas, inadequately documented systems, as we have already noted, are a real threat to the safe operation and maintenance of the process components and systems. A significant number of industrial problems result from working on or around systems that are not installed as the records (if available) show. Nevertheless, the question at this point is, "What do I do if I don't have the information?"

The regulations are specific. If the information has been lost, paragraph (d)(2)(ii) states that:

> *Where the original technical information no longer exists, such information may be developed in conjunction with the process hazard analysis in sufficient detail to support the analysis.*

In other words, you're going to have to perform hazard analysis without all the process safety information, developing it to the extent necessary (and possible) as a part of the hazard analysis. Does that make the hazard analysis harder? Absolutely.

Further, if the process is very old, paragraph (d)(3)(iii) states:

> *For existing equipment designed and constructed in accordance with codes, standards, or practices that are no longer in general use, the employer shall determine and document that the equipment is designed, maintained, inspected, tested, and operating in a safe manner.*

Or, more succinctly stated, the employer must prove that it's still safe. Appendix C to the regulation puts it this way:

> *For existing equipment designed and constructed many years ago in accordance with the codes and standards available at that time and no longer in general use today, the employer must document which codes and standards were used and that the design and construction along with the testing, inspection and operation are still suitable for the intended use. Where the process technology requires a design which departs from the applicable codes and standards, the employer must document that the design and construction is suitable for the intended purpose.*

To that end, a strong history of safe operation is useful but not conclusive. You're going to have to get some good engineering judgments to go with your safety record.

Finally, this section of regulation places the burden upon the employer to ensure and document that process equipment meets "accepted good engineering practices". Paragraph (d)(3)(ii) states:

> *The employer shall document that equipment complies with recognized and generally accepted good engineering practices.*

For an old process, this may be the most difficult of the requirements to meet. What constitutes "accepted good engineering practices" can be somewhat subjective (and may be different today than it was five years ago). Appendix C advises:

> *The information pertaining to process equipment design must be documented. In other words, what were the codes and standards relied on to establish good engineering practice. These codes and standards are published by such organizations as the American Society of Mechanical Engineers, American Petroleum Institute, American National Standards Institute, National Fire Protection Association, American Society for Testing and Materials, National Board of Boiler and Pressure Vessel Inspectors, National Association of Corrosion Engineers, American Society of Exchange Manufacturers' Association, and model building code groups....*
> *In addition, various engineering societies issue technical reports which impact process design. For example, the American Institute of Chemical Engineers has published technical reports on topics such as two phase flow for venting devices. This type of technically recognized report would constitute good engineering practice.*

Your best bet is to find an engineering firm with a track record in this area of compliance and see if they can assist *your* engineering staff in evaluating your process. Be sure to emphasize the word *assist*. If *your* engineering staff is not closely involved in the evaluation, you are at the mercy of the firm you hired. You fail, as well, to educate your own people on how this problem can be attacked and solved.

A Good Investment

It's pretty clear that valid and comprehensive process safety information creates a foundation on which everything else is built. Perhaps Appendix C summarizes best:

> Complete and accurate written information concerning process chemicals, process technology, and process equipment is essential to an effective process safety management program and to a process hazards analysis. The compiled information will be a necessary resource to a variety of users including the team that will perform the process hazards analysis as required under paragraph (e); those developing the training programs and the operating procedures; contractors whose employees will be working with the process; those conducting the pre-startup reviews; local emergency preparedness planners; and insurance and enforcement officials.

The time and money spent in ensuring that the foundation is well-laid will pay enormous dividends in future operations, maintenance, and training.

Process Hazard Analysis

Comprehensive process safety information, as we have noted, is requisite to competent hazard analysis. And hazard analysis is at the heart of process safety. Appendix C states it simply:

> A process hazard analysis (PHA), sometimes called a process hazard evaluation, is one of the most important elements of the process safety management program.... A PHA provides information which will assist employers and employees in making decisions for improving safety and reducing the consequences of unwanted or unplanned releases of hazardous chemicals.

It's adequately clear to most of us that if you don't know what you're dealing with, you can't protect against it. But, before we examine what the law says about process hazard analysis, we need to review what the term *hazard analysis* and its cousins mean.

Essential Definitions

In Chapter 4, we defined a *hazard* as a potential danger, usually of personnel injury or equipment damage, incurred as a result of engaging in or being near an activity. CCPS similarly defines the term in **Guidelines for Hazard Evaluation Procedures** (Second Edition with Worked Examples) as "An inherent physical or chemical characteristic that has the potential for causing harm to people, property, or the environment."

We also stated in the text that hazards are identified through the process of hazard analysis, referred to in some quarters as a *hazards and operability study*. We continued, stating that the objective of hazard study is to determine what dangers could result from an activity, process, or technology.

Hazard analysis seems to have many names, depending upon whose literature you read. But, the meanings vary little. Appendix C defines process hazard analysis in these terms:

> A PHA is an organized and systematic effort to identify and analyze the significance of potential hazards associated with the processing or handling of highly hazardous chemicals.

The CCPS's **Guidelines for Hazard Evaluation Procedures** [Second Edition with Worked Examples] uses slightly different terminology:

> **Hazard Evaluation:** *The analysis of the significance of hazardous situations associated with a process or activity. Uses qualitative techniques to pinpoint weaknesses in the design and operation of facilities that could lead to accidents.*
>
> **Hazard and Operability Analysis (HAZOP):** *A systematic method in which process hazards and potential operating problems are identified using a series of guide words to investigate process deviations.*

You can see that there's not much difference in the definitions. Regardless of nomenclature, we're addressing the same topic.

What Does It Do?

So despite what name we choose to give it, we need to determine what it does (or what it can do if it is appropriately used). We should ask, "Is this a process that I should understand and employ? Or is this an overburdensome environmental regulation that has no place in my business?" Let's check the logic.

The fundamental question of hazard analysis proceeds along these lines: "What substantial dangers to the operating personnel, the environment, and the general public are presented by this process, operation, or activity?"

Dissecting the Law

Once we answer that question, we need to know if the hazards that we have identified are *probable*. That's what risk analysis does. It extends the study with these inquiries: "How could such hazards become a reality and what is the probability of such an occurrence?" (Remember our graphic on risk management priorities from Chapter 4? Risk is the product of probability and consequences.)

Then, with the answers to the questions of hazard analysis and risk analysis in hand, design, construction, and operating teams can devise reasonable means for protecting against the hazards. We called this last process *risk management*—protecting against identified hazards which have a significant probability of occurring.

Obviously, then, process hazard analysis must be employed if you are ever to achieve comprehensive protection. In fact, the analysis process is part of an iteration of constant evaluation and improvement.

Once we agree that a PHA is necessary, we must then determine how comprehensive it needs to be. Let's look at the minimums. Here's what the law says:

1910.119(e) **(e) Process hazard analysis.**

(1) The employer shall perform an initial process hazard analysis (hazard evaluation) on processes covered by this standard. The process hazard analysis shall be appropriate to the complexity of the process and shall identify, evaluate, and control the hazards involved in the process. Employers shall determine and document the priority order for conducting process hazard analyses based on a rationale which includes such considerations as extent of the process hazards, number of potentially affected employees, age of the process, and operating history of the process. The process hazard analysis shall be conducted as soon as possible, but not later than the following schedule:

(i) No less than 25 percent of the initial process hazards analyses shall be completed by May 26, 1994;

(ii) No less than 50 percent of the initial process hazards analyses shall be completed by May 6, 1995;

(iii) No less than 75 percent of the initial process hazards analyses shall be completed by May 26, 1996;

(iv) All initial process hazards analyses shall be completed by May 26, 1997.

(v) Process hazards analyses completed after May 26, 1987 which meet the requirements of this paragraph are acceptable as initial process hazards analyses. These process hazard analyses shall be updated and revalidated, based on their completion date, in accordance with paragraph (e) (6) of this standard.

(2) The employer shall use one or more of the following methodologies that are appropriate to determine and evaluate the hazards of the process being analyzed:

(i) What-If;

(ii) Checklist;

(iii) What-If/Checklist;

(iv) Hazard and Operability Study (HAZOP);

(v) Failure Mode and Effects Analysis (FMEA);

(vi) Fault Tree Analysis; or

(vii) An appropriate equivalent methodology.

(3) The process hazard analysis shall address:

(i) The hazards of the process;

(ii) The identification of any previous incident which had a likely potential for catastrophic consequences in the workplace;

(iii) Engineering and administrative controls applicable to the hazards and their interrelationships such as appropriate application of detection methodologies to provide early warning of releases. (Acceptable detection methods might include process monitoring and control instrumentation with alarms, and detection hardware such as hydrocarbon sensors.);

(iv) Consequences of failure of engineering and administrative controls;

(v) Facility siting;

(vi) Human factors; and

(vii) A qualitative evaluation of a range of the possible safety and health effects of failure of controls on employees in the workplace.

(4) The process hazard analysis shall be performed by a team with expertise in engineering and process operations, and the team shall include at least one employee who has experience and knowledge specific to the process being evaluated. Also, one member of the team must be knowledgeable in the specific process hazard analysis methodology being used.

(5) The employer shall establish a system to promptly address the team's findings and recommendations; assure that the recommendations are resolved in a timely manner and that the resolution is documented; document what actions are to be taken; complete actions as soon as possible; develop a written schedule of when these actions are to be completed; communicate the actions to operating, maintenance and other employees whose work assignments are in the process and who may be affected by the recommendations or actions.

(6) At least every five (5) years after the completion of the initial process hazard analysis, the process hazard analysis shall be updated and revalidated by a team meeting the requirements in paragraph (e) (4) of this section, to assure that the process hazard analysis is consistent with the current process.

(7) Employers shall retain process hazards analyses and updates or revalidations for each process covered by this section, as well as the documented resolution of recommendations described in paragraph (e) (5) of this section for the life of the process.

Elements of the Law

This section of law prescribes seven major requisites. It requires: (1) that you perform a hazard analysis if your process is covered by this regulation; (2) that you use established methodologies when you perform the analysis; (3) that you address several critical elements of the process in the analysis; (4) that the analysis be performed by a qualified team; (5) that the employer establish a method for addressing findings and recommendations of the analysis team; (6) that the analysis be updated and validated at least every five years; and (7) that the analysis (and its iterations) be held as a permanent record.

Appendix C summarizes well:

> *The PHA focuses on equipment, instrumentation, utilities, human actions (routine and nonroutine), and external factors that might impact the process. These considerations assist in determining the hazards and potential failure points or failure modes in a process.*

Let's examine each major requirement in greater detail.

Prioritizing the Analyses

Paragraph (e)(1) establishes a timetable for accomplishing initial hazard analyses. As with most new regulations, it provides a period of time to catch up with the requirements of law.

Next, it charges employers with prioritizing which ones should be done first. Appendix C explains:

> *When the employer has a number of processes which require a PHA, the employer must set up a priority system of which PHAs to conduct first. A preliminary or gross hazard analysis may be useful in prioritizing the processes that the employer has determined are subject to coverage by the process safety management standard. Consideration should first be given to those processes with the potential of adversely affecting the largest number of employees. This prioritizing should consider the potential severity of a chemical release, the number of potentially affected employees, the operating history of the process such as the frequency of chemical releases, the age of the process and any other relevant factors.... The use of a preliminary hazard analysis would assist an employer in determining which process should be of the highest priority....*

Prioritizing makes sense. It allows you to address the greatest risks first.

Paragraph (e)(1) also implies (as it should) that the extent, rigor, and complexity of the hazard analysis should conform to the extent and complexity of the expected hazards. In other words, a complex process is probably going to require a complex hazard analysis. (Department of Energy documents appropriately call this a *graded approach*—applying principles or methods with a rigor that suits the risk or complexity of the activity.)

Matching Method to Process

Paragraph (e)(2) provides guidance on how hazard analysis should be performed. It delineates six specific analysis methodologies (as well as allowing any "appropriate equivalent methodology"), with the expectation that more than one method will probably be required. Appendix C elaborates:

> All PHA methodologies are subject to certain limitations. For example, the checklist methodology works well when the process is very stable and no changes are made, but it is not as effective when the process has undergone extensive change. The checklist may miss the most recent changes and consequently the changes would not be evaluated.

It is important, then, to choose the right tool for each task of the analysis—to customize the analysis using as many methods as necessary. Again, Appendix C:

> The application of a PHA to a process may involve the use of different methodologies for various parts of the process. For example, a process involving a series of unit operations of varying sizes, complexities, and ages may use different methodologies and team members for each operation. Then the conclusions can be integrated into one final study and evaluation.

The foregoing discussion shows hazard analysis to be a complex process. If you have not engaged in hazard analysis before, you may be intimidated by the names and the numbers of the methods. You have available, however, several resources for assistance. Each of the analysis methods is thoroughly discussed in CCPS's **Guidelines for Hazard Evaluation Procedures** [Second Edition with Worked Examples]. Also, nearly any engineering technical library will contain numerous other sources on analysis methodologies.

Analyzing the Right Things

The third requirement, stipulated in paragraph (e)(3), states, at a minimum, what the analysis must address:

- Known hazards of the process. (Have you thoroughly analyzed this process to determine what hazards it presents? Does industry experience back up your analysis?)
- Other hazards which are suggested by previous close calls or accidents. (Are there hazards that you haven't considered but that are indicated by events which have occurred in the past? Both *your* past and that of others who do similar work?)
- Protective engineering and administrative controls and how they might fail. (For the hazards which you have identified, are you reasonably protected by physical, procedural, and administrative barriers? Is the protection in-depth? What if a barrier fails?)
- Problems associated with facility location that could adversely affect the environment or the general public. (Is your facility sited such that, if you experience an event whose effects will go beyond your fence-line, the public and the environment will be reasonably protected?)
- Human factors. (Have you considered probable team member actions that could lead to an undesirable event? Is the design user-friendly and forgiving? Does your training provide assurance that your team members can handle their equipment during both normal and abnormal operating conditions?)

These questions are not new. You've encountered the same concerns in the first four parts of this book. Furthermore, the methodologies listed in the second paragraph all steer you, in one form or another, to the same questions.

Using Competent People

The next section of law, paragraph (e)(4), advises us that hazard analysis must be performed by competent people—adept in engineering, process operations, and, to a degree, in hazard analysis. Appendix C:

> The team conducting the PHA needs to understand the methodology that is going to be used.... The team leader needs to be fully knowledgeable in the proper implementation of the PHA methodology that is to be used and should be impartial in the evaluation. The other full or part time team members need to provide the team with expertise in areas such as process technology, process design, operating procedures and practices, including how the work is actually performed, alarms, emergency procedures, instrumentation, maintenance procedures, both routine and nonroutine tasks, including how the tasks are authorized, procurement of parts and supplies, safety and health, and any other relevant subject as the need dictates. At least one team member must be familiar with the process.

Clearly, your hazard analysis isn't going to be efficient (or complete) unless the people involved know what they're doing. If you've never done this before, you're well advised to acquire assistance from someone who has a proven track record, especially on your type of facility. But, as we said before, don't just hire it done. Have them lead a team of *your* people. You're going to have to do it again anyway, so teach your staff how to do it. Take some time to undergo formal training. It will save you time and money in the long run. (One prominent company that specializes in providing analysis methodology training is the Process Safety Institute, a division of JBF Associates, Inc., headquartered in Knoxville, TN.)

The hazard analysis, of course, is of little use if you don't do something with the results. So paragraph (e)(5) tells us to establish a system for acting on findings and recommendations, documenting how each concern was addressed. Some of the fixes are going to be easy, and some are going to be hard. Some will be short-term, and others are going to take a long time. Don't let them drop through the cracks. Remember, if you're operating your plant correctly, this isn't a drill simply to comply with the law. These are things you should be doing even without the law.

Keeping the Analysis Current

The sixth major requirement in this section of law establishes a five year analysis update—a mandatory re-evaluation. If little has changed, the analysis validation will not be nearly as consuming as the initial hazard analysis.

But, don't be lulled into a false sense of security. As we learned earlier in this text, hazards can (and frequently do) change—sometimes subtly over long periods of time. Do you recall our discussion of changes in the safe operating envelope from Chapter 4? We said that safety envelopes often change as a result of changing environments and circumstances. Just as the normal operating zone for a vehicle becomes more limiting as road conditions deteriorate, so will safety margins and boundaries for equipment operation under varying conditions. In Prince William Sound, retreat of Columbia Glacier confounded the original safety analysis. The hazards upon which the safety analysis was originally predicated slowly changed over a ten-year period beginning about 1979. Protection, however, was not modified to address changing hazards.

Note, also, that the five year requirement is a minimum. Significant changes in equipment, operating environment, or the operating staff require us to reanalyze for safety. This is part of *management of change*, the term used in OSHA law to address plant modifications. We'll learn more about that when we discuss paragraph (l).

Documenting the Analysis

The last of seven requirements in paragraph (e) states, as you would expect, that the hazard analysis is part of the documentation to prove process pedigree. You must keep a permanent record (for the life of the process) of the initial analysis, how it has been modified over the years, and how the recommendations of the analyses were resolved.

Based on Formal Safety Studies?

Safety studies (and protective actions) are predicated upon thorough hazard analyses. They lay the foundation for suitable protection. That's the purpose of this important element of process safety management.

At the Phillips complex, hazard analysis was incomplete. As a result, too many protective barriers were bypassed or just not there.

Are you protected? Can you answer "Yes" to the "Questions to Consider" at the end of Chapter 4. Are the design, construction, and operation of your plant or process based on formal safety studies? Do you have and periodically review those studies? Are they current? Are safety studies revised to reflect modifications to safety-related equipment?

Operating Procedures

Once hazards are identified, protective barriers must be erected to safeguard operating personnel, the environment, and the general public from recognized dangers. As we have already learned, conservative design and superior construction are the foundation for protection. But we also determined (in Chapters 3 and 4) that, regardless of design and construction, if the equipment is not operated in accordance with design constraints, the defense is incomplete. As a result, written directions for equipment operation—*operating procedures*—form an important part of the protective barrier defense.

We are not surprised, then, to find that the fourth major section of process safety management law addresses the need for creating and employing valid operating procedures in hazardous chemical processing facilities. For those of us who have studied (and practiced) systematic industrial operations, we reply, "How can you operate without them?"

The Man/Machine Interface

In Chapter 3 of the text, we identified valid policies and procedures as one of the eight essential elements of the strategy for operating success. We stated that an operating strategy will not succeed without clear operating policies and procedures. Legitimate policies and procedures form the interface between machines, processes, and their human controllers. Well-written procedures provide operators with proven guidelines for controlling their machinery.

We defined a procedure as a step-by-stop operating or maintenance instruction, the purpose of which is to govern the control or repair of a component, system, or process. We deemed procedures to be so important that, later in the text, we devoted an entire chapter—Chapter 9—to the subject of understanding and using procedures.

OSHA law ascribes similar importance to the subject of operating procedures and defines them in much the same way. Appendix C states:

> Operating procedures describe tasks to be performed, data to be recorded, operating conditions to be maintained, samples to be collected, and safety and health precautions to be taken.... [They] will include specific instructions or details on what steps are to be taken or followed in carrying out the stated procedures.

We already know what they are, so let's review the requirements of paragraph (f) on the subject of procedures and see if we find anything that doesn't make sense. Here's what the law says:

1910.119(f)

(f) **Operating procedures.**
(1) The employer shall develop and implement written operating procedures that provide clear instructions for safely conducting activities involved in each covered process consistent with the process safety information and shall address at least the following elements.
    (i) Steps for each operating phase:
        (A) Initial startup;
        (B) Normal operations;
        (C) Temporary operations;
        (D) Emergency shutdown including the conditions under which emergency shutdown is required, and the assignment of shutdown responsibility to qualified operators to ensure that emergency shutdown is executed in a safe and timely manner.
        (E) Emergency operations;
        (F) Normal shutdown; and,
        (G) Startup following a turnaround, or after an emergency shutdown.
    (ii) Operating limits:
        (A) Consequences of deviation; and,
        (B) Steps required to correct or avoid deviation.
    (iii) Safety and health considerations:
        (A) Properties of, and hazards presented by, the chemicals used in the process;
        (B) Precautions necessary to prevent exposure, including engineering controls, administrative controls, and personal protective equipment;
        (C) Control measures to be taken if physical contact or airborne exposure occurs;
        (D) Quality control for raw materials and control of hazardous chemical inventory levels; and,
        (E) Any special or unique hazards.
    (iv) Safety systems and their functions.
(2) Operating procedures shall be readily accessible to employees who work in or maintain a process.
(3) The operating procedures shall be reviewed as often as necessary to assure that they reflect current operating practice, including changes that result from changes in process chemicals, technology, and equipment, and changes to facilities. The employer shall certify annually that these operating procedures are current and accurate.

(4) The employer shall develop and implement safe work practices to provide for the control of hazards during operations such as lockout/tagout; confined space entry; opening process equipment or piping; and control over entrance into a facility by maintenance, contractor, laboratory, or other support personnel. These safe work practices shall apply to employees and contractor employees.

Steps for Major Operating Phases

Paragraph (f)(1)(i) mandates that, as a minimum, operating procedures must be developed for major operating phases of a process including normal process startups, operations, and shutdowns; response to abnormal conditions such as equipment casualties and emergencies (including emergency shutdowns); restoration startups following outages (i.e., recovery from shutdowns, usually associated with major maintenance, also known as *turnarounds*); and "temporary" operations.

The wording of this section of law is designed to emphasize (and differentiate between) normal and abnormal situations. Procedures are necessary for both. Appendix C explains:

> ...[I]n starting up or shutting down the process...different parameters will be required from those of normal operation. [The] operating instructions need to clearly indicate the distinctions between startup and normal operations such as the appropriate allowances for heating up a unit to reach the normal operating parameters. Also the operating instructions need to describe the proper method for increasing the temperature of the unit until the normal operating temperature parameters are achieved.

One term used in this paragraph—*temporary operations*—deserves special comment. It implies a situation which is abnormal or non-routine. Perhaps the best advice is that, prior to performing a complex or risky activity—especially one which is not frequently undertaken—the wise course is to think through the process and develop a procedure for performance. It will take some extra time in the beginning, but it is likely to save a great deal more time than "flying by the seat of your pants".

Operating Limits

Succinct procedural steps, though the meat of operating procedures, are usually insufficient for safe equipment control. The boundaries, or limits, of operation must also be communicated. As expected, then, the next requirement—paragraph (f)(1)(ii)—states that equipment and process operating limits should be clearly conveyed through operating procedures. Appendix C:

> Operating procedures will include specific instructions or details on what steps are to be taken or followed in carrying out the stated procedures.... [T]he operating procedures addressing operating parameters will contain operating instructions about pressure limits, temperature ranges, flow rates, what to do when an upset condition occurs, what alarms and instruments are pertinent if an upset condition occurs, and other subjects.

The term *operating limit* should be a familiar one for the student of systematic operations. You will recall from Chapter 4 that operating limits are limits for control of critical operating parameters which, if observed, ensure safe operation. They are normally specified by procedure and rely upon the intelligent action of operators to be effective.

We categorized operating limits as either *safe*, *normal*, or *abnormal* limits and illustrated the differences in Figure 4-2 (A Model Operating Envelope). (The Center for Chemical Process Safety uses slightly different terminology summarized in Appendix 12B of **Plant Guidelines for Technical Management of Chemical Process Safety** [Revised Edition].) We also found that, in literally every case study, operating limits were violated—sometimes intentionally as at Chernobyl. In fact, we identified degradation of operating limits and procedural non-compliance as two of the common contributors to accidents.

We should be little surprised, then, to find a requirement in law for conveyance of operating limits within procedures.

Safety and Health Considerations

Early in this section (and within the strategy for operating success) we established that procedures are an important protective barrier when correctly contrived, thoroughly taught, and conscientiously applied. Not only do they enhance efficient equipment and process operations, but, in so doing, they improve protection. In fact, in studying procedures (and every other element of the systematic approach), we perpetually return to what is by now becoming a mantra—*protection of the operating personnel,* the environment, and the general public. Efficiency and safety are not exclusive; rather, they are symbiotic.

Hence, paragraph (f)(1)(iii) requires that operating procedures for hazardous chemical process operations include warnings, precautions, and control measures associated with the process chemicals. Appendix C puts it this way:

> [O]perating instructions for each procedure should include the applicable safety precautions and should contain appropriate information on safety implications.

Safety Systems

Finally, paragraph (f)(1)(iv) requires that special emphasis be accorded within the procedures to *safety systems*. Though no further guidance is offered, it is reasonable that the importance and function of safety systems be accentuated in the procedural text. A word of warning, however: We must forever remember (as we learned in our study of the systematic approach) that procedures, no matter how well conceived and written, can *never* substitute for well trained people.

Accessibility

Paragraph (f)(2) addresses the need for accessibility of procedures. They won't be used and they can't be effective if they are not available.

Also, availability means a little more than just having the procedures there. If your team members can't comprehend them, they are essentially unavailable. Appendix C reminds us:

> If workers are not fluent in English then procedures and instructions need to be prepared in a second language understood by the workers.

At Bhopal, the procedures were written in English, the language of that nation's professionals. Originally, English was appropriate; but, as selection criteria degraded, team members were hired who did not speak English. The procedures, however, were never translated. And, we all know that, without usable procedures, we tend to operate based upon "tribal knowledge" and trial and error—a prescription for failed operations.

Review, Revision, and Certification

Procedural review and *certification* are required by paragraph (f)(3) once annually as a minimum. Beyond that, the regulation requires procedural review and *revision* "as often as necessary" to keep the procedures current with the operation.

The concept of keeping procedures up-to-date is fundamental to the systematic approach. Chapter 9 stated that procedures must not only be user-friendly, they must also be easy to revise when change is necessary. If the change process is excessively cumbersome, operating and maintenance technicians are apt simply to "work around" the procedure. This is a dangerous precedent to establish because it devalues **all** procedures. Also, unapproved changes made by technicians may have consequences that have not been considered.

Usually, procedure revision is required as a part of equipment or process modification. Appendix C explains:

> ...[O]perating procedures need to be changed when there is a change in the process as a result of the management of change procedures. The consequences of operating procedure changes need to be fully evaluated and the information conveyed to the personnel. For example, mechanical changes to the process made by the maintenance department (like changing a valve from steel to brass or other subtle changes) need to be evaluated to determine if operating procedures and practices also need to be changed.

Note that procedural revision alone may be inadequate. Some level of training is often necessary to familiarize the operators prior to process restart. Again, Appendix C:

> All management of change actions must be coordinated and integrated with current operating procedures and operating personnel must be oriented to the changes in procedures before the change is made. When the process is shutdown in order to make a change, then the operating procedures must be updated before startup of the process.

Yet, the training will be, at best, incomplete without the procedures which support the modification. In fact, as we learned in Chapter 9, well-written procedures are, next to excellent instructors, the best training tools. Appendix C similarly concludes that:

Operating procedures and instructions are important for training operating personnel.

We'll discuss training more thoroughly in our review of paragraph (g).

For Support and Maintenance

The title of paragraph (f) is "Operating Procedures". But, as we indicated in our study of the systematic approach, operators are not the only team members who need procedures. Paragraph (f)(4) calls for the establishment of "safe work practices...for control of hazards" during support and maintenance-related activities such as "lockout/tagout, confined space entry, opening process equipment or piping, and control over entrance into a facility".

Though this paragraph falls short of mandating maintenance procedures, paragraph (j)—mechanical integrity—does require them. As we stated in the introduction to Chapter 9, competent operators (and maintenance technicians) manage equipment within prescribed tolerances *in accordance with approved procedures*.

Accurate, Understandable, Up-to-Date

The need for operating procedures was thoroughly documented as an important element in our study of the strategy for operating success. At the Phillips complex, procedural deficiencies (at least in the area of settling leg isolation) contributed to the mistakes that resulted in the explosion and fire. Without reliable procedures, team members may be left to navigate through a problem without a map.

Obviously, the people at OSHA (and members of the chemical process industry) also consider valid procedures essential for process safety management. As Appendix C states:

> *The procedures need to be technically accurate, understandable to employees, and revised periodically to ensure that they reflect current operations.*

We can't say it any better than that.

Training

Process safety information, hazard analysis, operating procedures—one necessarily leads to the other in the development of successful process safety management. That idea aligns precisely with what we learned in our study of the systematic approach. In Chapter 9, we determined that when correctly developed, component and system procedures have a lineage that can be traced all the way back to the design basis of the equipment.

Yet, none of these matter—not process safety information, not hazard analysis, not operating procedures—unless there are alert, well-trained operators on-hand who are committed to operating and maintaining in accordance with the design specifications and operating constraints. (Remember Chernobyl?)

We emphasized in the text that operators who are taught the lineage and usefulness of procedures are far more likely to respect and employ them competently. In fact, we stated in Chapter 5 that the first principle of operation is to **operate and maintain equipment within the boundaries of the safe operating envelope.**

In Chapter 6, we went so far as to say that alert, well-trained operators are perhaps the most important element of the operating strategy. Operators are the final defense against failure and are ultimately responsible for safe and successful operations. Excellent design and construction must be supplemented with superior operating and maintenance instructions and then supported with the best-selected, best-trained people possible.

And we devoted three entire chapters (Chapters 6, 18, and 21) to the process of developing alert, well-trained operators, including topics such as the tenets of training, determination of job requirements, selection of candidates, development of the training program, selection and training of instructors, initial classroom training, on-the-job training, operator certification, and the advancement of operator knowledge and skills through a continuing training program.

Clearly, training is an important topic. Let's determine if OSHA law and the systematic approach are compatible. Here's what the law says:

1910.119(g)

(g) Training.

(1) Initial training.

(i) Each employee presently involved in operating a process, and each employee before being involved in operating a newly assigned process, shall be trained in an overview of the process and in the operating procedures as specified in paragraph (f) of this section. The training shall include emphasis on the specific safety and health hazards, emergency operations including shutdown, and safe work practices applicable to the employee's job tasks.

(ii) In lieu of initial training for those employees already involved in operating a process on May 26, 1992, an employer may certify in writing that the employee has the required knowledge, skills, and abilities to safely carry out the duties and responsibilities as specified in the operating procedures.

(2) Refresher training. Refresher training shall be provided at least every three years, and more often if necessary, to each employee involved in operating a process to assure that the employee understands and adheres to current operating procedures of the process. The employer, in consultation with the employees involved in operating the process, shall determine the appropriate frequency of refresher training.

(3) Training documentation. The employer shall ascertain that each employee involved in operating a process has received and understood the training required by this paragraph. The employer shall prepare a record which contains the identity of the employee, the date of training, and the means used to verify that the employee understood the training.

Establishing a Training Program

The requirements of paragraph (g) address three major topics: (1) initial training, (2) refresher training, and (3) documentation of training. We must admit, however, that, though implied, the law does not address (nor should it) the methods for developing an effective training *program*. That's up to you. But, don't get trapped in the "minimum compliance" perspective. You'll experience the same difficulties faced by Warrior & Gulf (Chapter 6). Where risk is high, thorough training is the only reasonable way to proceed.

As we learned in Chapter 6, there's a lot more to comprehensive instruction than delivery alone. The training function must include a philosophy (tenets of training); a method to determine *what* should be trained (job requirements and, subsequently, qualification standards); a way to select *who* should be trained (selection criteria); a process for choosing and training instructors (instructor selection and qualification); *how* the training should be performed (classroom? on-the-job?); *when* the training should be performed (planning and scheduling); how training should be advanced (continuing training); and, finally, how training should be evaluated.

Appendix C provides some insightful guidance for establishing a training program:

In establishing their training programs, employers must clearly define the employees to be trained and what subjects are to be covered in their training. Employers in setting up their training program will need to clearly establish the goals and objectives they wish to achieve with the training that they provide to their employees. The learning goals or objectives should be written in clear measurable terms before the training begins. These goals and objectives need to be tailored to each of the specific training modules or segments. Employers should describe the important actions and conditions under which the employee will demonstrate competence or knowledge as well as what is acceptable performance.

The foregoing is a very good description of a performance-based training program (including qualification standards) as discussed in Chapter 6.

Performance-based training usually means lots of on-the-job training. Appendix C emphasizes the need for comprehensive OJT:

Hands-on-training where employees are able to use their senses beyond listening, will enhance learning. For example, operating personnel, who will work in a control room or at control panels, would benefit by being trained at a simulated control panel or panels. Upset conditions of various types could be displayed on the simulator, and then the employee could go through the proper operating procedures to bring the simulator panel back to the normal operating parameters. A training environment could be created to help the trainee feel the full reality of the situation but, of course, under controlled conditions. This realistic type of training can be very effective in teaching employees correct procedures while allowing them to also see the consequences of what might happen if they do not follow established operating procedures. Other training techniques using videos or on-the-job training can also be very effective for teaching other job tasks, duties, or other important information. An effective training program will allow the employee to fully participate in the training process and to practice their skill or knowledge.

Note the significance accorded to realistic casualty and emergency training. As we stated in Chapter 15, unpracticed procedures are almost worthless in an emergency. Events often develop so rapidly that, without prior mastery of procedural steps, the procedure will be of little practical use. Emergency response training and testing is, therefore, necessary to evaluate the skills of operating personnel, resources staged for such events, and the procedures that have been developed to manage them.

If all of this sounds like a lot of work, you're right. But, remember what we learned in Chapter 3. The process of cultivating a team is time consuming and frustrating. Sometimes it seems that progress is more often lost than advanced. Yet experienced leaders recognize that most lasting changes in organizations take time, dedication, and persistence to plan and implement.

Proper training is not an exception. In fact, we would go so far as to suggest that you cannot meet the requirements of paragraph (g) without establishing a comprehensive training program.

Now, let's look at the standards for initial training, refresher training, and training documentation in more detail.

Initial Training

The initial training requirement—paragraph (g)(1)—is straightforward. As we would expect, it prescribes that anyone operating a hazardous chemical process must be trained to the procedural standards contained in paragraph (f). Appendix C elaborates:

All employees, including maintenance and contractor employees, involved with highly hazardous chemicals need to fully understand the safety and health hazards of the chemicals and processes they work with for the protection of themselves, their fellow employees and the citizens of nearby communities.... [A]dditional training in subjects such as operating procedures and safety work practices, emergency evacuation and response, safety procedures, routine and nonroutine work authorization activities, and other areas pertinent to process safety and health will need to be covered by an employer's training program.

We agree wholeheartedly. From our study of systematic operations, we concluded in Chapter 6 that the initial training phase must ensure that graduating operators not only have the skills to meet the challenges of routine operations, but also that they are prepared to manage abnormal or emergency situations.

Common sense dictates nothing less. How can you even attempt to operate a hazardous chemical process (or any other high risk venture) without first ensuring that the operating staff is smarter than the machines they intend to control? Remember the cardinal rule? **Humans must remain in control of their machinery at all times. Any time the machinery operates without the knowledge, understanding, and assent of its human controllers, the machinery is out of control.**

Grandfather Clause

Paragraph (g)(1)(ii) provides a grandfather clause for team members who were operating their processes prior to May 26, 1992. In other words, you may waive initial training for employees who have proven their ability through performance before the specified date. The employer must, however, certify in writing that they are competent in knowledge, skills, and abilities to safely operate.

Be careful. There are a couple of pitfalls here that you should avoid. If the team member in question was qualified on the process prior to the specified date, but has experienced a long break since last operating, you are making a mistake to give him (or her) a pass. At least you should develop for the team member a customized requalification process (sometimes termed a *partial requalification*) to bring them back up to par.

Further, if you have incorporated knowledge and performance requirements into the initial training qualification standard which you believe are not well understood (or performed) by those who you intend to grandfather, make sure these team members partially requalify in those areas.

Finally, refresher training applies to *everyone* who operates the process. No one is exempt—grandfathered or not.

Dissecting the Law

Refresher Training   In our recent study of implementing the systematic approach, we learned the value of *continuing training*—an on-going training process for team members who are already qualified (or certified) at an initial level of equipment operation. Further, we identified its objective as constant improvement of individual and team skills. Finally, we concluded that, in its absence, no operator (or operating team) ever reaches an elevated level of performance.

In Chapter 21, we learned that, for continuing training to be effective, it must be goal-directed, needs-based, taught by skilled instructors, formally conducted, and properly documented. We used as a model commercial air transport operations in which the phases and types of training included *initial* training, *transition* training, *upgrade* training, *differences* training, and *recurrent* training—all of which are designed to ensure that pilots are comprehensively prepared to handle their ships.

Refresher training is a different name for the same process. Although required by law only once every three years, our experience indicates that vital, on-going refresher training for high risk industries should be, at least, a monthly process. (We're more comfortable with formal weekly training; and, frankly, when you consider self-assessment, continuing training should be occurring informally every day. If that statement shocks you, it's time for a review of Chapter 22.)

Note that, in paragraph (g)(2), the law states that "The employer, in consultation with the employees involved in operating the process, shall determine the appropriate frequency of refresher training." Your team members have a pretty good idea of how often formal improvement training is necessary as long as they understand expected performance levels. (Here's another place where, if you're consulting your people, you should be taking credit for employee participation.)

What should be covered in refresher training? In Chapter 21, we suggested that continuing training should include problems identified during self-assessment, new equipment training, emergency preparedness, infrequent and abnormal conditions, lessons learned from mistakes (yours and others), emergent problems, operating proficiency, and recertification. Appendix C offers refresher training topic guidance:

> *Careful consideration must be given to assure that employees including maintenance and contract employees receive current and updated training. For example, if changes are made to a process, impacted employees must be trained in the changes and understand the effects of the changes on their job tasks (e.g., any new operating procedures pertinent to their tasks). Additionally, as already discussed the evaluation of the employee's absorption of training will certainly influence the need for training.*

Note that maintenance team members and contract employees—team members who are often not considered—must also be kept up-to-date, particularly regarding equipment changes.

A final comment on this topic: Don't neglect refresher training! Your team will either excel or languish based upon how you deal with this need. Remember, we wouldn't even consider neglecting practice for professional sports teams. How can we consider it for industrial teams?

Training Documentation   The final legal requirement of paragraph (g)(3) addresses the need for training documentation. "The employer shall prepare a record which contains the identity of the employee, the date of training, *and the means used to verify that the employee understood the training* [our emphasis]." In other words, you need proof that the training was effective. For knowledge requirements, a written or oral test is indicated. For skill requirements, a practical examination is appropriate. When the tests are oral, a valid examiner's signature is suitable.

You've got to stay on top of training records. It's little different than keeping a checkbook balanced—keep it current and it won't be a problem. As we stated in Chapter 6, the greatest burden (and often the weakest area) in training record management is keeping records current. To overcome that problem, the record-keeping process must be "user-friendly" and provide for easy review and revision as well as security from unauthorized access.

Periodic Evaluation   Periodic training evaluation is a process that is *not* prescribed in paragraph (g) as a part of the law. Yet, in its absence, training tends to degrade to a poor state. And we're not talking about a simple review of the training records. The issue here is whether people can *perform*. If they can't, training isn't working. Remember Rickover's sixth tenet of training:

Continual performance review of individuals and the entire operating team through on-the-job proficiency evaluation, inspection, and formal operational readiness testing.

Appendix C provides non-mandatory guidance on periodic evaluation:

Employers need to periodically evaluate their training programs to see if the necessary skills, knowledge, and routines are being properly understood and implemented by their trained employees. The means or methods for evaluating the training should be developed along with the training program goals and objectives. Training program evaluation will help employers to determine the amount of training their employees understood, and whether the desired results were obtained. If, after the evaluation, it appears that the trained employees are not at the level of knowledge and skill that was expected, the employer will need to revise the training program, provide retraining, or provide more frequent refresher training sessions until the deficiency is resolved. Those who conducted the training and those who received the training should also be consulted as to how best to improve the training process.

Training evaluation is an important part of every leader's job. Though not specified in this section of law, it remains fundamental to the systematic approach. And, to say that you don't have time is an admission of defeat.

A Final Thought

It's easy to neglect training when doing so seems to bear no immediate consequences. But inadequate training played a key role in literally every accident that we studied. We suspect that it was also a factor in the Phillips explosion and fire.

Don't leave this element out. Training is clearly a critical part of process safety management.

Contractors

Most industrial operations out-source some work, particularly in the areas of construction, modification, and specialized maintenance. Few have the resources afforded to our armed forces which must, of necessity, be self-contained. For an industrial organization to maintain a staff capable of managing all functions would be prohibitively expensive.

Therefore, occasions will arise in which contract employees—members of other organizations temporarily employed to accomplish limited objectives—must be engaged to assist the host company. Accordingly, the issue of this section of law is simply this: The host employer must apply reasonable measures to ensure that outsiders working in safety-related areas of hazardous chemical process operations and maintenance do so safely. Here's what the law says:

1910.119(h)

(h) **Contractors.**

(1) Application. This paragraph applies to contractors performing maintenance or repair, turnaround, major renovation, or specialty work on or adjacent to a covered process. It does not apply to contractors providing incidental services which do not influence process safety, such as janitorial work, food and drink services, laundry, delivery or other supply services.

(2) Employer responsibilities.

(i) The employer, when selecting a contractor, shall obtain and evaluate information regarding the contract employer's safety performance and programs.

(ii) The employer shall inform contract employers of the known potential fire, explosion, or toxic release hazards related to the contractor's work and the process.

(iii) The employer shall explain to contract employers the applicable provisions of the emergency action plan required by paragraph (n) of this section.

(iv) The employer shall develop and implement safe work practices consistent with paragraph (f) (4) of this section, to control the entrance, presence and exit of contract employers and contract employees in covered process areas.

(v) The employer shall periodically evaluate the performance of contract employers in fulfilling their obligations as specified in paragraph (h) (3) of this section.

(vi) The employer shall maintain a contract employee injury and illness log related to the contractor's work in process areas.

(3) Contract employer responsibilities.

(i) The contract employer shall assure that each contract employee is trained in the work practices necessary to safely perform his/her job.

(ii) The contract employer shall assure that each contract employee is instructed in the known potential fire, explosion, or toxic release hazards related to his/her job and the process, and the applicable provisions of the emergency action plan.

(iii) The contract employer shall document that each contract employee has received and understood the training required by this paragraph. The contract employer shall prepare a record which contains the identity of the contract employee, the date of training, and the means used to verify that the employee understood the training.

(iv) The contract employer shall assure that each contract employee follows the safety rules of the facility including the safe work practices required by paragraph (f) (4) of this section.

(v) The contract employer shall advise the employer of any unique hazards presented by the contract employer's work, or of any hazards found by the contract employer's work.

Which Contractors?

The first paragraph of this section—paragraph (h)(1)—makes it clear that these constraints do not apply to contractors not directly involved in process operations, maintenance, or safety. It is addressed, rather, to those engaged in "maintenance or repair, turnaround [outages], major renovation, or specialty work on or adjacent to a covered process." (Recall that a *covered process* merely means a hazardous chemical process covered by this regulation.)

Does that mean that contractors engaged in custodial work, food services, or laundry assistance needn't be apprised of the indigenous hazards on site? Of course not. As a part of emergency preparedness (and as a part of hazards communication) they require general employee training and sufficient communication to fulfil their roles safely. Remember that, if they are not treated like team members, they will not behave like team members.

Employer Responsibilities

Responsibility for contract employees is shared between the host employer and the contract employer. Therefore, the same chain of authority problems that we discussed in Chapter 16 regarding the operations/maintenance interface and the operations/testing interface will arise. As we indicated then, support agreements and policies should be established beforehand to ensure that the lines of authority and communication are clearly understood.

Paragraph (h)(2) mandates host employers to comply with six specific requirements. The first is to "obtain and evaluate information regarding the contract employee's safety performance and programs." Even without the law, this is a sensible thing to do. You certainly don't want poor performers on your job site if you can reasonably determine beforehand that they are apt to be unsafe. Further, the host contractor will probably not be exonerated of liability if contract employees cause injury or damage to the operating staff, the environment, or the general public. Appendix C provides this guidance:

> *Employers who use contractors to perform work in and around processes that involve highly hazardous chemicals, will need to establish a screening process so that they hire and use contractors who accomplish the desired job tasks without compromising the safety and health of employees at a facility.... For contractors whose safety performance on the job is not known to the hiring employer, the employer will need to obtain information on injury and illness rates and experience and should obtain contractor references.*

Nor is a review of the contractor's general record enough. The host employer should also ensure that specific qualifications and certifications for specialized work are current and valid. Appendix C:

> *Additionally, the employer must assure that the contractor has the appropriate job skills, knowledge and certifications (such as for pressure vessel welders). Contractor work methods and experiences should be evaluated.*

The second requirement of paragraph (h)(2) requires the employer to "inform contract employers of the known potential fire, explosion, or toxic release hazards related to the contractor's work and the process." In other words, contract employees need to know the dangers to which their working environment exposes them. Again, it makes no sense not to do this. Ignorance may be bliss, but knowledge, skills, and vigilance are what prevent accidents.

The third requirement is closely akin to the second. It states that the host employer must "explain to contract employers the applicable provisions of the emergency action plan." Not only do you want your contract employees to know what the hazards are, but, in case of emergencies, you want them to know what to do.

The fourth requirement compels the host employer to "develop and implement safe work practices…to control the entrance, presence and exit of contract employers and contract employees in covered process areas." Appendix C provides some guidance on this issue, emphasizing the process of work control:

> *Contract employees must perform their work safely. Considering that contractors often perform very specialized and potentially hazardous tasks such as confined space entry activities and nonroutine repair activities it is quite important that their activities be controlled while they are working on or near a covered process. A permit system or work authorization system for these activities would also be helpful to all affected employers. The use of a work authorization system keeps an employer informed of contract employee activities, and as a benefit the employer will have better coordination and more management control over the work being performed in the process area. A well run and well maintained process where employee safety is fully recognized will benefit all of those who work in the facility whether they be contract employees or employees of the owner.*

Performance evaluation is the fifth requirement of paragraph (h)(2). This paragraph requires host employers to "periodically evaluate the performance of contract employers in fulfilling their obligations." Again, even without the law, you're making a mistake if you're not overseeing contract employee work. If your own team members (who are thoroughly familiar with the environment and hazards) make mistakes, it is likely that your temporary team members will not do any better. In fact, they probably won't do as well.

Finally, paragraph (h)(2)(vi) requires host employers to "maintain a contract employee injury and illness log related to the contractor's work in process areas." Appendix C elaborates:

> *Maintaining a site injury and illness log for contractors is another method employers must use to track and maintain current knowledge of work activities involving contract employees working on or adjacent to covered processes. Injury and illness logs of both the employer's employees and contract employees allow an employer to have full knowledge of process injury and illness experience. This log will also contain information which will be of use to those auditing process safety management compliance and those involved in incident investigations.*

Contract Employer's Responsibilities

As we already stated, the responsibility for contract employee performance resides with both host and contract employers. Just as when a specialized military unit is temporarily attached to a foster-parent organization, a contract organization is answerable to its host. The host has overall responsibility; yet, the contract leaders must exercise immediate supervisory control and keep the host informed. It's a two-way street. Accordingly, paragraph (h)(3) prescribes four requirements for contract employers.

Training is the subject of the first requirement. Paragraph (h)(3)(i) states that contract employers must ensure that "each contract employee is trained in the work practices necessary to safely perform his/her job." That's not unreasonable. In fact, we would expect no less. If, as a contractor, I claim that my team members possess nuclear-grade welding skills, I had better make sure that my team members are skilled nuclear-grade welders (with the requisite certifications to prove it). The host employer has the responsibility to check my *bona fides* and frequently observe to make sure the specified job is being properly performed.

The second requirement coincides with paragraphs (h)(2)(ii) and (iii), mandating that the contract employer ensure (along with the host employer) that "each contract employee is instructed in the known potential fire, explosion, or toxic release hazards related to his/her job and the process, and the applicable provisions of the emergency action plan." In other words, the contract employer must ensure that, once informed of inherent job site hazards, those dangers are effectively communicated to contract employees. Again, we should expect no less.

Dissecting the Law

In the third requirement, the burden for contract employee training documentation is placed (as it should be) upon the contract employer. Note that training documentation requirements for host and contract employers are the same.

The fourth requirement of paragraph (h)(3) places immediate responsibility for supervising contract employees on their own leaders. "The contract employer shall assure that each contract employee follows the safety rules of the facility including the safe work practices required by paragraph (f)(4) of this section." The host employer will always bear final responsibility, but the contract employer is charged with detailed supervision and control.

The final requirement of paragraph (h)(3) mandates that contract employers inform host employers of "unique hazards" presented by the work. For example, if boiler cleaning is contracted, and the boiler cleaning team must use hazardous chemical agents to accomplish their work, the contract employer must so inform the host employer. Failure to inform deprives the host of information necessary to protect the operating staff, the environment, and the public from potential hazards—a lawful duty. Further, the contract employer is obligated by this paragraph to inform the host of new or unexpected hazards encountered as a part of work performance, a reasonably requirement.

You're Responsible! The bottom line of this segment of the regulation is that, as a host employer, you must ensure that everyone who works on your site—whether they belong to you or not—is informed and controlled.

That seems not to be the case in the Phillips 66 Houston Complex accident. Maintenance oversight and control was deficient. Closer teamwork and a more disciplined operations/maintenance interface well may have prevented that disaster.

Pre-Startup Safety Review

Before newly constructed (or significantly modified) industrial facilities are placed into operation, a final engineering and operational review is in order. The purpose of the review is to reasonably ensure that the equipment, procedures, and team members can efficiently function to accomplish the mission while protecting the operating staff, the environment, and the general public.

OSHA's process safety management standard recognizes the need for such an assessment—called a *pre-startup safety review*—in paragraph (i). Let's examine the requirements for pre-startup safety review and determine whether they are reasonable requirements. Here's what the law says:

1910.119(i) (i) Pre-startup safety review.
(1) The employer shall perform a pre-startup safety review for new facilities and for modified facilities when the modification is significant enough to require a change in the process safety management information.
(2) The pre-startup safety review shall confirm that prior to the introduction of highly hazardous chemicals to a process:
  (i) Construction and equipment is in accordance with design specifications;
  (ii) Safety, operating, maintenance, and emergency procedures are in place and are adequate;
  (iii) For new facilities, a process hazard analysis has been performed and recommendations have been resolved or implemented before startup; and modified facilities meet the requirements contained in management of change, paragraph (l).
  (iv) Training of each employee involved in operating a process has been completed.

What Is It? A **pre-startup safety review** is a specialized form of process safety review designed as a final check of equipment, procedures, and people prior to starting a new or significantly modified hazardous chemical process. It should, as the law indicates, reasonably verify that facilities and equipment are constructed in accordance with design specifications. Further, it must determine if supporting procedures are in place that, if followed, will assure safe operation and maintenance. (In other words, do the procedures support the design?) It must also evaluate whether requisite training for process operations (and maintenance) have been accomplished. Finally, the pre-startup review has to determine whether the hazard analysis is complete and safety questions resolved.

When Is It Required? As paragraph (i) states, a pre-startup safety review is necessary for new process startups and for startups following significant modification of components or systems. Appendix C states that:

> For existing processes that have been shutdown for turnaround, or modification, etc., the employer must assure that any changes other than "replacement in kind" made to the process during shutdown go through the management of change procedures.

The term *replacement in kind* means that repairs have been effected with materials and components which meet or exceed the original design specifications *and introduce no changes in protection*. Its meaning corresponds to our definition of *corrective maintenance* from Chapter 3. We stated that the purpose of corrective maintenance is to restore equipment to original operating specifications when parts or components fail. Unless new data indicates a flaw in the original protection analysis, no pre-startup review is necessary for replacement-in-kind repairs. The latest safety review should suffice.

Modification, you will recall, differs from maintenance. As we said in Chapter 3, modification changes the design specifications. If safety-related process equipment has been significantly modified, you must reevaluate protection. A pre-startup safety review is necessary.

How Extensive Should It Be? If a pre-startup safety review is required, you may wonder how extensive it should be. The answer is that it should be extensive enough to assure that protection is maintained.

Do you recall our earlier discussion of *graded approach*? The complexity of the process safety review will correlate well with the complexity and risk associated with the new process (or process modifications).

Check the Design In Chapter 3, we learned that reliable equipment and facilities—the second element in the strategy for operating success—result from conservative design, superior construction, and precise maintenance and modification. Naturally, then, design review is a fundamental step toward new process startup.

Sometimes, however, modification of an established facility is not subjected to the same scrutiny of design review, a potentially dangerous mistake. The systematic approach teaches us that modifications must be carefully engineered, reviewed, and tested to ensure that they don't introduce new hazards or negate other protective features. Always, the crucial question following modification is, "Has protection been maintained?"

Check the Procedures We also learned in Chapter 3 that legitimate policies and procedures form the interface between machines, processes, and their human controllers. In their absence, operators must "guess" at equipment characteristics, features and guidance necessary to run the equipment.

Consequently, a pre-startup safety review is incomplete without an assessment of the implementing procedures. Appendix C advises us that:

> The initial startup procedures and normal operating procedures need to be fully evaluated as part of the pre-startup review to assure a safe transfer into the normal operating mode for meeting the process parameters.

And, just as with design review, if a process modification has been introduced, the accompanying procedures must be validated prior to startup.

Check the Team Members Finally, a pre-startup safety review is deficient without an evaluation of the preparedness of team members to operate. Paragraph (i)(2)(iv) adjures the employer to ensure that, prior to startup, "Training of each employee involved in operating a process is completed." Appendix C adds:

> If the changes made to the process during shutdown are significant and impact the training program, then operating personnel as well as employees engaged in routine and nonroutine work in the process area may need some refresher or additional training in light of the changes.

Unfortunately, the requirement to check team member readiness is too often met with a simple training record audit. Wise leaders eventually learn, however, that record review is insufficient. Therefore, a well-devised *operational readiness review* is advisable prior to startup. A properly conducted readiness review tests individual and team skills through real-time performance checks, incorporating response to infrequent and abnormal operating conditions, emergency and casualty situations, and new equipment operation.

Dissecting the Law

Operational readiness reviews have been a part of miliary operations for decades. Many public and private sector organizations are also learning their value. The concept of pre-startup safety review is closely akin to readiness review when used to check readiness of people and procedures as well as design and material. It fits nicely into the thought process surrounding systematic industrial operations and is a critical part of process safety management.

Mechanical Integrity

As we recently stated, equipment reliability is founded, in great part, upon conservative design and superior construction. We also found that equipment can't remain reliable unless properly maintained and, when necessary, intelligently modified.

The components of reliability are addressed in paragraph (j) of the process safety management standard under the title *mechanical integrity*. Appendix C explains:

> *Equipment used to process, store, or handle highly hazardous chemicals needs to be designed, constructed, installed and maintained to minimize the risk of releases of such chemicals. This requires that a mechanical integrity program be in place to assure the continued integrity of process equipment.*

To ensure mechanical integrity, the law requires written maintenance procedures, training for those engaged in maintenance activities, equipment inspection and testing, and a deficiency identification/correction process. Let's examine mechanical integrity requirements in more detail. Here's what the law says:

1910.119(j)

(j) **Mechanical integrity.**
(1) Application. Paragraphs (j)(2) through (j)(6) of this section apply to the following process equipment:
  (i) Pressure vessels and storage tanks;
  (ii) Piping systems (including piping components such as valves);
  (iii) Relief and vent systems and devices;
  (iv) Emergency shutdown systems;
  (v) Controls (including monitoring devices and sensors, alarms, and interlocks) and,
  (vi) Pumps.
(2) Written procedures. The employer shall establish and implement written procedures to maintain the on-going integrity of process equipment.
(3) Training for process maintenance activities. The employer shall train each employee involved in maintaining the on-going integrity of process equipment in an overview of that process and its hazards and in the procedures applicable to the employee's job tasks to assure that the employee can perform the job tasks in a safe manner.
(4) Inspection and testing.
  (i) Inspections and tests shall be performed on process equipment.
  (ii) Inspection and testing procedures shall follow recognized and generally accepted good engineering practices.
  (iii) The frequency of inspections and tests of process equipment shall be consistent with applicable manufacturers' recommendations and good engineering practices, and more frequently if determined to be necessary by prior operating experience.
  (iv) The employer shall document each inspection and test that has been performed on process equipment. The documentation shall identify the date of the inspection or test, the name of the person who performed the inspection or test, the serial number or other identifier of the equipment on which the inspection or test was performed, a description of the inspection or test performed, and the results of the inspection or test.
(5) Equipment deficiencies. The employer shall correct deficiencies in equipment that are outside acceptable limits (defined by the process safety information in paragraph (d) of this section) before further use or in a safe and timely manner when necessary means are taken to assure safe operation.
(6) Quality assurance.
  (i) In the construction of new plants and equipment, the employer shall assure that equipment as it is fabricated is suitable for the process application for which they will be used.

(ii) Appropriate checks and inspections shall be performed to assure that equipment is installed properly and consistent with design specifications and the manufacturer's instructions.

(iii) The employer shall assure that maintenance materials, spare parts and equipment are suitable for the process application for which they will be used.

What Is Process Equipment?

As you can see, this area of law focuses predominantly upon the need for comprehensive maintenance, testing, and quality assurance for process equipment. But, where do you draw the line? What is process equipment and what isn't?

Paragraph (j)(1) identifies process equipment as pressure vessels and tanks; pumps, piping, and relief systems; instrument, alarm, and control systems; and emergency shutdown systems—in other words, all equipment (primarily installed equipment) used to route or control hazardous chemicals.

Sometimes, however, lists tend to rule out common sense. A short-sighted analyst might say, "If it's not on the list, I'm not including it as process equipment." So here's some guidance from the CCPS's **Guidelines** for Technical Management of Chemical Process Safety (Chapter 8, Process and Equipment Integrity):

> *Any equipment used to handle or process hazardous materials should be designed, built, installed, and maintained to control the risk of releases and other accidents. Of primary concern are those components that could release hazardous materials as a result of a single failure, or those components whose operation is critical during an equipment or system malfunction.*

We like the CCPS definition better than the list. Test your classification with a little common sense. Ask, "Is this component or system an important part of process control?" If so, it's process equipment. That means it should receive a high priority in care and maintenance.

Achieving Mechanical Integrity

Achieving mechanical integrity (or *process equipment integrity* as CCPS prefers) requires skilled team members designing, building, installing, inspecting, testing, repairing, replacing, and, if necessary, modifying process equipment so that its design functions remain intact. Appendix C provides some very useful guidance on what it takes to keep process equipment in top shape:

> *Elements of a mechanical integrity program include the identification and categorization of equipment and instrumentation, inspections and tests, testing and inspection frequencies, development of maintenance procedures, training of maintenance personnel, the establishment of criteria for acceptable test results, documentation of test and inspection results, and documentation of manufacturer recommendations as to meantime to failure for equipment and instrumentation.*

Master Equipment List

Please notice that the beginning of mechanical integrity is knowing what you own. Without first identifying and categorizing process equipment, you can never get a handle on maintaining it. So, anyone who has been in the maintenance business for very long will tell you that you need a *master equipment list*.

A **master equipment list**, as the name implies, is a complete list of equipment for which an operating and maintenance staff is responsible. The list creates an index upon which an equipment maintenance schedule can be built. Appendix C explains:

> *The first step of an effective mechanical integrity program is to compile and categorize a list of process equipment and instrumentation for inclusion in the program. This list would include pressure vessels, storage tanks, process piping, relief and vent systems, fire protection system components, emergency shutdown systems and alarms and interlocks and pumps.*

Categorizing Equipment

When thoughtfully devised, a master equipment list categorizes its constituent components and systems by importance. Importance is determined by the answers to questions such as, "If this component or system breaks, what dangers does it pose to the operating staff, the environment, and the general public? At what frequency does history show it to fail? How much does it cost to procure, install, and test? How often does it need to be inspected and serviced?"

Appendix C describes some factors affecting categorization:

> *For the categorization of instrumentation and the listed equipment the employer would prioritize which pieces of equipment require closer scrutiny than others. Mean time to failure of various instrumentation and equipment parts would be known from the manufacturers data or the employer's*

Dissecting the Law

experience with the parts, which would then influence the inspection and testing frequency and associated procedures. Also, applicable codes and standards such as the National Board Inspection Code, or those from the American Society for Testing and Material, American Petroleum Institute, National Fire Protection Association, American National Standards Institute, American Society of Mechanical Engineers, and other groups, provide information to help establish an effective testing and inspection frequency, as well as appropriate methodologies.

Preventive Maintenance

With a comprehensive master equipment list in hand, you are at the starting point for developing a good preventive maintenance program. You probably remember from Chapter 3 that the purpose of preventive maintenance is to detect and forestall degradation so that equipment maintains its original operating specifications. The idea is to find potential equipment problems before they develop into something more serious.

Preventive maintenance is accomplished primarily through the use of regularly scheduled inspections and tests. By observing and checking (most often using methods and tools prescribed in the manufacturer's technical manual), maintenance technicians can judge the state of components and systems, determining whether further care is needed.

Accordingly, paragraph (j)(4)—entitled "Inspection and Testing"—mandates that inspections and tests be performed at least at minimum frequencies prescribed by the manufacturer (more often if history proves necessary), employing "recognized good engineering practices", and documented to keep the pedigree of components and systems current.

But how do you know whether your preventive maintenance program is working? Breakdowns is a good indicator. When preventive maintenance is working, corrective maintenance is reduced. Therefore, team leaders need to look for maintenance trends. If certain components or systems are failing with regularity, perhaps they are in need of preventive maintenance rather than corrective (breakdown) maintenance.

Corrective Maintenance

Even with the best preventive maintenance, corrective maintenance will still be required. Components and systems sometimes fail unexpectedly or just plain wear out.

The process safety management standard addresses corrective maintenance in paragraph (j)(5) under the title "Equipment Deficiencies". It simply requires that "The employer shall correct deficiencies in equipment that are outside acceptable limits…before further use or in a safe and timely manner when necessary means are taken to assure safe operation."

As the law implies, all deficiencies don't warrant the same priority. In Chapter 14, we said that, though some discrepancies can (and should) be corrected at the time of discovery, others, for want of time, parts, or capability must wait. In those cases, deficiencies must be noted and reported for future scheduling and disposition. Therefore, in addition to knowledgeable technicians and leaders, an industrial complex must have in place a reliable and user-friendly system for reporting deficient conditions.

Obviously, a deficiency must be evaluated to determine the urgency with which it should be addressed. Some need to be repaired before the equipment is used again. Others can wait. Remember the categorization scheme in the master equipment list? That's one of the tools you should use to determine repair priorities.

Maintenance Training

Clearly, preventive and corrective maintenance will not be successful unless the team members assigned those tasks are both knowledgeable and skilled. As we would expect, the process safety management standard requires training (an "overview" as well as specific task and procedure training) for those engaged in maintenance. Appendix C offers this guidance:

> *Appropriate training is to be provided to maintenance personnel to ensure that they understand the preventive maintenance program procedures, safe practices, and the proper use and application of special equipment or unique tools that may be required. This training is part of the overall training program called for in the standard.*

But, don't forget maintenance training for your operators as well. As we indicated in Chapter 14, they are a critical link in the maintenance process. Before corrective repairs can ever take place, they must first be identified. Equipment operators, if well-trained, are usually in the best position to detect deficiencies. We concluded in Chapter 14 that an important tenet of sound maintenance is that material deficiencies must be recognized and reported in a timely manner so that they can be promptly repaired (or scheduled for repair when conditions permit.)

Written Procedures

The process safety management standard would be remiss if it did not address maintenance procedures as a part of mechanical integrity. We don't expect to operate without procedures nor should we expect to repair without them.

In Chapter 9, we said that when properly written and used, procedures form an important line of defense to ensure that equipment is operated and maintained within design specifications—within the boundaries of the safe operating envelope. In fact, good procedures are the foundation for effective training. Without them comprehensive qualification of operators and maintenance technicians is, at best, difficult.

Quality Assurance

The final requirement in the mechanical integrity section is for a quality assurance program. Appendix C explains:

> *A quality assurance system is needed to help ensure that the proper materials of construction are used, that fabrication and inspection procedures are proper, and that installation procedures recognize field installation concerns. The quality assurance program is an essential part of the mechanical integrity program and will help to maintain the primary and secondary lines of defense that have been designed into the process to prevent unwanted chemical releases or those which control or mitigate a release.*

Perhaps a word of caution is in order here. We believe in quality assurance and quality control departments. But the systematic approach (and process safety management) depends upon every team member upholding quality. As Dr. Deming so often reminded us, quality is everyone's job. It doesn't result solely from the efforts of a QA department. It can't be inspected in—it has to be built in.

In-Depth Defense

Mechanical integrity is certainly a critical part of process safety management. But, in the systematic approach, we learned that no one element of the strategy for operating success was sufficient for protection. All are necessary for an in-depth defense. Appendix C offers this excellent summary perspective on the role of mechanical integrity in relation to process safety management:

> *The first line of defense an employer has available is to operate and maintain the process as designed, and to keep the chemicals contained. This line of defense is backed up by the next line of defense which is the controlled release of chemicals through venting to scrubbers or flares, or to surge or overflow tanks which are designed to receive such chemicals, etc. These lines of defense are the primary lines of defense or means to prevent unwanted releases. The secondary lines of defense would include fixed fire protection systems like sprinklers, water spray, or deluge systems, monitor guns,…dikes, designed drainage systems, and other systems which would control or mitigate hazardous chemicals once an unwanted release occurs.*

At the Phillips complex, the primary defensive barrier failed when the maintenance team failed to appropriately isolate the Number 4 settling leg, resulting in an uncontrolled release to atmosphere. Unfortunately, fire suppression, the secondary line of defense, was not at full capability, a serious defect in protection. Accident investigators found that:

> *The fire protection system was not maintained in a state of readiness necessary to provide effective firefighting capability. Unknown to the fire chief, one of three emergency standby diesel-powered water pumps had been taken out of service, and another was not fully fueled, with the result that it ran out of fuel during firefighting activities. Further, electric cables supplying power to regular service fire pumps were not located underground, thereby exposing them to blast damage.*

There's no question—mechanical integrity is critical. But, as we have stated repeatedly, the best design and construction can never substitute for good people. All of the elements of the strategy must work together to provide an in-depth defense against hazards while, at the same time, efficiently accomplishing the mission.

Dissecting the Law

Hot Work Permit

As we learned in Chapter 16, equipment maintenance, particularly when performed on installed components and systems, requires close coordination between the operating and maintenance staffs. We stated that maintenance, modification, and testing are important team processes shared by operating and supporting teams—all of whom must cooperate closely to ensure that equipment is installed, repaired, restored, and operated within design specifications. Their efforts must be coordinated by the leaders of the cooperating agencies and, ultimately, by one leader responsible for the entire organization. We even referred to maintenance as a "team sport".

To help coordinate work control, most organizations use work authorization permits which, after having been reviewed and approved by operations, are issued to grant permission for specified work to commence. Appendix C explains it this way:

> *Nonroutine work which is conducted in process areas needs to be controlled by the employer in a consistent manner. The hazards identified involving the work that is to be accomplished must be communicated to those doing the work, but also to those operating personnel whose work could affect the safety of the process. A work authorization notice or permit must have a procedure that describes the steps the maintenance supervisor, contractor representative or other person needs to follow to obtain the necessary clearance to get the job started. The work authorization procedures need to reference and coordinate, as applicable, lockout/tagout procedures, line breaking procedures, confined space entry procedures and hot work authorizations. This procedure also needs to provide clear steps to follow once the job is completed in order to provide closure for those that need to know the job is now completed and equipment can be returned to normal.*

Paragraph (k) codifies one particular type of work authorization—the *hot work permit*. It is a work authorization granting permission to perform cutting, welding, grinding, or other high heat-producing processes in a specified area (usually at a specified time).

Hot work permits are designed to minimize the potential for fire, an event that frequently leads to significant damage to equipment and people. (In Chapter 15, we defined such an event as an *equipment casualty*.) The permit most often will require an inspection of the area in which the work is to be performed to determine that flammable materials have either been removed (or at least minimized), and that the work will be controlled to prevent fires. Also, a check of the atmosphere may be required to sample for flammable or explosive gases such as hydrogen.

Let's take a look at the law and see if we agree with its requirements. Here's what the law says:

1910.119(k)

(k) **Hot work permit.**

(1) The employer shall issue a hot work permit for hot work operations conducted on or near a covered process.

(2) The permit shall document that the fire prevention and protection requirements in 29 CFR 1910.252(a) have been implemented prior to beginning the hot work operations; it shall indicate the date(s) authorized for hot work; and identify the object on which hot work is to be performed. The permit shall be kept on file until completion of the hot work operations.

Fire Prevention and Protection

Rather than delineate each hot work requirement, this section of the law simply defers to OSHA Standard 1910.252(a) entitled "Fire Prevention and Protection". A quick review of that standard shows paragraph (a) to have four main subparagraphs—basic precautions, special precautions, welding and cutting containers, and confined spaces. The requirements are straightforward and probably familiar to most readers, so we will leave it for you to review on your own.

One part of "special precautions", however, deserves a moment of attention. Paragraph (a)(2)(iii) is a section on *fire watch*. It states that:

> *Fire watchers shall be required whenever welding or cutting is performed in locations where other than a minor fire might develop....*

Sometimes the job of fire watch can be rather boring. But a moment of inattentiveness can lead to a full-fledged fire. And once they get started, they're very hard to extinguish. Don't let your fire watches be lax. Teach them the importance of this seemingly inconsequential job.

Beyond Hot Work Permits

As we saw in the Phillips accident, explosions and fires are devastating casualties, usually accompanied by loss of life and injury. That's why hot work permits are so important. But, hot work permits are only part of the safe work permit picture. Safe electrical work, rigging jobs, confined space operations, and a host of other hazardous activities rest on the same concept and principles. Competent people from both maintenance and operations must evaluate the work that has been proposed and what difficulties it might present. Work authorization controls such as hot work permits or electrical safety checklists provide a forum for making sure you are protected.

Management of Change

Mechanical integrity may be seriously degraded (or completely lost) when changes in equipment or processes are introduced without appropriate analysis, review, approval, documentation, and training. Therefore, OSHA's process safety management standard mandates that changes to hazardous chemical processing equipment, systems, and procedures must be closely controlled. The requirements for change control are termed *management of change*.

The Definition of Change

Management of change (as described in the process safety management standard) is, for the most part, synonymous with the term *modification* as it was used in our recent study of systematic industrial operations. Both should be seen in contrast to maintenance or replacement in kind.

In Chapters 3 and 16, we drew a sharp distinction between the terms *maintenance* and *modification*. We said that maintenance refers to the restoration of components and systems to original design specifications. Modification, on the other hand, changes the design specifications. We said that, as a result, modifications must be carefully engineered, reviewed, and tested to ensure that they don't introduce new hazards or negate other protective features.

The non-mandatory guidance of Appendix C provides this comprehensive definition and explanation:

> *To properly manage changes to process chemicals, technology, equipment and facilities, one must define what is meant by change. In this process safety management standard, change includes all modifications to equipment, procedures, raw materials and processing conditions other than "replacement in kind." These changes need to be properly managed by identifying and reviewing them prior to implementation of the change. For example, the operating procedures contain the operating parameters (pressure limits, temperature ranges, flow rates, etc.) and the importance of operating within these limits. While the operator must have the flexibility to maintain safe operation within the established parameters, any operation outside of these parameters requires review and approval by a written management of change procedure.*

Appendix C goes on to explain that changes in process equipment and technology can occur in many different ways:

> *Management of change covers...changes in process technology and changes to equipment and instrumentation. Changes in process technology can result from changes in production rates, raw materials, experimentation, equipment unavailability, new equipment, new product development, change in catalyst and changes in operating conditions to improve yield or quality. Equipment changes include among others change in materials of construction, equipment specifications, piping pre-arrangements, experimental equipment, computer program revisions and changes in alarms and interlocks. Employers need to establish means and methods to detect both technical changes and mechanical changes.*

The Significance of Change

Some changes may not seem to warrant evaluation. For example, a simple increase in process feed flow might seem a trifling detail. But if the increase exceeds design specification, it may have serious effects. You probably recall from our study of the systematic approach that changes to design specifications are of particular concern since the protective scheme devised to fend against hazards is based upon the original design specifications. It takes a mature team to recognize that something which, on the surface, appears inconsequential may, in fact, have a major impact on operations.

Dissecting the Law

Temporary and Inadvertent Changes

Perhaps the most insidious of changes are those that occur inadvertently or are installed as "temporary". For example, a component, jumper, bypass, or interim alteration may be installed as a "jury rig" to expedite production. The reasoning goes something like this: "Yes, I know we're supposed to do an engineering evaluation by procedure, but it won't be in place long enough to matter." (If you have much experience in industrial operations, you know exactly what we're talking about!) Of course, if the temporary change doesn't cause an immediate problem, it will then be found years later by someone tracing systems using the as-built drawings which, for some unexplained reason, don't show the change!

Appendix C has this to say about temporary changes:

> *Temporary changes have caused a number of catastrophes over the years, and employers need to establish ways to detect temporary changes as well as those that are permanent. It is important that a time limit for temporary changes be established and monitored since, without control, these changes may tend to become permanent. Temporary changes are subject to the management of change provisions. In addition, the management of change procedures are used to insure that the equipment and procedures are returned to their original or designed conditions at the end of the temporary change. Proper documentation and review of these changes is invaluable in assuring that the safety and health considerations are being incorporated into the operating procedures and the process.*

Clearly, vigilance on the part of operators, maintenance technicians, engineers, and team leaders is necessary to ensure that changes to design specifications and protective features—whether administrative, procedural, or installed safeguards—do not "sneak" into the process. Notice also that it's not enough simply to have proper change controls in place to manage *proposed* temporary changes. Team members must also be able to discern undocumented changes that were installed long ago and have gone undetected.

With this improved understanding of change management, let's check the regulatory requirements. Here's what the law says:

1910.119(l)

(l)**Management of change**

(1) The employer shall establish and implement written procedures to manage changes (except for "replacements in kind") to process chemicals, technology, equipment, and procedures; and, changes to facilities that affect a covered process.

(2) The procedures shall assure that the following considerations are addressed prior to any change:

 (i) The technical basis for the proposed change;
 (ii) Impact of change on safety and health;
 (iii) Modifications to operating procedures;
 (iv) Necessary time period for the change; and,
 (v) Authorization requirements for the proposed change.

(3) Employees involved in operating a process and maintenance and contract employees whose job tasks will be affected by a change in the process shall be informed of, and trained in, the change prior to start-up of the process or affected part of the process.

(4) If a change covered by this paragraph results in a change in the process safety information required by paragraph (d) of this section, such information shall be updated accordingly.

(5) If a change covered by this paragraph results in a change in the operating procedures or practices required by paragraph (f) of this section, such procedures or practices shall be updated accordingly.

Codifying the Change Process

The first requirement of this regulation simply mandates that you have a formal, standardized procedure for managing equipment or process changes. But what should the procedure address? What tools should it provide? Appendix C provides some instructive ideas:

> *Employers may wish to develop a form or clearance sheet to facilitate the processing of changes through the management of change procedures. A typical change form may include a description and the purpose of the change, the technical basis for the change, safety and health considerations, documentation of changes for the operating procedures, maintenance procedures, inspection and testing, P&IDs, electrical classification, training and communications, pre-startup inspection, duration if a temporary change, approvals and authorization. Where the impact of the change is minor and well understood, a check list reviewed by an authorized person with proper communication to others who are affected may be sufficient. However, for a more complex or significant design change, a hazard evaluation procedure with approvals by operations,*

maintenance, and safety departments may be appropriate. Changes in documents such as P&IDs, raw materials, operating procedures, mechanical integrity programs, electrical classifications, etc., need to be noted so that these revisions can be made permanent when the drawings and procedure manuals are updated. Copies of process changes need to be kept in an accessible location to ensure that design changes are available to operating personnel as well as to PHA team members when a PHA is being done or one is being updated.

For detailed instruction and examples, we recommend that you refer to the guidance contained in **Plant Guidelines for Technical Management of Chemical Process Safety** [Revised Edition].

Analyzing Proposed Changes

Since modification implies a change to original equipment design, it usually requires an engineering study in advance of installation to ensure that personnel and equipment protection is maintained. Therefore, your management of change procedure must delineate how proposed equipment and process changes will be reviewed and approved.

Also, don't forget that, after installation, acceptance testing is often necessary to validate the modification. So, your procedure should specify administrative steps to determine if changes (as well as all systems affected by them) are functioning as predicted. There's too much at stake to allow this to be a loose process.

Altering Procedures and Drawings

Coincidental to installing changes, affected process safety information (such as as-built drawings) and operating/maintenance procedures must also be altered. Therefore, paragraphs (l)(4) and (l)(5) require that the change process include procedure changes and process safety information changes if necessary.

Training for Changes

Finally, the standard would be incomplete if it didn't also address training. Accordingly, paragraph (l)(3) requires suitable training to ensure that those affected by the change have the knowledge and skills necessary to safely operate and maintain their equipment.

Training, of necessity, should utilize the graded approach—with complexity and comprehensiveness matched to risk. If the change is minor, nothing more than a personal review of the change synopsis may be necessary. On the other hand, if the change is complex, having great impact on actions, more thorough training will obviously be required.

In devising the training, don't forget to consider infrequent and abnormal conditions. Appendix C reminds us that:

Training in how to handle upset conditions must be accomplished as well as what operating personnel are to do in emergencies such as when a pump seal fails or a pipeline ruptures.

Complex and Difficult

Before we leave this important section on management of change, we should remember that change management isn't an easy process. It takes a lot of time and it requires a well-skilled staff, working in conjunction with competent team members and leaders. Nevertheless, you can't achieve consistent mechanical integrity unless you establish and control the management of change process.

Incident Investigation

In our discussion of implementing the systematic approach, we identified abnormal event investigation (Chapter 20) as a principal consideration. We stated that the ability (and commitment) to investigate abnormal events is an important ingredient in implementing the strategy for operating success. Through aggressive investigation, the causes of things that went wrong (or things that almost went wrong) can be determined and others within the industrial team inoculated against similar errors.

Accordingly, OSHA's process safety management standard recognizes the need for incident investigation to identify causes of underlying events, using the lessons to educate team members in an effort to avoid repetition. Appendix C elaborates:

Incident investigation is the process of identifying the underlying causes of incidents and implementing steps to prevent similar events from occurring. The intent of an incident investigation is for employers to learn from past experiences and thus avoid repeating past mistakes.

Paragraph (m) establishes seven fundamental requirements for incident investigation. Let's review them. Here's what the law says:

1910.119(m) (m) **Incident investigation.**

(1) The employer shall investigate each incident which resulted in, or could reasonably have resulted in a catastrophic release of highly hazardous chemical in the workplace.

(2) An incident investigation shall be initiated as promptly as possible, but not later than 48 hours following the incident.

(3) An incident investigation team shall be established and consist of at least one person knowledgeable in the process involved, including a contract employee if the incident involved work of the contractor, and other persons with appropriate knowledge and experience to thoroughly investigate and analyze the incident.

(4) A report shall be prepared at the conclusion of the investigation which includes at a minimum:

    (i) Date of incident;

    (ii) Date investigation began;

    (iii) A description of the incident;

    (iv) The factors that contributed to the incident; and,

    (v) Any recommendations resulting from the investigation.

(5) The employer shall establish a system to promptly address and resolve the incident report findings and recommendations. Resolutions and corrective actions shall be documented.

(6) The report shall be reviewed with all affected personnel whose job tasks are relevant to the incident findings including contract employees where applicable.

(7) Incident investigation reports shall be retained for five years.

What to Investigate

In Chapter 20, we stated that every abnormal event is worthy of investigation. Yet, some require an in-depth, formal investigation while others can be adequately addressed with a simple, informal analysis conducted by an on-duty operator. When, then, is a formal investigation required? There is no perfectly objective answer. A good rule of thumb, however, is that a formal investigation should be conducted if an occurrence led to (or potentially could have caused) serious equipment damage or personnel injury.

Paragraph (m), however, does not mandate that every event worthy of being called an incident or an abnormal event must be formally investigated. Rather, it states that "The employer shall investigate each incident which resulted in, or could reasonably have resulted in a catastrophic release of highly hazardous chemical in the workplace." Appendix C offers this additional guidance:

> *The incidents for which OSHA expects employers to become aware and to investigate are the types of events which result in or could reasonably have resulted in a catastrophic release. Some of the events are sometimes referred to as "near misses", meaning that a serious consequence did not occur, but could have.*

It's usually clear when an event causes catastrophic damage. In those circumstances, there's little doubt that an investigation is warranted. The difficulty is in recognizing and investigating near misses. If an event results in no damage, leaders may be tempted to brush off the event without investigating. Certainly there is incentive to do so, because investigations are painful, expensive, and may cause you to learn things that you don't want to acknowledge. That, however, is a short-sighted view. Remember what we learned in Chapter 20. Most accidents are not first-time events. They have usually been preceded by a number of near misses—events which, under less conservative circumstances, may have resulted in damage to people or equipment. By investigating near misses, facility personnel are made aware of dangerous conditions, allowing an opportunity to adjust behavior or correct conditions.

Facing facts in the near term is usually far less expensive and painful than allowing an uncorrected problem to continue, perhaps leading to injury or loss of life later. It takes a mature organization to recognize that.

How Soon to Investigate

In Chapter 20, we learned that an investigation can only be as good as the evidence that supports it. And we all know by now that, if evidence (including witness statements) is not gathered and preserved soon after an event, the ability of an investigating team to determine what really happened will diminish as time passes.

Subsequently, the process safety management standard is very specific on how soon an investigation must start: "An incident investigation shall be initiated as promptly as possible, but not later than 48 hours following the incident." In other words, begin the investigation as soon as you recognize the event to be an incident worthy of formal investigation. With few exceptions, forty-eight hours should be enough time.

The Investigation Team

When constituting an investigation team, the standard mandates that the team shall consist of "at least one person knowledgeable in the process involved, including a contract employee if the incident involved work of the contractor, and other persons with appropriate knowledge and experience to thoroughly investigate and analyze the incident." Obviously, you have a wide latitude in team composition. The spirit of the law is simply to involve team members who will perform a competent, quality investigation.

Don't neglect the use of subject matter experts—people knowledgeable in the process—as the law indicates. We said in Chapter 20 that investigators may not have all the technical knowledge necessary to conduct the investigation alone, and thus may require assistance from subject matter experts—leaders and technicians currently qualified on the equipment and processes under investigation.

If you are to avoid untimely delays (and violations of this section of the standard), you must be prepared to investigate promptly. You've got to be able to pull an investigative team together quickly. Appendix C provides wise guidance:

> *Employers need to develop in-house capability to investigate incidents that occur in their facilities. A team needs to be assembled by the employer and trained in the techniques of investigation including how to conduct interviews of witnesses, needed documentation and report writing. A multi-disciplinary team is better able to gather the facts of the event and to analyze them and develop plausible scenarios as to what happened, and why. Team members should be selected on the basis of their training, knowledge and ability to contribute to a team effort to fully investigate the incident. Employees in the process area where the incident occurred should be consulted, interviewed or made a member of the team. Their knowledge of the events form a significant set of facts about the incident which occurred.*

We agree that a multi-disciplinary team is valuable. In Chapter 20, we said that without a thorough understanding of the event contributors, successful resolution is unlikely. Individuals alone are limited in perspective and, therefore, limited in their abilities to thoroughly identify contributors. Multiple perspectives are necessary. Multiple perspective is achieved through including in the investigation and resolution phases people from different backgrounds and disciplines.

We must remember, however, that investigation isn't an isolated activity conducted by a sequestered team. We learned in Chapter 20 that the practice of investigating and reporting abnormalities should be taught and expected as a job responsibility for all equipment operators and their leaders. Investigation is everyone's responsibility. It should begin as soon as an anomalous situation is discovered. Appendix C concurs:

> *The cooperation of employees is essential to an effective incident investigation.*

Accordingly, investigation—both formal and informal—must be fostered as an individual and team trait.

The Focus of Investigation

We stated in Chapter 20 that fact-finding is the investigation's purpose—to recount and clarify the details of the event in order to learn and prevent recurrence. We emphasized that the investigation must be (and must be perceived as) an objective, fact-finding process or future investigations will be jeopardized. Appendix C:

> *The focus of the investigation should be to obtain facts, and not to place blame. The team and the investigation process should clearly deal with all involved individuals in a fair, open and consistent manner.*

The Center for Chemical Process Safety views the investigation in a similar fashion, admonishing us that "incidents should be viewed as opportunities to improve management systems rather than as opportunities to assign blame."

If, on the other hand, investigations take on the aura of "witch hunts", people—the primary source of observation and reporting—will withdraw, clam up, and participate only to the point that coercion demands. So, be very careful to establish a climate of openness, objectivity, and team-mindedness in your investigations.

Report and Response

Paragraph (m)(4) requires that, following completion of the investigation, a report including, among other things, a description of the event, contributing factors, and recommendations for correction or improvement be compiled and issued. Obviously, to be valuable, the report should be accurate, objective, and timely.

If the investigation has been competently performed, the underlying causes will have been identified. It is upon those causes that the actions for correction and improvement should be focused. Accordingly, paragraph (m)(5) requires that the employer "establish a system to promptly address and resolve the incident report findings and recommendations."

Dissemination of Findings

A well-conducted investigation has little value unless the findings are intelligently disseminated. In Chapter 20, we stated that accidents are all the more tragic if they are not used to educate others in the mistakes that were made. Therefore, dissemination of *lessons learned* is a vital part of investigation.

Paragraph (m)(6) then mandates that findings be disseminated, stating that "The report shall be reviewed with all affected personnel whose job tasks are relevant to the incident findings including contract employees where applicable." Note that the law only requires that you share with those for whom you are responsible. But, don't be short-sighted. Others may have vital need of your lessons. If so (and with due regard for legal considerations), please share with others!

Report Retention Requirement

The final requirement of paragraph (m) is that investigative reports be retained (as a minimum) for five years.

Record requirements always seem onerous. Actually, though, five years isn't very long in the life of an industrial complex. If you are serious about improvement, a comprehensive incident investigation file (covering the entire life history of the process if possible) will be exceptionally useful in evaluating operating and maintenance trends.

Training Incident Investigators

In our study of the systematic approach, we concluded that incident investigation was a necessary implementing element. It is, therefore, no less important for achieving competent process safety management. And, if you're going to do it right, your investigators need to be trained. A good place to start is with the Center for Chemical Process Safety's discourse on incident investigation entitled **Guidelines for Investigating Chemical Process Incidents**.

Emergency Planning and Response

In Chapter 2, we identified inadequate emergency preparedness as a common contributor to accidents. We defined *emergency preparedness* as the readiness of an individual or an organization to successfully combat or control abnormal conditions. We also stated that emergency preparedness is the result of anticipating potential hazards (probable events capable of causing unacceptable damage), and then planning the means and the methods to counter them. Finally, we devoted all of Chapter 15 to the study of combatting emergencies and casualties, one of the twelve vital operating skills.

Not surprisingly, we find that OSHA believes emergency preparedness to be an important part of process safety management as well, requiring in paragraph (n) that employers running a hazardous chemical process covered under this standard to "establish and implement an emergency action plan."

Though the regulation appears to be short, it actually is one of the more complex requirements of the process safety management standard, referring to "Employee Emergency Plans and Fire Protection Plans" (OSHA 1910.38) and "Hazardous Waste Operations and Emergency Response" (OSHA 1910.120). Here's what the law says:

1910.119(n)

(n) Emergency planning and response. The employer shall establish and implement an emergency action plan for the entire plant in accordance with the provisions of 29 CFR

1910.38(a). In addition, the emergency action plan shall include procedures for handling small releases. Employers covered under this standard may also be subject to the hazardous waste and emergency response provisions contained in 29 CFR 1910.120(a), (p) and (q).

Must I Have an Emergency Plan?

If you operate a hazardous chemical process covered by the process safety management standard, you are required to have an emergency plan. The minimum standards for your plan are delineated in OSHA Standard 1910.38, the general regulation on emergency plans. Paragraph (a)(1)—scope and application—states that its requirements apply to "all emergency action plans required by a particular OSHA standard." It further states that the plan "shall be in writing…and shall cover those designated actions employers and employees must take to ensure employee safety from fire and other emergencies." Appendix C explains:

> *Employers at a minimum must have an emergency action plan which will facilitate the prompt evacuation of employees [following] an unwanted release of highly hazardous chemicals. This means that the employer will have a plan that will be activated by an alarm system to alert employees when to evacuate, and that employees who are physically impaired will have the necessary support and assistance to get them to the safe zone as well. The intent of these requirements is to alert and move employees to a safe zone quickly. Delaying alarms or confusing alarms are to be avoided.*

If, in addition, your process operation falls within the purview of OSHA Standard 1910.120 (HAZWOPER) regarding hazardous waste operations and emergency response, life becomes more complicated. Paragraphs (a), (p), and (q) of that standard address, respectively, the scope and application of Standard 120; emergency response for personnel operating at treatment, storage, and disposal facilities subject to the Resources Conservation and Recovery Act of 1976 (RCRA); and emergency response to hazardous substance releases. You need to review those sections of law thoroughly.

The emergency response plan and training, obviously, must be more comprehensive for people who must do more than simply evacuate an area. Appendix C offers the following guidance:

> *Preplanning for releases that are more serious than incidental releases is another important line of defense to be used by the employer. When a serious release of a highly hazardous chemical occurs, the employer through preplanning will have determined in advance what actions employees are to take. The evacuation of the immediate release area and other areas as necessary would be accomplished under the emergency action plan. If the employer wishes to use plant personnel such as a fire brigade, spill control team, a hazardous materials team, or use employees to render aid to those in the immediate release area and control or mitigate the incident, these actions are covered by 1910.120, the Hazardous Waste Operations and Emergency Response (HAZWOPER) standard. If outside assistance is necessary, such as through mutual aid agreements between employers or local government emergency response organizations, these emergency responders are also covered by HAZWOPER. The safety and health protection required for emergency responders are the responsibility of their employers and of the on-scene incident commander.*

Clearly, the requirements of OSHA Standard 1910.120 (HAZWOPER) for emergency response are stricter and they take precedence over 1910.38. We recommend that you review the two regulations closely to determine which applies to you.

What Must It Contain?

OSHA Standard 1910.38(a)(2) delineates six elements which, as a minimum, an emergency plan must contain: emergency escape procedures and escape route assignments; procedures for employees who must remain to combat an emergency; procedures to account for team members after an emergency evacuation has been initiated; assigned rescue and medical duties for designated team members; a formal method for reporting fires and emergencies; and names of contact personnel who can further explain the plan and its requirements. Alarm systems, evacuation plans, and emergency response training are also addressed.

Should it apply to you, OSHA Standard 1910.120(p)(8)(ii) includes several other required plan elements. Together, they include: pre-emergency planning and coordination with outside parties; personnel roles, lines of authority, training, and communication; emergency recognition and prevention; safe distances and places of refuge; site security and control; evacuation routes and procedures; decontamination procedures; emergency medical treatment and first aid; emergency alerting and response procedures; critique of response and follow-up; and PPE [personal protective equipment] and emergency equipment. Paragraph (p)(8) also addresses emergency response training and procedures for handling emergency incidents.

You can see that, in general, the two lists of elements are similar, but the 1910.120 list is much more specific.

Appendix C helps with the following guidance:

> *Each employer must address what actions employees are to take when there is an unwanted release of highly hazardous chemicals.... Employers will need to decide if they want employees to handle and stop small or minor incidental releases. Whether they wish to mobilize the available resources at the plant and have them brought to bear on a more significant release. Or whether employers want their employees to evacuate the danger area and promptly escape to a preplanned safe zone area, and allow the local community emergency response organizations to handle the release. Or whether the employer wants to use some combination of these actions. Employers will need to select how many different emergency preparedness...lines of defense they plan to have and then develop the necessary plans and procedures, and appropriately train employees in their emergency duties and responsibilities and then implement these lines of defense.*

Emergency Response Training

In Chapter 15, we learned that emergency response has to be practiced. Without drills, emergency response actions aren't going to progress smoothly.

Accordingly, the process safety management standard—both in OSHA Standard 1910.38 and 1910.120—requires emergency response training. Paragraph (p)(8)(iii) of Standard 120 provides an excellent description of what the training must accomplish:

> *Training for emergency response employees shall be completed before they are called upon to perform in real emergencies. Such training shall include the elements of the emergency response plan, standard operating procedures the employer has established for the job, the personal protective equipment to be worn and procedures for handling emergency incidents.*

OSHA Standard 1910.38 focuses primarily upon individual training. But, as we learned in the systematic approach, emergency response is exceptionally dependent upon how the *team* reacts. Therefore, team training is essential. Appendix C emphasizes the need for team training and close coordination with outside resource centers, stressing the need for practice:

> *Responders may be working under very hazardous conditions and therefore the objective is to have them competently led by an on-scene incident commander and the commander's staff, properly equipped to do their assigned work safely, and fully trained to carry out their duties safely before they respond to an emergency. Drills, training exercises, or simulations with the local community emergency response planners and responder organizations is one means to obtain better preparedness. This close cooperation and coordination between plant and local community emergency preparedness managers will also aid the employer in complying with the Environmental Protection Agency's Risk Management Plan criteria.*

A Worthwhile Investment

In the Phillips event, emergency actions weren't precise. The warning signal apparently wasn't heard by all, and evacuation, as evidenced by the number dead within the blast zone, progressed too slowly. Similarly, fire fighting response wasn't in top form because of out-of-service and unfuelled equipment. Both deficiencies could have been identified in advance with proper casualty and emergency drills.

No question, emergency planning and response is a difficult and complicated part of the systematic approach (or process safety management). It requires, for most operations, a full-time staff and constant upgrade training for those who must be prepared to respond (which is, essentially, everyone). If you're just getting started in the development of an emergency response program, you will be well-served to seek help.

Compliance Audits

In Chapter 22, we stated that without proper evaluation, team improvement—and ultimately product or service improvement (which is the real purpose of evaluation)—cannot be achieved. We learned that evaluation accomplishes at least five purposes:
- Verifies that policies and procedures are properly interpreted and implemented;
- Evaluates the effectiveness of policies, procedures, and standards;
- Provides objective, timely, and reliable performance data for achievements as well as deficiencies;

- Provides recommendations for improvement where appropriate; and,
- Provides opportunities to teach, correct, and reinforce proper work practices and principles.

In our study of the systematic approach, we divided evaluation into two categories—outside evaluation and self-assessment. We applied the term *outside evaluation* to assessment performed by any person or agency foreign to our own organization. *Self-assessment*, on the other hand, is the term we used to describe all assessment performed by our own people—everyone (at every organizational level) evaluating their own and their team's performance. Both categories included tests and audits as methods to assist in evaluation.

In the same vein, *compliance audits* are codified in paragraph (o) of the process safety management standard to establish a minimum measure of process safety performance. Though called by a different name—*process safety audit*—the Center for Chemical Process Safety defines the audit (in **Plant** Guidelines for Technical Management of Chemical Process Safety [Revised Edition] with these words:

> *In-depth process safety audits are a management device to ensure that a proper level of safety and property protection is maintained despite the evolutionary changes which occur during operation of a facility.*

Let's see what the law has to say on the subject of compliance audits.

1910.119(o)

(o) Compliance audits.

(1) Employers shall certify that they have evaluated compliance with the provisions of this section at least every three years to verify that the procedures and practices developed under the standard are adequate and are being followed.

(2) The compliance audit shall be conducted by at least one person knowledgeable in the process.

(3) A report of the findings of the audit shall be developed.

(4) The employer shall promptly determine and document an appropriate response to each of the findings of the compliance audit, and document that deficiencies have been corrected.

(5) Employers shall retain the two (2) most recent compliance audit reports.

What is an Audit?

In Chapter 22, we defined an audit of industrial operations to be the formal examination of performance and the records associated with all aspects of facility operations. We stated that the records evaluated during audits include training and qualification records, process data, operating logs and data record sheets, tagouts and tagout logs.

In contrast to a financial audit, we learned that a comprehensive performance audit must proceed beyond record review. It should include—and, if possible, focus on—real-time observations and assessment of individual and team performance as the team engages in routine process operations, response to abnormal conditions, and reaction to casualties and emergencies. Appendix C offers this guidance:

> *An effective audit includes a review of the relevant documentation and process safety information, inspection of the physical facilities, and interviews with all levels of plant personnel. Utilizing the audit procedure and checklist developed in the preplanning stage, the audit team can systematically analyze compliance with the provisions of the standard and any other corporate policies that are relevant. For example, the audit team will review all aspects of the training program as part of the overall audit. The team will review the written training program for adequacy of content, frequency of training, effectiveness of training in terms of its goals and objectives as well as to how it fits into meeting the standard's requirements, documentation, etc. Through interviews, the team can determine the employee's knowledge and awareness of the safety procedures, duties, rules, emergency response assignments, etc. During the inspection, the team can observe actual practices such as safety and health policies, procedures, and work authorization practices. This approach enables the team to identify deficiencies and determine where corrective actions or improvements are necessary.*

When to Audit

The minimum frequency of compliance auditing is established in paragraph (o)(1) at three years. We must emphasize again that the law prescribes *minimum* standards. We find that the best industries, though they also incorporate formal audits and assessments, actually engage in on-going compliance auditing as part of an inculcated self-assessment process, perpetually checking themselves against the standards of the law and, more importantly, their own achievement standards. By finding and fixing at more frequent intervals, problems are not allowed to take root and grow to large proportions. It's cheaper in the long run.

Selecting the Audit Team

In **Industrial Facility Self-Assessment**, we state that no amount of procedural documentation can substitute for an expert observer. The experience, knowledge, and judgment of a well-qualified observer is the **most** important part of any assessment program.

Compliance audits are no exception. The selection of the audit team will, for the most part, determine how accurate, incisive, and comprehensive (and, therefore, how valuable to you) the audit will be. Guidance from Appendix C reinforces that point:

> *The selection of effective audit team members is critical to the success of the program. Team members should be chosen for their experience, knowledge, and training and should be familiar with the processes and with auditing techniques, practices and procedures. The size of the team will vary depending on the size and complexity of the process under consideration. For a large, complex, highly instrumented plant, it may be desirable to have team members with expertise in process engineering and design, process chemistry, instrumentation and computer controls, electrical hazards and classifications, safety and health disciplines, maintenance, emergency preparedness, warehousing or shipping, and process safety auditing. The team may use part-time members to provide for the depth of expertise required as well as for what is actually done or followed, compared to what is written.*

Of course, as with nearly all other teams, the team leader will set the tone (and the standards) by which the audit will be conducted. So, be sure to start with a firm foundation. Appendix C reminds us that:

> *The audit should be conducted or led by a person knowledgeable in audit techniques and who is impartial towards the facility or area being audited.*

The size and composition of the team will, in all probability, correspond to the proportions and complexity of the process to be audited. For a small, established process, it's possible that the team could even consist of a single person. ("The compliance audit shall be conducted by at least one person knowledgeable in the process.") Appendix C states that:

> *Employers need to select a trained individual or assemble a trained team of people to audit the process safety management system and program. A small process or plant may need only one evaluation of the design and effectiveness of the process safety management system and a field inspection of the safety and health conditions and practices to verify that the employer's systems are effectively implemented.*

More probably, the audit will require substantially more participants.

Perhaps more important than all other aspects of team selection, however, is auditor attitude. As we stated in Chapter 22, every evaluator must understand that the primary objective of evaluation is to facilitate continued improvement through identifying and correcting deficiencies, recognizing and measuring success, and evaluating the effectiveness of implemented changes. If the audit becomes an exercise in power and ego, the response is apt to be unfavorable.

Planning the Audit

Evaluation of any sort requires some planning. In Chapter 22, we determined that an evaluation strategy required:
- Understanding and defining the purpose and objectives of evaluation;
- Delineating the scope and methods of evaluation;
- Identifying and allocating the resources necessary to implement the program;
- Determining the means of reporting and communicating evaluation results;
- Developing a means for tracking and evaluating corrective actions and improvements; and,
- Making provisions for analyzing program effectiveness.

Audit preparation is constrained by essentially the same consideration. Appendix C states that:

> *The essential elements of an audit program include planning, staffing, conducting the audit, evaluation and corrective action, follow-up and documentation.... Planning in advance is essential to the success of the auditing process. Each employer needs to establish the format, staffing, scheduling and verification methods prior to conducting the audit. The format should be designed to provide the lead auditor with a procedure or checklist which details the requirements of each section of the standard. The names of the audit team members should be listed as part of the format as well. The checklist, if properly designed, could serve as the verification sheet which provides the auditor with the necessary*

information to expedite the review and assure that no requirements of the standard are omitted. This verification sheet format could also identify those elements that will require evaluation or a response to correct deficiencies. This sheet could also be used for developing the follow-up and documentation requirements.

Responding to the Audit

We also learned in Chapter 22 that evaluation and self-assessment are not ends unto themselves. Rather, they are only starting points for additional action—whether deficiencies are found or not. Remember, the ultimate objective is improvement.

A compliance audit is only a data grinder. The quality of what comes out depends upon what goes in and who processes it. Assuming that the audit has been comprehensively performed, it should provide some very valuable insights. What you do with the information after the audit is what really counts. Appendix C elaborates:

Corrective action is one of the most important parts of the audit. It includes not only addressing the identified deficiencies, but also planning, followup, and documentation. The corrective action process normally begins with a management review of the audit findings. The purpose of this review is to determine what actions are appropriate, and to establish priorities, timetables, resource allocations and requirements and responsibilities. In some cases, corrective action may involve a simple change in procedure or minor maintenance effort to remedy the concern. Management of change procedures need to be used, as appropriate, even for what may seem to be a minor change. Many of the deficiencies can be acted on promptly, while some may require engineering studies or indepth review of actual procedures and practices. There may be instances where no action is necessary and this is a valid response to an audit finding. All actions taken, including an explanation where no action is taken on a finding, needs to be documented as to what was done and why.

Finally, remember what we learned in Chapter 22 regarding attitude toward the audit. We stated that evaluation is worthless if those evaluated are unwilling to listen and act on the results, or (worse) if evaluators use the process just to exercise authority. The *right attitude* toward evaluation is essential.

Tracking Actions

In **Industrial Facility Self-Assessment**, we said that facility observation is valuable if only from the perspective of providing on-the-spot feedback to managers, supervisors, and employees. But without a recording and tracking system, the observation process is incomplete. Many deficient conditions cannot be immediately corrected. They must be identified and entered into a system of evaluation and prioritization so that facility resources can be efficiently used.

With this realization, the law requires in paragraph (o)(4) that employers "promptly determine and document an appropriate response to each of the findings…and document that deficiencies have been corrected." Appendix C explains:

It is important to assure that each deficiency identified is addressed, the corrective action to be taken noted, and the audit person or team responsible be properly documented by the employer. To control the corrective action process, the employer should consider the use of a tracking system. This tracking system might include periodic status reports shared with affected levels of management, specific reports such as completion of an engineering study, and a final implementation report to provide closure for audit findings that have been through management of change, if appropriate, and then shared with affected employees and management. This type of tracking system provides the employer with the status of the corrective action. It also provides the documentation required to verify that appropriate corrective actions were taken on deficiencies identified in the audit.

Even if your intentions are favorable, valuable findings and solutions sometimes get lost in the day-to-day war of operational activities. Without a way to track actions and responsibilities, they will never be acted upon.

Audit Retention Requirements

In our discussion of incident investigation, we stated that report retention was important for tracking trends. In compliance auditing, the same is true. Accordingly, paragraph (o)(5) requires employers to "retain the two…most recent compliance audit reports." Assuming compliance audits are performed at the minimum frequency of every three years, record retention and storage should not be difficult. Moreover, we would hope that the audit report, findings, and corrective action plan is stored on computer disk for ease of access.

Dissecting the Law

Evaluation Resources Obviously, compliance auditing is a costly part of the process safety management standard. As with so many other of the requirements, it will pay you well to improve your knowledge.

For a thorough study of compliance auditing (and the self-assessment process) we recommend a review of two publications. The first is our own **Industrial Facility Self-Assessment: A Guide for Observation Program Planning and Implementation**. This study provides an in-depth look at what it takes to develop and implement a comprehensive self-assessment program. The second publication which is of great assistance is, of course, **Plant Guidelines for Technical Management of Chemical Process Safety** (Revised Edition). In particular, Appendix 13D (Example of Safety and Property Protection Procedures) provides excellent guidance for planning and performing a process safety (or compliance) audit.

Trade Secrets

The last regulatory paragraph of Standard 119—paragraph (p)—is titled "Trade Secrets". A *trade secret* is defined to be "any confidential formula, pattern, process, device, information or compilation of information that is used in an employer's business, and that gives the employer an opportunity to obtain an advantage over competitors who do not know or use it." Here's what the law says:

1910.119(p) **(p) Trade secrets.**

(1) Employers shall make all information necessary to comply with the section available to those persons responsible for compiling the process safety information (required by paragraph (d) of this section), those assisting in the development of the process hazard analysis (required by paragraph (e) of this section), those responsible for developing the operating procedures (required by paragraph (f) of this section), and those involved in incident investigations (required by paragraph (m) of this section), emergency planning and response (paragraph (n) of this section) and compliance audits (paragraph (o) of this section) without regard to possible trade secret status of such information.

(2) Nothing in this paragraph shall preclude the employer from requiring the persons to whom the information is made available under paragraph (p) (1) of this section to enter into confidentiality agreements not to disclose the information as set forth in 29 CFR 1910.1200.

(3) Subject to the rules and procedures set forth in 29 CFR 1910.1200(i) (1) through 1910.1200(i) (12), employees and their designated representatives shall have access to trade secret information contained within the process hazard analysis and other documents required to be developed by this standard.

What Does That Mean? This final paragraph of the process safety management standard simply says that an employer may not jeopardize hazardous chemical process safety by limiting access to safety-related information—either through hazard analysis, procedures, or evaluation—on the grounds that the information is a trade secret. In other words, employees (or others engaged in evaluating process operations) must have all safety-related information available for process operations and evaluation. But, as the law also clearly states, the employer may protect trade secret information by requiring binding agreements to be signed by those who will have access to the information.

Conclusion

We began this supplement with the goal of examining process safety management in light of the systematic approach to industrial operations, the theme of the main text. What we have learned is that there is little, if any difference, in their purposes. Both have been developed to improve efficiency and to protect the operating staff, the environment, and the general public through a disciplined approach to equipment control. Though the 14 elements of law presented in OSHA 1910.119 codify only minimum standards, compliance with those elements will go a long way toward preventing chemical process accidents involving highly hazardous chemicals.

Part VI

Glossary/Index

Glossary/Index

Abnormal Operating Limit	A temporary limit which is usually more restrictive than the normal operating limit and that is imposed to compensate for an abnormal operating condition. [60]
Abnormality	A condition or event which deviates notably from desired or expected conditions. [103, 154]
Acceptance Testing	Testing of new or modified equipment after it has been received and installed. [182]
Accident Precursor	Warning signs of impending danger in equipment operation. [23, 103]
Administrative Prerequisite	A requirement, usually placed upon those who control equipment, specifying the qualifications, certifications, and conditions of operators for operating equipment. [59]
Audit	Formal examination or verification, primarily of records. [240]
Blank Flange	Plates which are inserted into bolted piping connections to physically block fluid flow; also known as slip blinds or pancake flanges. [187]
Calibration	The process of aligning a component or system to a known operating standard. [64]
Caution Tag	A warning tag, frequently yellow in color, indicating unusual or abnormal operating precautions that must be observed when operating the subject equipment. [189]
Certification	The examination process by which an authority validates an operator's knowledge and skill. [89]
Close Call	An event or condition which almost became an accident; a near miss. [28, 158]
Configuration Control	The process of operating, maintaining, and testing equipment and processes in accordance with the limits and practices imposed by design. [101]
Continuing Training	An on-going training process for team members who are already qualified (or certified) at an initial level of equipment operation; the purpose of continuing training is constant improvement of individual and team skills. [231]
Corrective Maintenance	Maintenance actions designed to restore equipment to its original operating specifications. [38]
Critical Parameter	An important operating property of a process, system, or component that must be controlled to avoid failure. [8, 40, 103, 123]
Critique	A formal proceeding similar to a grand jury, preliminary hearing, or congressional hearing for the purpose of recounting and clarifying the details of an event in order to learn from it and to prevent recurrence. [225]
Danger Tag	A warning tag, often red in color, installed on the operating controls and power sources of equipment warning that the equipment, if operated, could result in injury to personnel or serious damage to machinery. [189]
Data Record Sheet	A preprinted form designed for recording the current values of important equipment operating parameters (e.g., temperatures, pressures, tank levels, voltages, etc.) [141]
Diagnostic Testing	Troubleshooting tests designed to determine what is wrong with equipment. [182]
Double Valve Isolation	The use of two valves in line to prevent the escape of fluid from a leak path. [187]
Education	The process of transmitting and acquiring thoughts, models, pictures, and ideas. [236]
Energy Isolating Device	A mechanical device that physically prevents the transmission or release of energy. [187]
Engineered Safeguard	An equipment design feature that physically protects against unsafe acts or conditions. [58]
Equipment Casualty	An event that, if it occurs, will probably cause significant damage to the operating personnel, the environment, or the general public. [104, 167]

Equipment Emergency	An event that, if it occurs and is combatted by prompt immediate action, is unlikely to result in serious damage to operating personnel, the environment, or the general public. [104, 167]
Equipment Margin	Safety margin incorporated into operating limit development to account for inexact and delayed component response. [63]
Emergency Preparedness	The readiness of an individual or organization to successfully combat and control abnormal conditions. [24, 167]
Failure	The condition in which a material, component, or system ceases to meet its specified performance requirements. [40, 62]
Failure Zone	The region of the safe operating envelope in which a condition for a given parameter is known to be unacceptably harmful to equipment, people, or the environment. [40, 62]
Fit-for-Duty	The state of being mentally and physically prepared to handle assigned routine and emergent equipment operating challenges; having the necessary skills and knowledge to perform during routine and emergency situations, as well as the proper mental and physical state to utilize them. [18, 46, 89]
Functional Testing	Routine equipment checks designed to verify operability. [182]
Graded Approach	A prioritized approach to hazard analysis wherein the extent, rigor, and complexity of the analysis corresponds to the extent and complexity of the expected hazards. [301, 315, 323]
Hazard	A potential danger, usually of personnel injury or equipment damage, incurred as a result of engaging in or being near an activity. [56]
Hazardous Energy	Any type or amount of energy that will cause injury to a person or damage equipment. [185]
Hazards Analysis	A process of determining the dangers of equipment operation and the consequences of failure. [40, 56]
Hot Work Permit	A work authorization granting permission to perform cutting, grinding, welding, or other high heat-producing operation with an associated risk of fire or explosion. [281, 320, 321]
Immediate Actions	Actions taken in response to an equipment casualty or emergency, the steps of which are necessary to place the equipment in a safe and stable condition. [172]
Independent Verification	The process of separately proving that a task has been correctly performed. [8, 9, 103, 127]
Industrial "Defensive Driving"	An attitude or state of awareness wherein professional operators anticipate and recognize the formation of potentially dangerous conditions and take preemptive actions to forestall threatening situations. [35]
Initial Training	The first phase of training which ensures that graduating operators have the skills necessary to conduct routine operations and manage abnormal or emergency situations. [88]
Initiating event	An accident event which, by itself, has little or no consequence, but, in the context of accident conditions, starts the accident chain. [5]
Inspection Tour	Usually a walking tour of an operating space during which a cognizant operator checks equipment for abnormalities. [157]
Limiting Control Setting	The setpoint at which an automatic protective feature activates. [58]
Limiting Condition of Operation	A condition which requires a limit more restrictive than the normal operating limit. [61, 65]
Lockout	The placement of a lockout device on an energy isolating device to indicate that the energy isolating device and the equipment being controlled may not be operated until the lockout device is removed. [188]
Lockout Device	A device that utilizes a positive means such as a lock to hold an energy isolating device in the safe position. [188]
Management of Change	Also see Modification. Closely controlling changes to processing equipment, systems, and procedures to avoid introduction of new hazards or avoid negating design safety features. [321–323]
Margins of Safety	Safety buffers employed to preclude or protect against equipment failure in the presence of identified conditions. [40]

Master Equipment List	A comprehensive list of equipment for which an operating team is responsible. [317]
Minor Operating Error	An operating error which constitutes (or leads to) a minor mistake or delay of action beyond normal reaction time. [64]
Modification	A maintenance action that changes equipment from its original design specifications. [38, 177]
Narrative Log	A chronological written account of the events which occur at operating stations or control centers in an industrial operation, military venture, or commercial enterprise requiring a legal record of occurrences. [143]
Needs-Based Training	Training determined by analysis to be relevant to the needs of the end user. [233]
Normal Operating Limit	A routine limit for controlling an operating parameter during normal conditions. [60]
On-the-Job Training	The process of explaining a task to a student, showing the student how to perform the task, and letting the student perform and explain the task, all under the observation and control of the instructor. [88, 104, 193]
Operating Band	The area of tolerance between upper and lower operating limits. [124]
Operating Limit	A restriction of operation imposed for control of a critical operating parameter. [58]
Operating Map	Usually a two-dimensional representation of a few important operating boundaries for a process or piece of equipment. [66]
Operating Margin	Safety margin incorporated into normal operating limit calculations to account for reasonable and expected changes that occur during (or as a result of) operation. [64]
Operating Team	The team of machinery operators and leaders who physically control the operating equipment necessary to perform an activity or facility mission. [43]
Operational Readiness Review	Evaluation of the preparedness of team members to operate. Involves testing of individual and team skills through real-time performance checks of response to infrequent and abnormal operating conditions, emergency and casualty situations, and new equipment operation. [315–316]
Operator Aid	A chart, graph, or other aid (usually not a part of an operating or maintenance procedure) which is useful to operating technicians or maintenance technicians in performance of their tasks. [117]
Policy	Written (or sometimes verbal) guidance describing a plan or general principles for performance (as opposed to step-by-step direction for equipment operation or maintenance). [41, 115]
Post-Maintenance Testing	A verification process designed to validate the effectiveness of repairs and ensure that a component or system has been restored to its design configuration. [180, 181]
Post-Task Debriefing	The process of evaluating task performance after completion. [114]
Pre-Shift Meeting	A meeting held in advance of turnover to transmit the overall status of equipment to an oncoming work group. [207]
Pre-Startup Safety Review	A specialized form of process safety review designed as a final check of equipment, procedures, and people before starting a new or significantly modified hazardous process. [314–316, 322]
Pre-Task Briefing	A preparatory meeting conducted before performing an operating, maintenance, or testing task; a meeting designed to ensure the safe and efficient execution of the task. [102, 111]
Preventive Maintenance	Maintenance actions designed to detect and forestall degradation so that equipment maintains its original design specifications. [38]
Primary Cause	A central or principle cause of a problem or event. [228]
Procedure	An equipment operating or maintenance instruction, the purpose of which is to govern operation or maintenance of a component or system. [40, 115]
Process Safety Information	The body of recorded design, construction, and operating information associated with hazardous chemical processing components and systems, including the chemicals themselves. [294]
Process Safety Management	The management of chemical process safety. For chemical processes, this is the equivalent of what we call the systematic approach to industrial operations. [287 ff., 291 ff.]

Glossary/Index

Production pressure	Organizational pressure for task completion which outweighs serious consideration of consequences. [24]
Proficiency	Current ability to operate components and systems. [235]
Qualification	The training process which prepares an operator to perform independently. [89]
Qualification Standard	A composite of written statements of the knowledge and skill requirements that an operator candidate must achieve in order to complete qualification. [84]
Reasonable Risk-Minimization	An understanding and application of the concept that few (if any tasks) have zero risk. Hence, reasonable and cost-effective means of risk minimization are employed based upon good engineering judgment. [36]
Recertification	The process of periodically demonstrating to a licensing authority the requisite knowledge and skills required by law or policy. [235]
Reliability	The ability of equipment and facilities to consistently fulfil their functions (especially without unnecessary cost and complexity). [36]
Rework	Having to do something again because the task was incorrectly performed the first time. [236]
Risk	The product of probability of an undesirable event multiplied by the potential consequences of the event. [56]
Risk Analysis	The process of analyzing the risk of failure. [40, 56]
Risk Management	The process of determining (and implementing) methods of minimizing and protecting against risk. [40, 57]
Safe Operating Envelope	A graphic or written representation (usually both) of the boundaries and margins of safety established for critical parameters of machinery design, construction, operation, and maintenance. [39, 65]
Safe Operating Limit	The maximum limit beyond which a parameter may not be permitted to proceed if avoidance of failure is to be guaranteed; also known as safety limit, technical specification, or operational safety requirement. [60]
Safety Analysis	The process of identifying operating hazards, determining the risks associated with each hazard, and erecting protective barriers and boundaries of design, construction, operation, and maintenance. [55]
Scan Pattern	A systematic habit of glancing at (and critically evaluating) important operating patterns. [125]
Secondary Cause	A cause which contributed to a problem or an event but in a far less significant way than a primary cause. [228]
Selection Criteria	The minimum attributes and characteristics deemed necessary to successfully complete a qualification program and perform the duties and responsibilities required by an operating position. [20, 84]
Self-Assessment	The process by which individuals (at every organizational level) and teams evaluate their own performances by formal and informal means including the processes of testing, auditing, and coaching. [236]
Skill	The ability to perform a task. [236]
Supporting Team	A team that performs in support of an operating team; support teams include such organizations as maintenance, training, engineering, quality assurance, safeguards and security, procurement, and human resources. [44]
Systematic Industrial Operations	The concept of controlling equipment and processes in a step-by-step fashion, minimizing risk in reasonable, cost-effective ways, with the objective of safe, efficient operation. [33]
Tagout	The placement of a tagout device (Danger Tag) on an energy isolating device to indicate that the energy isolating device and the equipment being controlled may not be operated until the tagout device is removed. [189]
Theory-to-Practice	A phrase which characterizes the process of instilling theoretical understanding of a task or process and then reinforcing it through practical performance and testing. [235]

Training Plan	A policy document which describes the training philosophy, the paths, and the methods by which candidate operators must be trained and evaluated to meet specified job requirements. [84, 195]
Turnover	The exchange of information between work units and the subsequent relief of one work unit by the other; often called *shift change*. [119, 205]
Uncertainty Margin	That margin added by the designer to account for uncertainties regarding the onset of failure for a critical parameter. [62]
Zone of Normal Operations	That area of the safe operating envelope which prescribes the region of acceptable routine operations; the area of the safe operating envelope left over after safety margin has been applied against the failure zone. [40, 63]

A

Abnormal condition margin, 65
Abnormalities, 103, 154–60
Abnormal operating limits, 60–61
Acceptance tests, 182
Accident precursors, 23–24, 103
Accidents
   Air Florida Flight 90 crash (1982), 147–60
   Big Bayou Canot Bridge accident (1993), 77–83
   Challenger accident (1986), 217–23
   Chernobyl Atomic Power Station accident (1986), 69–76
   Continental Express accident (1991), 199–205
   Eastern Airlines Flight 401 crash (1972), 3–10
   Electric Light Division utility crew electrocution, 107–10
   Exxon-Valdez grounding (1989), 12–29
   Phillips 66 chemical complex explosion and fire (1989), 276–85
   Three-Mile Island-2 accident (1979), 161–76
   Titanic, sinking of (1912), 131–35
   Union Carbide fatal gas release (Bhopal, India, 1984), 95–101
Administrative prerequisites, 59
Air Florida Flight 90 crash (1982), 147–60
Alcohol, 18
Alertness, 17–18, 46, 89–90
Attitude, 46, 90
Audits, 240

B

Bhopal, India, Union Carbide fatal gas release at, 95–105
Big Bayou Canot Bridge Accident (1993), 77–83
Blank flange, 187
Boundaries, 38–40

C

Calibration, 64
Cardinal rule of operation, 74–75
Casualties, 167–76
Caution tags, 189–90
Certification, 89
Chain of insignificant events, 4–5
Challenger accident (1986), 217–23
Chemical energy, 186
Chernobyl Atomic Power Station accident (1986), 69–76
Close calls, 28, 158–59
Coaching, 240, 265. *See also* Instructors; Leadership
Command responsibility, 178
Communication, 50, 135
   Challenger accident and, 222
   Eastern Airlines Flight 401 crash and, 8
   emergency, 138–39
   Exxon-Valdez grounding and, 18–19
   face-to-face, 136–37
   of plant information, 103
   radio/telephone, 137
   Titanic sinking and, 134–35
   verbal, 136–37
   written, 137–38
Complex industrial failure, 5–6
Compliance audits, 328–32
Computer-controlled equipment, 126
Configuration control, 101, 103, 158
Construction characteristics, 57–58
Continental Express accident (1991), 199–205
Continuing training, 87, 89, 231–37
Continuous verification, 128
Corporate-initiated evaluation, 245
Corrective maintenance, 38
Critical parameters, 40, 58, 103
   Eastern Airlines Flight 401 crash and, 8
   monitoring, 123–29
Critical tasks, verification of, 22
Critiques, 225–27

D

Danger tags, 189–90
Data record sheets, 141–43
   turnovers and, 208, 209
   See also Documentation; Logs
Defensive driving, 35
Depot maintenance, 179
Design characteristics, 57–58
Diagnostic tests, 182
Disciplined operations, 35–36
Documentation, 113, 229. *See also* Data record sheets; Logs
Double valve isolation, 187

E

Eastern Airlines Flight 401 crash (1972), 3–10
Education, 236. *See also* Training
Electrical energy, 185
Electric Light Division, utility crew electrocution (1994), 107–10
Emergencies, 167–76
   notification of, 139
   OSHA Regulation 1910.119 and, 326–28
   Phillips 66 chemical complex explosion and, 280
   Three-Mile Island-2 accident and, 161–66
Emergency preparedness, 113, 167, 234
   Exxon-Valdez grounding and, 24–25
   Union Carbide fatal gas release and, 101
Energy hazards, isolation of, 185–91
Energy isolating device, 187
Engineered safeguards, 58
Environmental anomalies, 155
Equipment
   casualty, 104, 167
   control, 95–102
   deficiencies, 154, 180
   emergency, 104, 167. (*See also* Emergencies)

limitation, 21
maintenance
   Exxon-Valdez grounding and, 21–22
   Union Carbide fatal gas release and, 100
   See also Maintenance
margin, 62–64, 63
monitoring, 20–21
operating characteristics, 7–8
   Exxon-Valdez grounding and, 20
   Union Carbide fatal gas release and, 97–98, 100
reliability of, 36–38
status, 20–21
Evaluations, 49, 87, 239–53
Event reports, 229
Evidence
   analyzing, 227–28
   gathering and preserving, 224–25
Evolution, 119, 120
Exxon-Valdez, grounding of (1989), 12–29

F

Face-to-face communication, 136–37
Facilities, reliable, 36–38
Facility-initiated evaluation, 244
Facility maintenance, 179
Failure, 40, 62
Failure limit, 62
Failure zone, 40, 62
Fatigue, 17–18, 46, 90
Final conditions, 112, 117
Final reports, 229
First-line evaluation, 243
Fit-for-duty, 18, 46, 89
Fitness, 46, 89–90
   Exxon-Valdez grounding and, 17–18
Functional tests, 182

G

Graded approach, 301, 315, 323

H

Hazard, 56
Hazardous energy, 185
   sources of, 185–86
Hazards analysis, 40, 56, 112
Hot work permits
   OSHA Regulation 1910.119 and, 320, 321
   Phillips 66 chemical complex explosion and, 281
Hydraulic energy, 186

I

Immediate actions, 172
Implementation, 255–69

Independent verification, 103, 127–30
   Challenger accident and, 222
   Eastern Airlines Flight 401 crash and, 8–9
Informal verification, 129–30
Initial conditions, 20, 112
Initial training, 88
Initiating event, 5
Inspection tours, 157
Instructors
   OJT, 194
   qualification of, 85–86
   selection of, 85
   See also Coaching; Leadership
Investigations, 223–30
   Challenger accident and, 217–23
   OSHA Regulation 1910.119 and, 323–26
Invulnerability, sense of, 22–23
Ionizing radiation, 186

J

Job and task analysis, 83
Job requirements, determination of, 83–84

L

Leaders, evaluation of, 250
Leadership, 26–27, 47–49, 139, 252
   acceptance by subordinates, 257–58
   Challenger accident and, 222–23
   Eastern Airlines Flight 401 crash and, 9
   Titanic sinking and, 134
   See also Coaching; Instructors
Lesson materials, 86
Lessons learned
   continuing training and, 235
   dissemination of, 229–30
Limiting control setting, 58
Lockout, 188
Lockout device, 188
Logs, 143–45
   Continental Express accident and, 204
   turnovers and, 208, 209
   See also Data record sheets; Documentation

M

Maintenance, 58, 104, 177–81
   corrective, 38
   Exxon-Valdez grounding and, 21–22
   preventative, 38
   Three-Mile Island-2 accident and, 165
   Union Carbide fatal gas release and, 97–98, 100
Management of change, 321–23
Management oversight, 27–28
Margins of safety, 40
Master equipment list, 317

MBWA (Management by wandering around), 240
Mechanical energy, 186
Minor operating error, 64
Mission, 41–42
Modification, 38, 58, 104, 177
Monitoring, 123–26
Multiple-point verification, 128

N

Narrative logs, 143–45
Needs-based training, 233
Night-time accidents, 17–18
    Big Bayou Canot Bridge accident, 77–83
    Chernobyl Atomic Power Station accident, 73–76
    Eastern Airlines Flight 401 crash, 3–10
    Exxon-Valdez grounding, 12–29
    Three-Mile Island-2 accident, 161–76
    Titanic sinking, 131–35
    Union Carbide fatal gas release (Bhopal, India, 1984), 95–101
    See also Challenger accident
Non-ionizing radiation, 186–87
Normal operating limits, 60

O

On-the-job training (OJT), 88, 104, 193–97
Operating
    band, 124
    boundaries, 38–40
    discipline, 52–53
    limits, 28–29, 58–65
        Chernobyl Atomic Power Station Accident and, 73–75
        Union Carbide fatal gas release and, 99
    maps, 66–67
    margin, 62–64
    organization, 43
    philosophy, 34–36
    principles, 74–76
    skills, 46, 101–4
    strategy, 33–34
    structure, 41–45
    team, 43–44
Operational readiness review, 315–16
Operator
    aids, 117
    alert/well-trained, 45–47, 77–91. *See also* Alertness; Training
    maintenance, 179
OSHA Regulation 1910.119, 288
    application, 291–92
    compliance audits, 328–32
    contractors, 311–14
    emergency planning and response, 326–28
    employee participation, 292–94
    hot work permit, 320–21
    incident investigation, 323–26
    management of change, 321–23
    mechanical integrity, 316–19
    operating procedures, 303–7
    pre-startup safety review, 314–16
    process hazard analysis, 299–303
    process safety information, 294–99
    trade secrets, 332
    training, 307–11
Oversight, 90

P

Phillips 66 chemical complex explosion and fire (1989), 276–85
Plant information, communication of, 103
Pneumatic energy, 186
Policies, 40–41, 115
Post-maintenance tests, 180, 181–82
Post-task debriefing, 114
Preliminary reports, 229
Pre-shift meetings, 209
Pre-shift tours, 209–10
Pre-startup safety review, 314–16
Pre-task briefings, 102, 111–14
    Electric Light Division accident and, 107–10
Preventive maintenance, 38
Primary causes, 228
Procedural non-compliance, 118, 121
    Continental Express accident and, 204
    Exxon-Valdez grounding and, 20
    Phillips 66 chemical complex explosion and, 281
    Three-Mile Island-2 accident and, 165
    Union Carbide fatal gas release and, 100–101
Procedures, 40–41, 115–22
Process failure, 276–85
Process safety information, 294
    OSHA Regulation 1910.119, 294–99
Process safety management, 287–89
    OSHA Regulation 1910.119, 291–332
Production pressure, 24
Proficiency, 235
Pro-shift meeting, 207

Q

Qualification, 89
Qualification standards, 84–85

R

Radio/telephone communication, 137
Reasonable risk-minimization, 36
Recertification, 235
Regulatory evaluation, 245–46
Reliability, 36–38
Reports, 229
    evaluation, 251

Responsibility, 177–79
Risk
   analysis, 40, 56, 112
   management, 40, 57

S

Safe operating envelope, 60, 65–67
Safe operating limits, 60, 62
Safety
   analysis, 55
      margins, 39–40
Sage operating envelope, 39–40
Scan pattern, 125
Secondary causes, 228
Selection criteria, 20, 84
Self-assessment, 51, 236, 249
Shift changes, 19–20
Site-initiated evaluation, 244–45
Site maintenance, 179
Skill, 236
Sound energy, 187
Staffing, 25–26
   Union Carbide fatal gas release and, 99
Status reports, 120
Superior construction, 37
Supporting teams, 44–54
Systematic industrial operations, 33

T

Tagout, 189
Tags, warning, 189–91
Team feeling, 53
Teamwork, 28, 42–44, 46–47, 49–54
Testing, 104, 177, 182–84, 240
Theoretical knowledge, 45
Theory-to-practice, 235
Thermal energy, 186
Three-Mile Island-2 accident (1979), 161–76
Titanic, sinking of (1912), 131–35
Tools and materials checklist, 112
Training, 83, 120, 264–66
   Big Bayou Canot Bridge accident and, 77–83
   continuing, 87, 89, 231–37
   deficient, 7–8, 20
   developing a program, 84–87
   Eastern Airlines Flight 401 crash and, 7–8
   for emergencies and casualties, 169–72
   evaluation, 87
   initial, 88
   needs-based, 233
   on-the-job, 88, 104, 193–97
   OSHA Regulation 1910.119 and, 307–11
   phases of, 87–89
   plans, 84, 195
   pre-test, 183
   procedures, 41, 45–46
   records, 86
   schedules, 86
   See also Coaching; Education
Turnovers, shift, 104, 119, 205–13
   Continental Express accident and, 199–205
   Exxon-Valdez grounding and, 19–20

U

Uncertainty margin, 62
Union Carbide fatal gas release (Bhopal, India, 1984), 95–101

V

Verification
   continuous, 128
   independent, 22
   informal, 129–30
   multiple-point, 128

W

Warning signs, 23–24, 103

Z

Zone of normal operations, 40, 63, 64

About the Author

A 1972 graduate of the United States Military Academy at West Point, Hop Howlett has been developing and presenting special skills training programs in a variety of technical arenas for more than two decades. His military service began with the Berlin Brigade in West Berlin, Germany. He was later trained and certified in the States as an explosive ordnance disposal officer. As the unit commander of a bomb squad in the Washington, DC area, he and his personnel provided support to the U.S. Secret Service for protection of the president and foreign heads of state against terrorist violence.

Following military service, Hop was certified by Westinghouse Electric Corporation as a nuclear plant engineer and shift supervisor for reactor plant operations at the Department of Energy's Naval Reactors Facility, part of the U.S. Navy's Nuclear Propulsion Program. As an operator, shift supervisor, and facility training manager, he amassed nearly a decade of hands-on experience in naval nuclear plant operations, maintenance, training, and emergency preparedness. He also served as a site security and anti-terrorist consultant for the Idaho National Engineering Laboratory.

In June of 1988, Hop founded TECHSTAR, a multi-service consulting firm, specializing in the development of custom training programs and materials for industrial and nuclear facilities. His extensive industrial background includes in-depth evaluation, consultation, and assistance at Department of Energy and commercial nuclear facilities throughout the United States.

In addition to his technical expertise in industrial operations, Hop is also a sought-after leadership instructor. His seminars in leadership and management have been attended by thousands of business men and women throughout the country.

Hop is the author of a number of books, articles, and seminar guides on leadership and industrial management. Popular TECHSTAR titles include *The Industrial Operator's Handbook, Managing People: The Art of Leadership,* and *Industrial Facility Self-Assessment.*